Rock Mechanics and Rock Engineering

Rock Mechanics and Rock Engineering

Volume 2: Applications of Rock Mechanics – Rock Engineering

Ömer Aydan

Department of Civil Engineering, University of the Ryukyus, Nishihara, Okinawa, Japan

CRC Press
Taylor & Francis Group
Boca Raton London New York

CRC Press is an imprint of the
Taylor & Francis Group, an **informa** business

A BALKEMA BOOK

Cover photograph description: Eurasia subsea rock tunnel beneath the Bosporus in Istanbul

Published by:
CRC Press/Balkema
Schipholweg 107C, 2316 XC Leiden, The Netherlands

First issued in paperback 2023

Volume 1
ISBN 13: 978-0-367-42162-5 (hbk)
DOI: https://doi.org/10.1201/9780367822293

Volume 2
ISBN-13: 978-1-03-265429-4 (pbk)
ISBN 13: 978-0-367-82230-9 (ebk)
ISBN 13: 978-0-367-42165-6 (hbk)
DOI: https://doi.org/10.1201/9780367822309

Two-volume set
ISBN 13: 978-0-367-02935-7 (hbk)
DOI: https://doi.org/10.1201/9780429001239

Library of Congress Cataloging-in-Publication Data
Applied for

Contents

Preface

Rock is the main constituent of the crust of the Earth, and its behavior is the most complex one among all materials in geosphere to be dealt by humankind. Furthermore, it contains various discontinuities, which make the thermo-hydro-mechanical behavior of rocks more complex. These simply require higher level of knowledge and intelligence in the Rock Mechanics and Rock Engineering (RMRE) community. Furthermore, the applications of the principles or rock mechanics to mining, civil and petroleum engineering fields, as well as earthquake science and engineering, are diverse, and it constitutes rock engineering. Recently, the International Society for Rock Mechanics (ISRM) added "Rock Engineering" in 2017 to its name while its acronym remains ISRM.

Rock mechanics is concerned with the theoretical and applied science of the mechanical behavior of rock and rock masses, and it is one of branches of mechanics concerned with the response of rock and rock masses to their physical-chemical environment. *Rock engineering* is concerned with the application of the principles of mechanics to physical, chemical and electromagnetic processes in the uppermost part of the Earth and the design of the rock structures associated with mining, civil and petroleum engineering. This book is intended to be a fundamental book for younger generations and newcomers, as well as a reference book for experts specialized in rock mechanics and rock engineering.

The book is divided into two volumes, due to the wide spectra of rock mechanics and rock engineering, titled *Rock Mechanics and Rock Engineering: Fundamentals of Rock Mechanics* and *Rock Mechanics and Rock Engineering: Applications of Rock Mechanics – Rock Engineering*. In the first volume, the fundamental concepts, theories, analytical and numerical techniques and procedures of rock mechanics and rock engineering, together with some emphasis on new topics, are described as concisely as possible while keeping the mathematics simple.

The second volume is concerned with the applications of rock mechanics and rock engineering in practice. It ranges from classical rock classifications, the response and stability of surface and underground structures, to model testing, monitoring, excavation techniques and rock dynamics. Particularly, earthquake science and engineering, vibrations and nondestructive techniques are presented as a part of rock dynamics.

Although *Rock Mechanics and Rock Engineering* consists of two volumes, each volume is complete in its content, and it should serve the purposes of educators, students, experts as well as practicing engineers. It is strongly hoped that these two volumes would fulfill the expectations and would serve further advances in rock mechanics and rock engineering.

Chapter 1

Introduction

Rock engineering is concerned with the applications of the principles of rock mechanics in practice. These applications involve the construction of transportation facilities such as tunnels, high-cut rock slopes, foundations of large bridges, nuclear power plants, dams, storage of oil and natural gases in caverns in civil engineering, exploitation of natural sources such as metallic minerals, coals in the form of open-pits or underground mines in mining engineering, extraction of oil and gas in petroleum engineering and utilization of geothermal energy.

Some of the practical applications of empirical, analytical and numerical methods involve the evaluation of deformation and stress state of surface, semi-underground and underground rock engineering structures in the short and long term, as well as under static and dynamic loading conditions. Some fundamental examples of applications to surface and underground structures are explained with the consideration of practical conditions.

The model testing technique in rock mechanics and rock engineering has been an important tool for engineers for understanding the response of rock engineering structures as well as obtaining design parameters if the similitude law is properly established for a given structure. With the development of numerical methods, response and stability of rock engineering structures could be evaluated under very complicated initial and boundary conditions as well as rock mass behavior. Nevertheless, model testing is still a useful yet powerful technique to have an insightful view of what is taking place with regard to a given structure under the given conditions. They may also provide a clear visual yet quantitative picture of the phenomenon, which may be quite difficult to evaluate by numerical methods.

From earlier times, many rock classifications have been proposed, and some of them provide quantitative characterization of rock masses with or without their applications to certain rock engineering structures. These classifications and their utilization for estimating the mechanical properties of rock masses are presented.

Model tests have been used in engineering for thousands of years, and it is still widely used in many engineering applications. The similitude law used in model testing is described, and some specific examples are given. Principles of various model testing techniques under static and dynamic conditions are explained, and various specific examples of model tests are described in order to illustrate their use in rock mechanics and rock engineering.

Humankind has devised many excavation techniques to create underground openings, construct foundations of rock engineering structures and pass through steep valleys. Blasting is still the most commonly used excavation technique, and its principles are explained. Besides the excavation, the positive and negative characteristics of the blasting technique are presented. Furthermore, the principles of machine-based excavation techniques and expansive chemical agents are described.

Vibrations in rock mechanics and rock engineering result from different processes such as blasting, machinery, impact hammers, earthquakes, rockburst, bombs (including missiles), traffic, winds, lightning, weight drop and meteorites. Vibrations may also be induced by impact hammers, blasting with small explosives, Tunnel Boring Machine (TBM), and they may be used to evaluate wave velocity characteristics of rocks and rock masses for assessing the rock mass properties for design purposes. The related chapter describes the devices for measuring vibrations and their utilization for characterization of rock mass conditions such as existence of weak/fracture zones, cavities and their properties. Furthermore, they are used to infer some yielding or loosening around rock structures. Various field examples of its utilization in rock mechanics and rock engineering are presented.

Degradation of rock masses subjected to atmospheric conditions and/or gas/fluids percolating through rocks and rock discontinuities is quite well-known. This issue is explained, and the causes of degradation such as the alteration of minerals, weakening of particle bonds and/or solution of particles are explained. Furthermore, the effects of degradation processes on the properties of rocks and rock engineering structures are described.

The monitoring of rock mass movements, as well as of their responses to various environmental conditions, has become quite common in the construction of various rock engineering structures. This chapter is devoted how to measure deformation responses utilizing direct and space-borne optical and laser techniques as well as variations of various parameters using the multiparameter monitoring technique, which may include temperature, water level, acoustic emissions, electric potential, infrared imaging technique. Various laboratory and field examples are described.

Earthquakes are often encountered in many long-term rock engineering projects. Therefore, understanding the behavior of rock mass during shaking as well as various rock engineering structures during earthquakes is of great importance. Chapter 10 is devoted to the science and engineering aspects of earthquakes and their effect on rock engineering structures. In addition, some specific examples are given for evaluating ground motions caused by earthquakes on the basis of principles of rock mechanics and how to design rock engineering structures against the motions caused by earthquakes. Furthermore, the possibility of earthquake prediction on the basis of principles of rock mechanics is studied and discussed.

Chapter 2

Applications to surface rock engineering structures

2.1 Cliffs with toe erosion

2.1.1 Analytical approach

As pointed out in the previous section, the toe erosion of rock cliffs results in overhanging rock blocks. If the overhanging part of the cliff is continuously connected to the rest of rock mass, these rock blocks may be modeled as cantilever beams. However, depending upon the erosion type, their configuration may change from a rectangular prism to triangular prism. If the bending theory is employed, one can easily derive the following set of equations by assuming that cliffs are subjected to gravitational and seismic loads as illustrated in Figure 2.1 for a unit thickness.

Equivalent beam thickness

$$h = h_b \left[1 - (1-\alpha)\frac{x}{L} \right] \tag{2.1}$$

Shear force

$$Q = V_o - (1+k_v)\gamma h_b x \left[1 - (1-\alpha)\frac{x}{2L} \right] \tag{2.2}$$

Bending moment

$$M = M_o + V_o x - (1+k_v)\gamma h_b x^2 \left(\frac{1}{2} - (1-\alpha)\frac{x}{6L} \right) \tag{2.3}$$

Bending stress at the outer fiber

$$\sigma = k_h \gamma h_b L \left(\frac{1+\alpha}{2} \right) + 6\frac{M}{h^2} \tag{2.4}$$

where

$$\alpha = \frac{h_s}{h_b}, \ V_o = (1+k_v)\gamma h_b L \left(1 - \frac{(1-\alpha)}{2} \right) \tag{2.5a}$$

$$M_o = -(1+k_v)\gamma h_b L^2 \left(\frac{1}{2} - \frac{(1-\alpha)}{3} \right) \tag{2.5b}$$

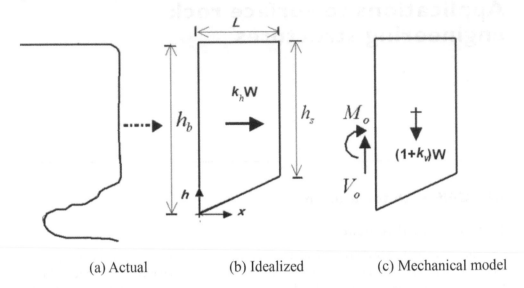

(a) Actual   (b) Idealized   (c) Mechanical model

Figure 2.1 Modeling of overhanging cliffs

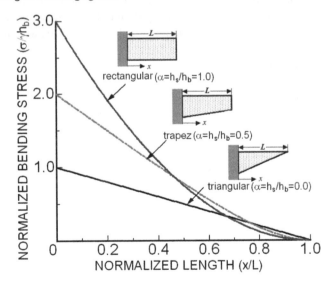

Figure 2.2 Comparison of distribution of bending stress at the outermost fiber of beam with different configurations

h_b, h_s, γ and L are beam height at the base and at the far end, unit weight of rock mass and erosion depth, respectively. k_h, and k_v are horizontal and vertical seismic coefficients.

   Figure 2.2 shows the bending stress distributions along the outermost fiber of the beam for different geometrical configurations. The severest condition occurs when the beam has a rectangular shape and the value of the bending stress is much higher for the rectangular configuration. Tensile stress is also the largest at the base of the cantilever beam. As discussed by Aydan and Kawamoto (1992), the cantilevers fail immediately once the tensile stress exceeds the tensile strength of rock mass. Furthermore, the seismic loads in addition to gravitational load would make the cliffs more vulnerable to failure during earthquakes.

While the consideration of seismic loads is based on the seismic coefficient method in this section, one may assess any amplification from a response analysis if the frequency content of earthquake waves is of great importance for a given earthquake record using the following formula for the natural frequency (first mode) of cantilever beams as a first approximation:

$$f_1 = \frac{1.875^2}{2\pi} \sqrt{\frac{EI}{mL^4}} \qquad (2.6)$$

where L is erosion depth, E is elastic modulus, m is mass per unit length, and I is the inertia moment of area.

2.1.2   Numerical analyses

Finite element method (FEM) is one of the powerful numerical techniques to analyze the stability of the cliffs. This technique was employed by Kawamoto et al. (1992) to back-analyze the failure of an overhanging cliff at Echizen along the Japan seashore, which killed 15 people in 1984. One of the main purposes of the finite element method was to check the application limits of the bending theory presented in the previous section.

The stress state of an overhanging block having different shapes was analyzed using a two-dimensional elastic finite element method under gravitational and seismic loading to illustrate stress changes during the erosion process and compare with estimations from the bending theory of cantilever beams presented in previous section for material properties given in Table 2.1. The reason for using elastic finite element analysis was that, when cantilever structures start to rupture in tension in brittle rock mass, it immediately results in total failure.

Figure 2.3 shows the finite element model for simulating the erosion process for a rectangular cliff model with a height of 10 m for three erosion depths (4 m, 8 m and 12 m), together with assumed boundary conditions. The horizontal stress distributions along the vertical sections, namely ES1 (4 m), ES2 (8 m) and ES3 (12 m) for a rectangular shape under gravitational loading, are shown in Figure 2.4. As noted from the figure, the stress distribution is not linear as expected from the bending theory of cantilevers.

Figure 2.5(a) shows the horizontal stress distribution near the top surface. As expected, tensile stress develops parallel to the top surface of the cliff model, and the amplitude of tensile stress increases as the erosion depth increases. When the ratio of erosion depth to overhanging beam thickness exceeds 1, it is also interesting to note that the maximum tensile stress occurs near the surface projection of the erosion tip as expected from the bending theory of cantilevers. However, the value of tensile stress is lower than that obtained from

Table 2.1 Properties used in elastic finite element analyses

Unit Weight (kN m⁻³)	Elastic Modulus (GPa)	Poisson's Ratio
25	10	0.25

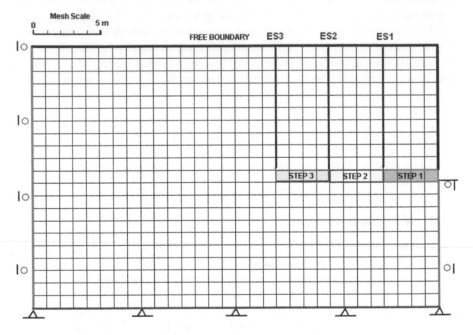

Figure 2.3 Finite element mesh and assumed boundary conditions

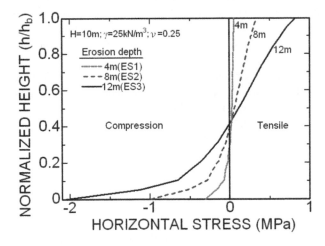

Figure 2.4 Horizontal stress distribution at the base of cantilever beam for various erosion depths computed by FEM

cantilever theory. This is considered to be due to the difference between the finite element model and cantilever models. Furthermore, the vertical stress shown in Figure 2.5(b) in the vicinity of the erosion depth is very high and compressive. However, the amplitude of the compressive stress is about 2.5 times the maximum tensile stress. This simply implies that the possibility of the failure of overhanging blocks is much more likely in tension than in

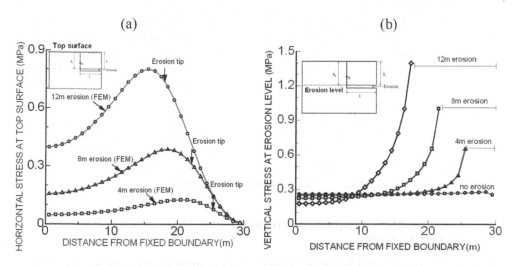

Figure 2.5 (a) Horizontal stress distribution near the top surface the FEM model for various erosion depths, (b) vertical stress distribution at the erosion level of the cliff for various erosion depths

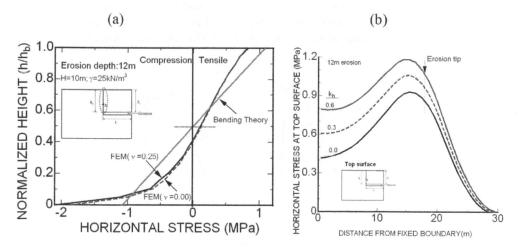

Figure 2.6 (a) Horizontal stress distribution at the base of cantilever beam for various erosion depths computed by FEM, (b) horizontal stress distribution near the top surface of the cliff for various values of horizontal seismic load coefficient

compression when the actual ratio (10–20) of the compressive strength of rocks to their tensile strength is taken into account.

Horizontal stress distributions along the vertical section emanating from the erosion tip computed from the finite element method for two different Poisson's ratios (0.00 and 0.25) are compared with the distribution from the bending theory of cantilever beams as shown in Figure 2.6(a). The stress distributions are not influenced by the variation of Poisson's ratio. The stress distribution is not linear as expected from the bending theory of cantilever beams. However, the distribution of tensile part is relatively linear as compared with compressive stresses. The maximum value of tensile stress computed from the finite element method is about 75% of that computed from the bending theory. Nevertheless, the work done is the same for both computational methods. Furthermore, the amplitude of the compressive part is about 2.5 times the maximum tensile stress. Once again, this simply implies that the possibility of the failure of overhanging blocks is much more likely in tension rather than in compression when the actual ratio (10–20) of the compressive strength of rocks to their tensile strength is taken into account.

The effect of seismic loads on the stress state of cliffs is simulated using the seismic coefficient method. Figure 2.6(b) shows the horizontal stress distribution near the top surface of the model for horizontal seismic load coefficients of 0.0, 0.3 and 0.6. As expected, the stress components increase as the seismic load coefficient increases. However, this increase will be linear, which may be inferred from the theoretical formulation.

2.2   The dynamic response and stability of slopes against wedge sliding

The authors have advanced the method of stability assessment proposed by Kovari and Fritz (1975) for wedge failure of rock slopes under different loading conditions and confirmed its validity through experiments (Kumsar *et al.*, 2000). Aydan and Kumsar (2010; Aydan, 2017) extended to evaluate sliding responses of rock wedges under dynamic loading conditions under submerged conditions with viscous resistance. Let us consider a wedge subjected to dynamic and water loading as shown in Figure 2.7. One can easily write the following

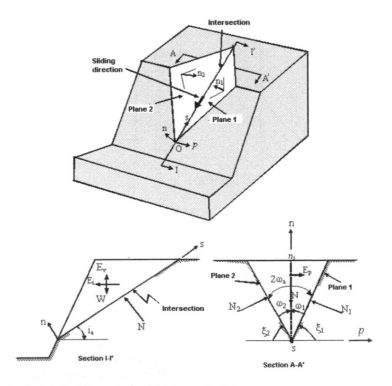

Figure 2.7 Illustration of mathematical model for wedge failure

dynamic equilibrium conditions for the wedge during sliding motion on two basal planes in a coordinate system *Osnp* shown in Figure 2.7.

$$\sum F_s = (W - E_v)\sin i_a - E_i \cos i_a - S = m\frac{d^2 s}{dt^2} \tag{2.7a}$$

$$\sum F_n = (W - E_v)\cos i_a - E_i \sin i_a - N = m\frac{d^2 n}{dt^2} \tag{2.7b}$$

$$\sum F_p = -N_1 \cos \omega_1 - N_2 \cos \omega_2 - E_p = m\frac{d^2 p}{dt^2} \tag{2.7c}$$

where $N = N_1 \sin \omega_1 + N_2 \sin \omega_2$, W is weight of wedge, E_v is dynamic vertical load, E_i is dynamic force in the direction of intersection line, and Ep is dynamic load perpendicular to intersection line. Other parameters are shown in Figure 2.7. Although the dynamic vectorial equilibrium equation is written in terms of its component, they correspond to a very general form for wedge sliding along the intersection line while being in contact with two basal planes. Furthermore, the earthquake force is decomposed to its corresponding components in the chosen coordinate system.

One can easily obtain the following identity from Equation (2.7c) by assuming that there are no motions upward and perpendicular to the intersection line:

$$N_1 + N_2 = \left[(W - E_v)\cos i_a - E_i \sin i_a\right]\lambda_i - E_p\lambda_p \tag{2.8}$$

where

$$\lambda_i = \frac{\cos \omega_1 + \cos \omega_2}{\sin(\omega_1 + \omega_2)}, \quad \lambda_p = \frac{\sin \omega_1 - \sin \omega_2}{\sin(\omega_1 + \omega_2)} \tag{2.9}$$

If the resistance is assumed to obey the Mohr-Coulomb criterion (Aydan and Ulusay, 2002; Aydan *et al.*, 2008) one may write the following:

$$T = (N_1 + N_1)\mu, \quad \mu = \tan \phi \tag{2.10a}$$

Following the initiation of sliding, the friction angle can be reduced to the kinetic friction angle as given here:

$$\mu = \tan \phi_r \tag{2.10b}$$

where ϕ_r are residual cohesion and friction angle. Under frictional condition, it should be noted that normal force ($N_1 + N_1$) cannot be negative (tensile). If such a situation arises, normal force ($N_1 + N_1$) should be set to 0 during computations. Let us introduce the following parameters:

$$\eta_v = \frac{E_v}{W} = \frac{a_v}{g}, \quad \eta_i = \frac{E_i}{W} = \frac{a_i}{g}, \quad \eta_p = \frac{E_p}{W} = \frac{a_p}{g} \tag{2.11}$$

where a_v, a_i, a_p are acceleration components resulting from dynamic loading.

The following dynamic equilibrium equation must be satisfied during the sliding motion of the wedge:

$$S = T \qquad (2.12)$$

If the relations given by equations are inserted in Equation (2.7a), one can easily obtain the following differential equation:

$$\ddot{s} = \frac{d^2 s}{dt^2} = g\left[(1-\eta_v)A + \eta_i B + \eta_p C\right] \qquad (2.13)$$

where

$$A = \left(\sin i_a - \cos i_a \mu \lambda_i\right), \; B = \left(\cos i_a + \sin i_a \mu \lambda_i\right); \; C = \mu \lambda_p$$

Since dynamic loads are very complex in the time domain, the solution of Equation (2.13) is possible only through numerical integration methods. The time-domain problems in mechanics are generally solved by finite difference techniques. For this purpose, there are different finite difference schemes. In this article, the solution of Equation (2.13) based on the linear acceleration finite difference technique (i.e. Aydan and Ulusay, 2002; Aydan et al., 2008). One can write the velocity ($\dot{s}$) and displacement of wedge (s) for a time step $n + 1$ as follows:

$$\dot{s}_{n+1} = \dot{s}_n + \frac{\ddot{s}_n}{2}\Delta t + \frac{\ddot{s}_{n+1}}{2}\Delta t \qquad (2.14)$$

$$s_{n+1} = s_n + \frac{\dot{s}_n}{1}\Delta t + \frac{\ddot{s}_n}{3}\Delta t^2 + \frac{\ddot{s}_{n+1}}{6}\Delta t^2 \qquad (2.15)$$

Provided that resulting dynamic shear force exceeds the shear resistance of the wedge at time ($t = t_i = i\Delta t$), one can easily incorporate the variation of shear strength of discontinuities from peak state ($\mu = \tan \phi_p$) to residual state ($\mu = \tan \phi_r$).

2.3   Complex shearing, sliding and buckling failure of an open-pit mine

2.3.1   A limiting equilibrium method

Ulusay et al. (1995) proposed a limiting equilibrium analysis method for a failure mechanism of an open-pit mine in Eastern Turkey as shown in Figure 2.8. The interslice forces were evaluated using the approach of Aydan et al. (1992). It is assumed that failure takes place through shearing of intact layers at the back of the slope and sliding along a bedding plane. For the pit-floor layer, there may be three possible modes: MODE 1: compressive failure, MODE 2: buckling failure, and MODE 3: combined compressive and buckling failure (Aydan et al., 1996b).

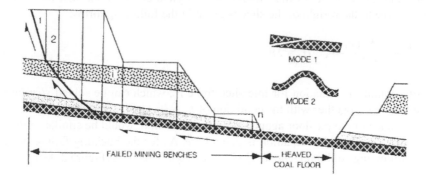

Figure 2.8 Failure mechanism

Source: Proposed by Ulusay *et al.* (1995)

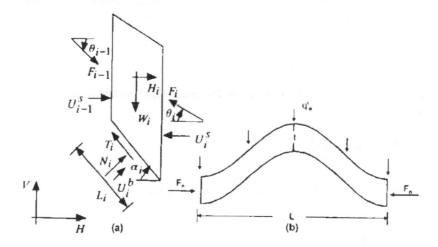

Figure 2.9 Mechanical models for stability assessment

For the sliding and shearing part, the force system acting on a typical block may be modeled as shown in Figure 2.9(a). Note that a lateral force is also assumed to act in order to consider the lateral stresses. The equilibrium equations for the chosen coordinate system can be written as:

s-direction

$$-T_i + W_i \sin\alpha_i + H_i \cos\alpha_i + F_{i-1} \cos(\alpha_i - \theta_{i-1}) - F_i \cos(\alpha_i - \theta_i)$$
$$+ (U_{i-1}^s - U_i^s)\cos\alpha_i = 0 \tag{2.16}$$

n-direction

$$N_i + U_i^b - W_i \cos\alpha_i + H_i \sin\alpha_i + F_{i-1} \sin(\alpha_i - \theta_{i-1}) - F_i \sin(\alpha_i - \theta_i)$$
$$+ (U_{i-1}^s - U_i^s)\sin\alpha_i = 0 \tag{2.17}$$

Assuming that the rock obeys the Mohr-Coulomb yield criterion and that the ratio of the horizontal force to the weight of the slice is given in the following forms:

$$T_i = \frac{c_i L_i + N_i \tan \varphi_i}{SF}, \quad H_i = \lambda W_i \tag{2.18}$$

One easily obtains an equation for interslice force F_i, which can be solved *step by step* to obtain the force F_n together with the condition of $F_0 = 0$.

The resistance of the pit-floor against compressive failure would be similar to the thrust-type faulting. Therefore, no equation is given here. As for the buckling failure of the coal seam, the following nonhomogeneous differential equation holds (Figure 2.9(b)):

$$\frac{d^2 u}{dx^2} + \frac{F_n}{EI} u = \frac{q_{o'}}{2EI} x(L - x) \tag{2.19}$$

where E is elastic modulus, I is second areal inertia moment, u is displacement, and $q_{o'}$ is effective distributed load. Solution of the preceding equation is:

$$u = A \cos kx + B \sin kx + \frac{q_{o'}}{4EIk^2}(Lx - x^2 + \frac{2}{k^2}); \quad k^2 = \frac{F_n}{EI} \tag{2.20}$$

If $u = 0$ and $du/dx = 0$ at the ends of the layer, the integration constants A and B are obtained as follows:

$$A = -\frac{q_{o'}}{F_n k^2}, \quad B = -\frac{q_{o'} L}{F_n k} \tag{2.21}$$

Assuming that $du/dx = 0$ at $x = L/2$, the critical buckling load is obtained as:

$$F_n = EI \left(\frac{8.99}{L}\right)^2 \tag{2.22}$$

Introducing $I = bt^3/12$ and $F_n = \sigma_o bt$, the critical axial stress for buckling is obtained as follows:

$$\sigma_o^{cr} = 6.735E \left(\frac{t}{L}\right)^2 \tag{2.23}$$

where t is layer thickness, and L is span. An application of the preceding approach is shown in Figure 2.10. Figure 2.10(a) was obtained from force $F_n = \sigma_o bt$ by considering the combined shearing and sliding failure for $SF = 1$ by varying lateral stress coefficient λ.

Figure 2.10(b) was obtained from buckling analysis by assuming that E/σ_C is 65 (continuous line) and 26 (broken line). Since it is more likely that the peak strength values hold, the slope may become unstable and the pit-floor fails in compression provided that the lateral stress coefficient is 0.13 for a uniaxial strength of 596 kPa. As for buckling failure, the lateral stress coefficient failure should be greater than 0.1 and less than 0.13 in view of the actual range of L/t at the time of failure. The uncovered span of the lignite seam was 113 m at the time failure. If the thin gyttja formation just above the lignite seam near the toe of the slope is neglected, the effective span is about 153 m. For a 5 m thick lignite seam, the value of L/t for compressive failure is found to be 21.95 from Figure 2.10(b). If the effect of the gyttja

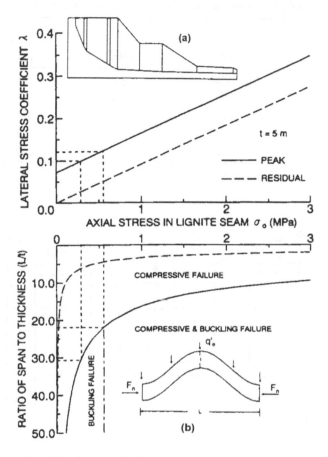

Figure 2.10 Computed stability chart for the lignite seam

formation on L/t ratio is taken into account, the L/t is 22.6. On the other hand, if its effect is neglected, L/t is 30.6. Considering these numbers, it has been contemplated that the lignite seam would likely be buckling rather than failing in compression. If the thickness of the seam involved in failure is greater than 5 m, the possibility of failure by buckling increases more rapidly as compared with that by compression.

2.3.2 Discrete finite element analyses

The discrete finite element method proposed by Aydan-Mamaghani (Aydan *et al.*, 1996a; Mamaghani *et al.*, 1999) was chosen to simulate the failure process of the slope. This method is based entirely on the finite element method and can simulate very large deformations of jointed media. Material properties used in the analyses are given in Table 2.2. Figure 2.11 shows the finite element mesh and boundary conditions.

First, a series of elastic analyses was carried out by varying lateral stress coefficient κ to see the magnitude of the axial stress of lignite seam at the location adjacent to the sliding benches. Figure 2.12 shows the relation between lateral stress coefficient κ and the axial stress in the lignite seam. As seen from the figure, the lateral stress coefficient must be greater than 0.58 and less than 0.78 to cause the buckling of the seam. Otherwise, the lignite

Table 2.2 Geomechanical parameters used in discrete finite element method (DFEM) analysis

Unit	λ	μ	c	ϕ
	(MPa)	(MPa)	(kPa)	(°)
Loam, marl, blue clay	408	5.4	60	25
Lignite	710	9.4	161	33
Base layer	670	38	161	33
Weak clay	3.2	1.1	23	6
Fracture plane	3.2	1.1	23	20

λ, μ are Lamé coefficients; c is cohesion; ϕ is internal friction angle.

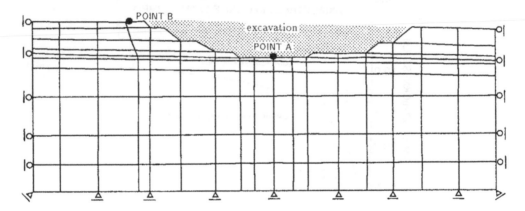

Figure 2.11 Finite element mesh used in analyses

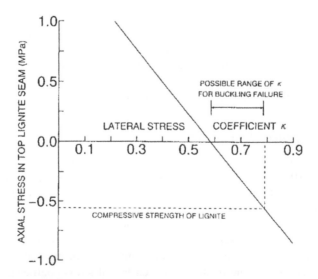

Figure 2.12 Computed stability chart for buckling failure

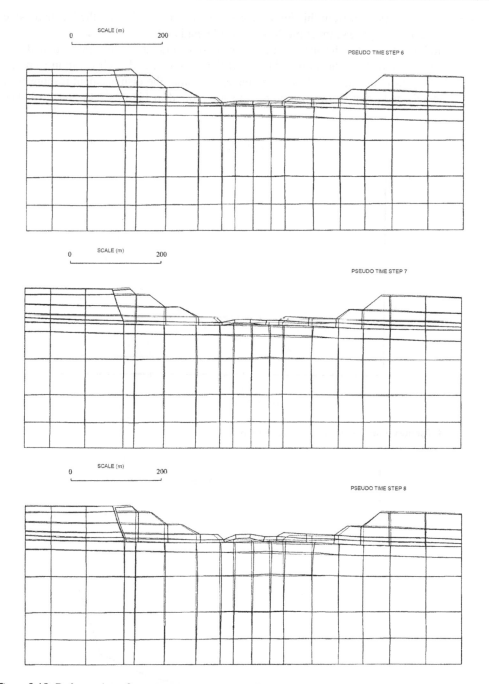

SCALE (m)
0   200

PSEUDO TIME STEP 6

SCALE (m)
0   200

PSEUDO TIME STEP 7

SCALE (m)
0   200

PSEUDO TIME STEP 8

Figure 2.13 Deformed configurations at various pseudo time steps

seam must fail in compression, which is contradictory to field evidence. The results further indicate that if the lateral stress coefficient is less than 0.58, the axial stress in the seam may be tensile. By setting the lateral stress coefficient κ as 0.7, an elasto-plastic analysis was carried out. Figure 2.13 shows the deformed configuration of the open-pit for each respective pseudo time step. As seen from this figure, the sliding of the benches on the left-hand side and buckling of the lignite seam at pit-floor are well simulated.

It should be noted that if the analysis becomes non-convergent in finite element analysis, this may be taken as the indication of failure of the structure, and each iteration step can be regarded as pseudo time step. With this concept in mind, the displacement responses of open-pit at selected point shown in Figure 2.13 are plotted in Figures 2.14 and 2.15. The

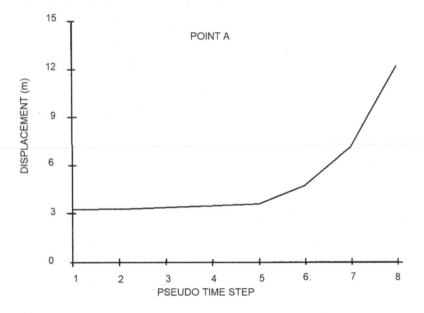

Figure 2.14 Pseudo time step vs displacement for point A

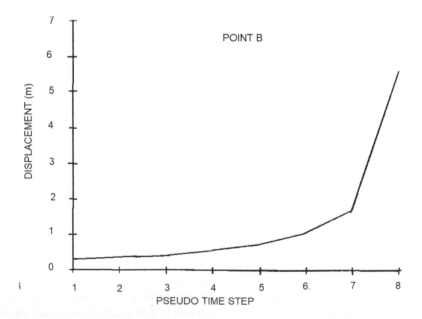

Figure 2.15 Pseudo time step vs displacement for point B

displacement response of point A, which is located at the center of the pit-floor, corresponds to the heaving of the floor. The heaving of the floor proceeds at a constant rate up to pseudo time step 6 and thereafter increases with an increasing rate, and the floor buckles.

The displacement response of point B, which is selected at the rear top of the sliding benches, corresponds to the horizontal displacement of the sliding body. This response is very similar to the measured response shown in Figure 2.15.

2.3.3   Estimation of postfailure deformation

The method used for estimating postfailure motions of the failed body is based on the earlier proposals by Aydan *et al.* (2006, 2008), Aydan and Ulusay (2002) and Tokashiki and Aydan (2010, 2011). Let us consider a landslide body consisting of N number of blocks sliding on a slip surface as shown in Figure 2.16. If interslice forces are assumed to be nil as assumed in the simple sliding (Fellenius-type) model, one may write the following equation of motion for the sliding body:

$$\sum_{i=1}^{n}(S_i - T_i) = \bar{m}\frac{d^2s}{dt^2} \tag{2.24}$$

where $\bar{m}, s, t, n, S_i$ and T_i are total mass, travel distance, time, number of slices, shear force and shear resistance, respectively. Shear force and shear resistance may be given in the following forms, together with Bingham-type yield criterion:

$$S_i = W_i(1+\frac{a_H}{g})\sin\alpha_i, \ T_i = c_iA_i + (N_i - U_i)\tan\phi_i + \eta W_i\left(\frac{ds_i}{dt}\right)^b \tag{2.25}$$

where W_i, A_i, N_i, U_i, α_i, α_v, α_H, c_i, ϕ_i, η and b are weight, basal area, normal force, uplift pore water force, basal inclination, vertical and horizontal earthquake acceleration, cohesion, friction angle of slice i, Bingham-type viscosity and empirical coefficient, respectively. If normal force and pore water uplift force related to the weight of each block as given here:

$$N_i = W_i(1+\frac{a_V}{g})\cos\alpha_i, \ U_i = r_uW_i \tag{2.26}$$

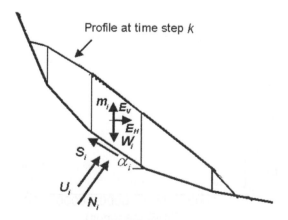

Figure 2.16 Mechanical model for estimating postfailure motions

One can easily derive the following equation with the use of Equations 2.25–2.26:

$$\frac{d^2 s}{dt^2} + \eta \left(\frac{ds}{dt}\right)^b - B(t) = 0 \tag{2.27}$$

where

$$B(t) = \frac{g}{\bar{m}}\left(\sum_{i=1}^{n} m_i (\sin \alpha_i (1 + \frac{a_H}{g}) - (\cos \alpha_i (1 + \frac{a_V}{g}) - r_u) \tan \phi_i) - \frac{c_i A_i}{g}\right) \tag{2.28}$$

In the derivation of Equation (2.28), the viscous resistance of the shear plane of each block is related to the overall viscous resistance in the following form:

$$\eta \bar{m} g \left(\frac{ds}{dt}\right)^b = \sum_{i=1}^{n} \eta m_i g \left(\frac{ds_i}{dt}\right)^b \tag{2.29}$$

Equation (2.29) can be solved for the following initial conditions together with the definition of the geometry of basal slip plane.
At time $t = t_0$:

$$s = s_0 \text{ and } v = v_0 \tag{2.30}$$

There may be different forms of constitutive laws for the slip surface (i.e. Aydan et al., 2006, 2008; Aydan and Ulusay, 2002). The simplest model is elastic-brittle plastic to implement. If this model is adopted, the cohesion will exist at the start of motion, and it will disappear thereafter. Therefore, cohesion component introduced in Equation (2.29) may be taken as nil as soon as the motion starts. Thereafter, the shear resistance will consist of mainly frictional component together with some viscous resistance.

The method just explained was applied to the Kitauebaru landslide in Okinawa (Japan) involving bedding plane and fault plane. This method utilizes the Bingham-type visco-plastic yield criterion. Although the assumed geometry of the open-pit mine is slightly different from the actual one, it was applied to the failure in the Kışlaköy open-pit mine (Ulusay et al., 2019). Figure 2.17 shows the displacement response during failure. The actual displacement

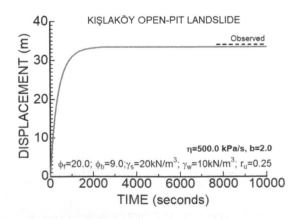

Figure 2.17 Displacement response of the mass center of the failed body

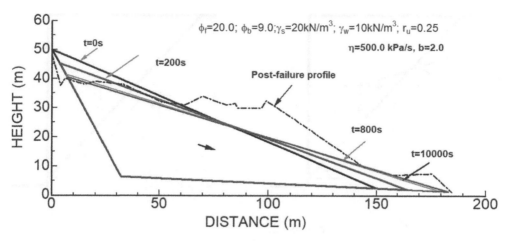

Figure 2.18 Deformed configurations of the failed body

of the failed body was about 34 m. The estimated displacement of the failed body is about 33 m. The material properties are shown in Figures 2.17 and 2.18, which are based on those given in Table 2.2. Figure 2.18 shows the deformation configuration of the failed body in space with time. Despite some difference between the assumed and actual geometries of the failed body, the estimations are very close to the actual ones.

2.4 Dynamic response of reinforced rock slopes against planar sliding

As the deformation of rock slope occurs mainly due to slippage along the failure surface, a dynamic limiting equilibrium method developed originally by Aydan and Ulusay (2002) and elaborated by Aydan *et al.* (2008) and Aydan and Kumsar (2010) was used to simulate the slip of the model slope on the failure surface (Aydan *et al.*, 2018). Figure 2.19 shows a view of the mechanical model for the dry condition with the consideration of the experimental fact; that is, the unstable part moves like a monolithic body irrespective of layered or single body. One can easily write the following limiting equilibrium equations for *s*- and *n*-directions, respectively, as follows:

$$W \sin \alpha + E \cos \alpha - T \cos (\alpha \pm \beta) - S = m \frac{d^2 s}{dt^2} \qquad (2.31\text{a})$$

$$W \cos \alpha + E \sin \alpha - T \sin (\alpha \pm \beta) - N = m \frac{d^2 n}{dt^2} \qquad (2.31\text{b})$$

where *W, E, T, S, N* and *m* are weight, seismic load, force provided by rock bolts/rock anchors, shear and normal forces and mass of the unstable body, respectively. α and β are the inclination of failure plane and rock bolts/rock anchors from the horizontal. *s* and *n* are the amount of shear and normal displacement of the unstable body. (−) is signed used when the acute angle between the bolt force and horizontal is positive, which is denoted as Case 1. (+) sign is used the acute angle between the bolt force and horizontal is negative, which is denoted as Case 2.

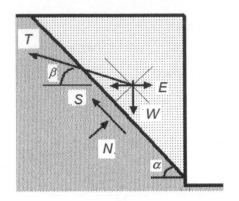
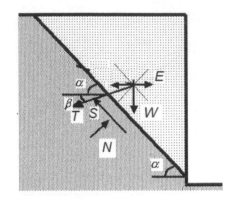

(a) Case 1: $\alpha - \beta$ (b) Case 2: $\alpha + \beta$

Figure 2.19 Mechanical model for reinforced rock slope

Let us assume that the inertia force for n-direction during sliding is negligible and the resistance of the failure plane is purely frictional as given here:

$$\left| \frac{S}{N} \right| = \tan(\phi) \tag{2.32}$$

One can easily obtain the following equation for the rigid body motion of the sliding rock burst body:

$$m \frac{d^2 s}{dt^2} = A_w + A_E - A_T \tag{2.33}$$

where

$$A_w = W\left(\sin\alpha - \cos\alpha \tan\phi\right)$$
$$A_E = E\left(\cos\alpha + \sin\alpha \tan\phi\right)$$
$$A_T = T\left(\cos\left(\alpha \pm \beta\right) + \sin\left(\alpha \pm \beta\right)\tan\phi\right)$$

As the earthquake or shaking-induced force E will be proportional to the mass of the sliding body, it can be related to ground shaking in the following form:

$$E = \frac{a_g(t)}{g} W \tag{2.34}$$

where a_g is base acceleration, and g is gravitational acceleration.

The reinforcement effect (T) of rock bolts is due to resolved components and the dowel effect. The bar/tendon may be pre-tensioned at the time of installation. Therefore, the total acting force (T_t) may be contemplated as the sum of pretension force (T_p) and deformation-induced force (T_d) as given here:

$$T_t = T_p + T_d \tag{2.35}$$

The deformation-induced axial force in a rock bolts during the shearing process may be given in the following form (Aydan, 2018):

$$T_d = B\delta^b \tag{2.36}$$

where b and B are empirical constants, and δ is the displacement of rock bolt. It is experimentally well-known that the rock bolts crossing a discontinuity plane are bent during shearing and that there is an effective length of rock bolts mobilized during the shearing process. Thus, the extension of rock bolt would be given by:

$$\delta = \ell - \ell_o = \sqrt{\left(\ell_o + \delta_h\right)^2 + \delta_v^2} - \ell_o \tag{2.37}$$

where ℓ_o, δ_h and δ_v are the effective length of rock bolt mobilized at the failure plane and horizontal and vertical movement of the sliding unstable body. The force in the rock bolt can then be obtained by inserting the extension value from Equation (2.37) into Equation (2.36).

The small amount of deformation may cause the yielding of rock bolts/rock anchors. In such cases, their axial force may be assumed to be equivalent to their yielding value. If there is a hardening type of response, then the following type constitutive may be adopted:

$$T_t = T_Y + H\delta \tag{2.38}$$

where T_Y and H are yielding strength and hardening modulus, respectively.

The mathematical model described in this section can be used for both unreinforced and reinforced rock slopes. For the unreinforced case, the resistance provided by rock bolts/rock anchors is neglected.

An application of the dynamic limiting equilibrium method is shown in Figure 2.19. Three different values are used for the kinetic (residual) friction angle. In this particular simulation, if the slip stops, the peak friction angle is used for the initiation of slip in the next cycle. Although it is not reported here, the slip becomes much larger if the friction angle is assumed to be equal to residual value once the slip is initiated. The values used in computations are also shown in Figure 2.20. The initiation of the slip occurred slightly at a higher friction angle than that determined from tilting tests.

Figure 2.21 compares the slip responses for three different values of residual friction angle. When the residual friction angle is equal to the peak friction angle, which corresponds to perfectly plastic behavior, the amount of slip is quite smaller than those for a lower residual friction angle. When the residual friction angle is 0.625 times the peak friction angle, it corresponds to the kinetic friction angle determined from tilting tests. For this particular situation, the slip is largest. When the residual friction angle is 0.725 times the peak friction angle, the slip at the first stage is equal to that measured in the experiment as seen in Figure 2.21 It is very likely that if the peak friction angle is reduced as a function of the slip cycles, it is quite possible to get better estimations of the measured responses. It

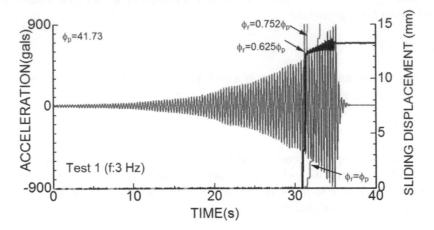

Figure 2.20 Comparison of measured slip response with the estimated responses for different values of residual friction angle

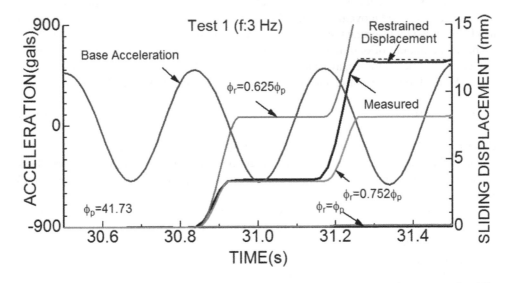

Figure 2.21 Expanded comparison of measured slip response with the estimated responses for different values of residual friction angle, shown in Figure 2.20

should be noted that the slip of the potentially unstable block is restrained to 12.3 mm, and the displacements exceeding this value are not considered.

The theoretical approach is applied to model tests shown in Figure 2.19 by selecting that the friction angle is 39 degrees. The computed results are shown in Figure 2.22. The reinforcement effect (T_B) of rock bolts are normalized by the weight (W_B) of the unstable body. As noted from the figure, the computed results are quite similar to experimental results both quantitatively and qualitatively. However, the computations indicate that the yielding should started a bit later than the measured results. The discrepancy may result from the complexity

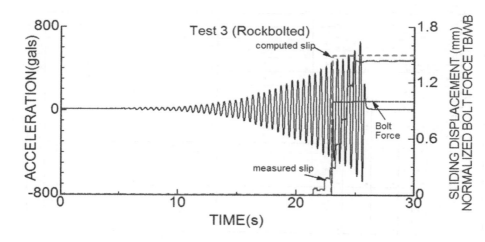

Figure 2.22 Comparison of computed responses with measured responses

of actual frictional behavior of the sliding surface. Nevertheless, the theoretical model is capable of modeling the dynamic response of the support system.

2.5   Bridge foundations

2.5.1   Back-analysis of the constitutive law parameters of the foundation rock

The application involves a back-analysis of the constitutive law parameters of the foundation rock of the pier 3P of the Akashi suspension bridge in western Japan (Aydan, 2016, 2018). The rock consists of Kobe tuff, which is a relatively soft rock. The diameter of the foundation was 40 m and its height was 80. Following the lowering of the caisson foundation to the sea bottom, it was filled with concrete, which increased the load on the foundation. First the filling of the inner ring was completed, and then the outer ring was filled with some time lag. The deformation of the ground was measured during the filling stages. It was required to obtain the time-dependent characteristics of foundation formation by considering the loading associated with the construction procedure. The constitutive law of the foundation rock was assumed to be of Kelvin type. The problem was considered to be an axisymmetric problem; the finite element mesh used in the back analyses is shown in Figure 2.23. The elastic modulus and viscosity coefficient of Kelvin model of foundation rock were 833 MPa and 3.3 GPa · day , respectively (Figures 2.24 and 2.25). Figure 2.26 compares the computed response with the measured response for the loading condition shown in the same figure. Figure 2.27 shows the displacement of the pier for about 4 years. The expected creep displacement is about 108 mm.

2.5.2   Settlement and stress state and circular rigid foundation

The example is concerned with settlement and stress state beneath foundations subjected to surcharge loads through a relatively rigid foundation with a diameter of 3 m, and it is modeled as an axisymmetric problem. Figure 2.28 shows the computed settlement and pressure

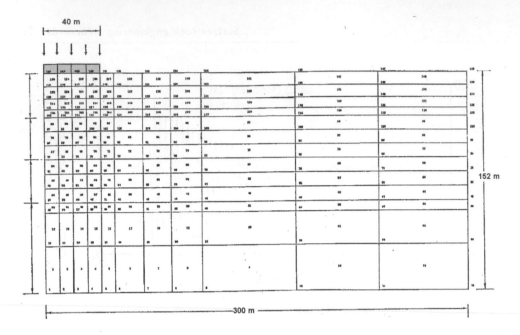

Figure 2.23 Finite element mesh used in the back-analyses

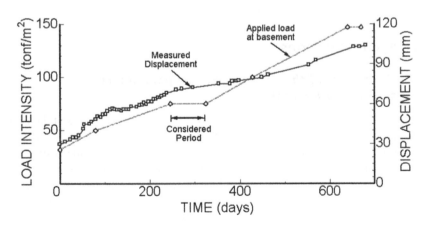

Figure 2.24 Time response of applied load and measured displacement

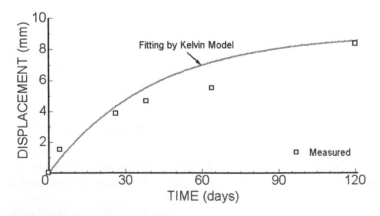

Figure 2.25 Back-analysis of measured displacement by Kelvin model

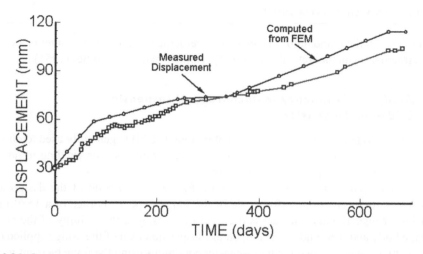

Figure 2.26 Comparison of measured and computed displacement

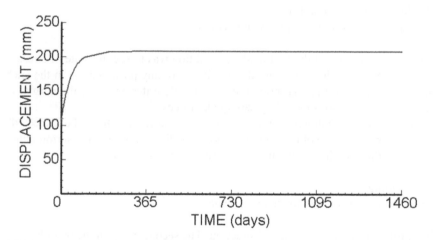

Figure 2.27 Estimated creep displacement of the pier

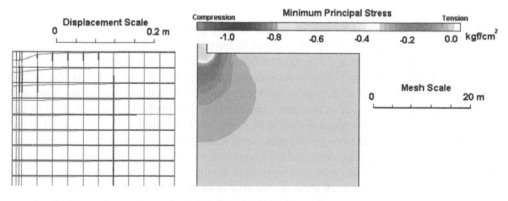

(a) Deformed configuration   (b) Minimum principal stress contours

Figure 2.28 Settlement and pressure contours beneath a circular foundation

contours beneath the foundation. The estimated settlement and pressure contours are generally in agreement with theoretical solutions by Timoshenko and Goodier (1951).

2.5.3   Analysis of tunnel-type anchorage of suspension bridge under tension

Numerical methods such as the finite element method (FEM) is generally used to check the local straining, stresses and local safety factors in linear analyses or yielding in nonlinear analyses. A specific example is described here.

The analyses reported in this subsection are for the examination of the design and to predict the post-construction performance of the tunnel-type anchorage of a 1570 m long suspension bridge. The analyses are carried out to investigate the stability of the concrete anchorage body and surrounding rock upon the suspension load of the bridge applied on the anchorage by considering two loading intensity conditions using the finite element method:

1   The design loading condition
2   Loading three times the design loading condition

The purpose of considering the first loading condition was to see the state of stress in concrete and in the surrounding rock and the possibility of any plastification in the anchorage and the rock for a given material properties and a yield criterion under the normal loading (i.e. design load) conditions including earthquake forces.

The second loading condition, on the other hand, was to see the effect of unexpectedly high loads upon the performance of the anchorage body and the response of the surrounding rock.

For each loading condition, two possible cases are investigated:

1   Full bonding
2   No tension slit just under the loading plane

The first case corresponds to an actual situation. The second case, on the other hand, visualizes a tensile crack just under the loading plane in a concrete body. By this, it is intended to see the effect of such a tensile crack on the stability of the concrete anchorage body and surrounding rock.

The rock at the site under consideration is mainly granite. However, there is an almost vertical dioritic volcanic intrusion that seems to have disturbed the surrounding rock, and it is highly weathered. The main portion of the anchorage is situated in granitic rock. The granitic rock is classified as *CH* class rock in the classification of DENKEN (Ikeda, 1970). The remaining part of the anchorage is located in the dioritic volcanic intrusion (dyke) and is classified as *DH* class rock in the same classification. The geological cross section of the rock mass along the anchorage is shown in Figure 2.29. The material properties used in the analyses were determined from *in-situ* shear tests and plate-bearing tests in other near construction sites with similar geology and are given in Table 2.3. The material properties listed in Table 2.3 is highly conservative and represents the lowest values of the respective tests.

Poisson's ratio for every rock class is assumed to be 0.2. The elastic modulus, cohesion and tensile strength is 0.6 times the undisturbed rock mass

The anchorage is modeled as an axisymmetric body considering the geological formations shown in Figure 2.29. Besides the modeling of geologic formations mentioned in the

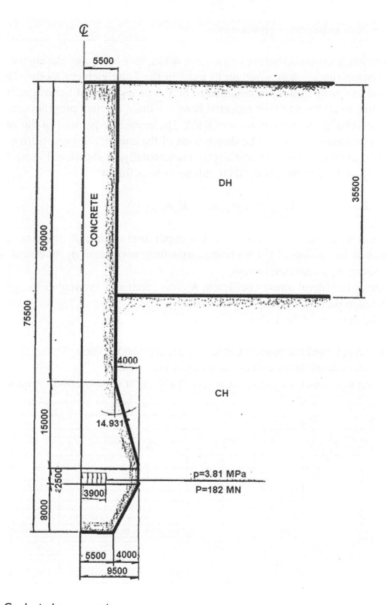

Figure 2.29 Geological cross section

Table 2.3 Material properties

Rock Class	Esb MPa	Es MPa	Ed MPa	Cohesion MPa	φ (°)	σc MPa	σt MPa	γ (kN m⁻³)
Soil-like	80	75	150	0	35	0	0	19
DH	80–150	180	350	0.1	37	0.4	0.04	20.5
				0.07	32	0.25	0.02	19.0
CL	150–300	330	560	0.5	40	2.14	0.21	22.0
				0.4	40	1.54	0.15	20.0
CM	300–600	600	1200	0.8	40	3.43	0.34	23.5
				0.7	40	3.00	0.30	21.0
CH	600–1200	1280	2400	1.2	45	5.79	0.58	24.5
				1.0	45	4.83	0.48	22.0
CONCRETE		25000		4.97	45	24	15	24

previous section, a loosening zone of 2 m wide, which may be caused during the excavation of the anchorage tunnel, was assumed to exist in the finite element model. The material properties of loosening zones are assumed to be 0.6 times those of the respective geological formations. In all the analyses reported herein, a finite element program considering an elasto-plastic behavior of rock mass was used. The element type used in the analyses is a 4-noded isoparametric element. The dimensions of the analysis domain were taken as two times the total anchorage length vertically and horizontally, as shown in Figure 2.30.

The gravity in the analyses, was taken into account as follows:

$$\gamma^* = \gamma \sin \alpha, \, \sigma_z = \gamma^* h, \, \sigma_r = K_o \sigma_z, \, \sigma_\theta = K_o \sigma_z \tag{2.39}$$

where γ is unit weight or rock or concrete, h is depth form surface, K_o is lateral initial-stress coefficient, α is inclination of the anchorage axis from horizontal, σ_z is vertical stress, σ_r is radial stress, and σ_θ is tangential stress.

The value of the lateral stress coefficient K_o was taken as 1 in view of the ground-stress measurements in the near vicinity of the anchorage site.

The results in terms of the following items:

1   Maximum principal compressive stress in concrete and in rock
2   Maximum principal tensile stress in concrete and in rock
3   Minimum safety factor against shear failure (SFS) for concrete and for rock

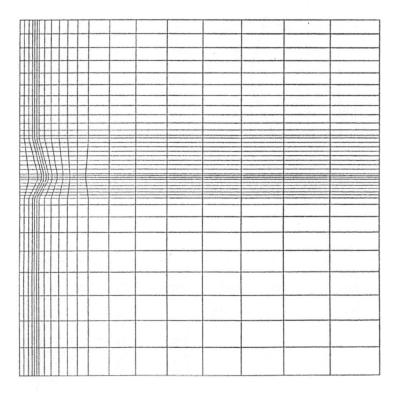

Figure 2.30 Finite element mesh

Minimum safety factor against shearing (SFS) is defined as:

$$SFS = \frac{c\cos\phi + \dfrac{\sigma_1 + \sigma_3}{2}\sin\phi}{\dfrac{\sigma_1 - \sigma_3}{2}} \tag{2.40}$$

Note that the sign of the compressive stress is taken as negative (−).

4   Minimum factor of safety against tensile failure (SFT) for concrete and for rock,

which is defined as follows:

$$SFT = \frac{\sigma_t}{\sigma_1 \; or \; \sigma_3} \tag{2.41}$$

First the calculated results are presented for the ordinary design load condition for fully bonded and with no-tension slit cases and are compared and summarized. Figure 2.31 shows

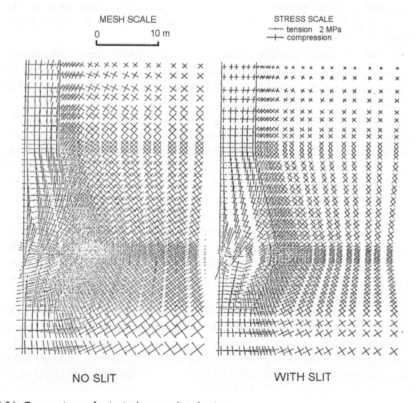

Figure 2.31 Comparison of principal stress distributions

Table 2.4 Comparison of maximum compressive and tensile stresses

Material Type	Fully Bonded		No-tension Slit	
	Max. Comp. Stress MPa	Max. Ten. Stress MPa	Max. Comp. Stress MPa	Max. Ten. Stress MPa
Concrete	3.59	1.24	6.46	0.35
D_H	0.44	–	0.44	–
C_H	2.34	–	2.34	–
D_H^*	0.44	0.004	0.44	0.004
C_H^*	1.217	–	1.221	–

principal stress distributions. As noted from the stress distributions, the magnitude of compressive principal stresses are higher in the fully bonded case than those in the no-tension slit case as the existence of the tension slit tends to create higher tensile stresses, which result in the reduction in the magnitude of initial *in-situ* compressive stresses. As expected, the no-tension slit relieves the concrete block just below the loading plate, which manifests itself as the principal stress directions become more vertical as compared with the fully bonded case. The maximum compressive stress and maximum tensile stress are summarized in Table 2.4. When this table is carefully examined, except concrete, there seems to be no difference between two cases. However, as noted from the overall principal stress distribution, there is a remarkable change in the overall distributions.

On the other hand, the comparison of the safety factors may be more relevant as they will reflect the local changes more clearly. Figures 2.32 and 2.33 show the safety factor distributions against shearing and tensile failures, respectively. The minimum safety factors are listed in Table 2.5.

The existence of no-tension slit has a very remarkable effect on the safety factor of concrete against tensile failure as it results in lower tensile stresses in concrete. On the contrary, the safety factors for rock formations tend to decrease in magnitude as expected. The safety factor in rock formation CH is about 5. The Gauss point at which these values are observed is very near the surface and is next to the anchorage body, as the given value for this rock formation corresponds to a very blocky rock mass with a very low value of tensile strength. In conclusion, it may be stated that the stress state at this point is not a representative for the overall anchorage body, and it could have no effect on the stability of the anchorage body.

Next, we summarize and discuss the calculated results for the case of three times the design loading condition. The principal stress distributions are shown in Figure 2.34. Tables 2.6 and 2.7 compare the maximum compressive and tensile stresses and minimum safety factors in concrete and rock formations.

Although the values of safety factors and maximum compressive and tensile stresses differ from the design load cases, the overall behavior and conclusions are same. The minimum safety factor distributions in concrete and rock are shown in Figures 2.35 and 2.36.

When the minimum compressive stresses given in Table 2.8 for two loading conditions are compared, it seems that the increase in the intensity of load has almost no effect on the increment of compressive stress in rock. As pointed out in previous sections, the superimposed initial state of stress causes such an impression. When the superimposed *in-situ* stresses are

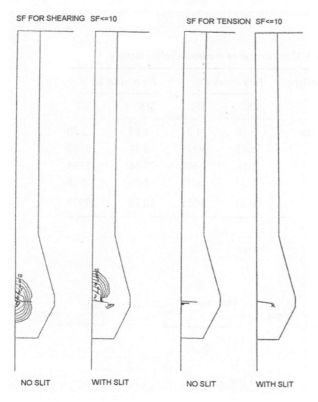

Figure 2.32 Comparison of safety factor distributions (concrete)

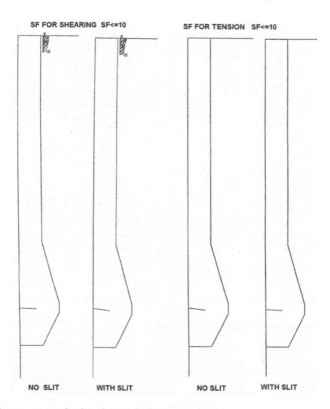

Figure 2.33 Comparison of safety factor distributions (rock)

Table 2.5 Comparison of minimum safety factors

Material Type	Fully Bonded		No-tension Slit	
	SFS	SFT	SFS	SFT
Concrete	2.96	1.03	2.69	2.70
D_H	15.43	9999	15.21	9999
C_H	21.12	9999	20.66	9999
D_H^*	5.34	5.51	5.25	5.18
C_H^*	10.29	9999	10.26	9999

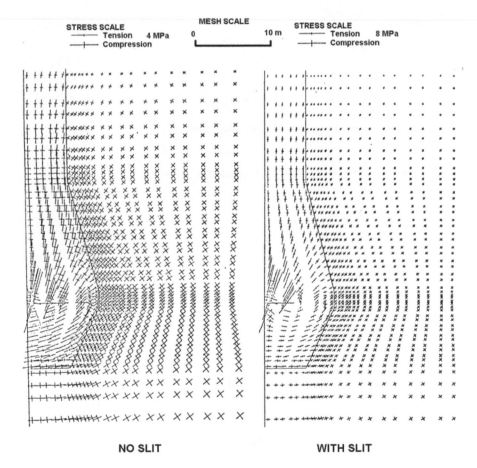

Figure 2.34 Comparison of principal stress distributions

Table 2.6 Comparison of maximum compressive and tensile stress

Material Type	Fully Bonded		No-tension Slit	
	Max. Comp. Stress MPa	Max. Ten. Stress MPa	Max. Comp. Stress MPa	Max. Ten. Stress MPa
Concrete	8.62	5.85	17.24	3.21
D_H	0.44	0.015	0.019	–
C_H	2.34		2.34	–
D_H^*	0.44	0.0296	0.44	0.031
C_H^*	1.262	–	1.275	–

Table 2.7 Comparison of minimum safety factors

Material Type	Fully Bonded		No-tension Slit	
	SFS	SFT	SFS	SFT
Concrete	0.375	0.23	1.11	0.40
D_H	4.72	2.17	4.65	2.12
C_H	6.76	9999	6.60	9999
D_H^*	1.34	0.54	1.31	0.53
C_H^*	3.13	9999	3.12	9999

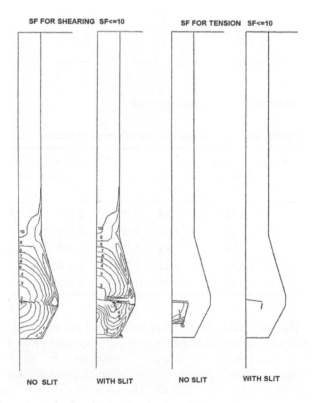

Figure 2.35 Comparison of safety factor distributions (concrete)

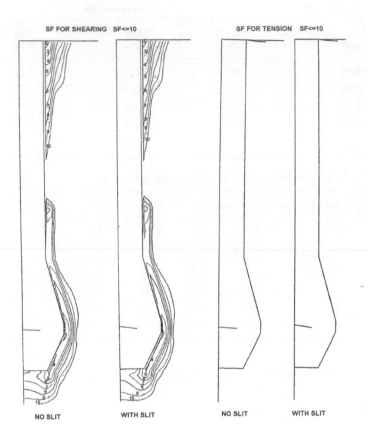

SF FOR SHEARING   SF<=10   SF FOR TENSION   SF<=10

NO SLIT   WITH SLIT   NO SLIT   WITH SLIT

Figure 2.36 Comparison of safety factor distributions (rock)

Table 2.8 Comparison of stress increment in rock for two loading conditions

Element No	Average Int. Stress	With Int. Stress		$\sigma_1^{3T} / \sigma_1^{IT}$	Without Int. Stress		$\sigma_1^{3T} / \sigma_1^{IT}$
		IT	3T		IT	3T	
324	0.966	1.16	1.38	1.190	0.194	0.414	2.13
344	0.940	1.13	1.34	1.185	0.190	0.400	2.11
324	0.920	1.16	1.45	1.250	0.240	0.530	2.21
324	0.890	1.10	1.33	1.210	0.210	0.440	2.10
324	0.860	1.07	1.28	1.190	0.210	0.420	2.00

excluded from the calculated results, the effect of such a load increase will be apparent. To show this, a calculation was carried out using the material properties given in Table 2.5.

The FEM analyses indicated that the anchorage of a suspension bridge based on the pure gravitational and frictional concept is highly safe for the given geometry and material properties, which are determined very conservatively. The anchorage is even safe for the loads applied three times the design load. When the rock has a few discontinuities and a relatively high tensile strength, the occurrence of tensile crack in the concrete just under the loading plane does not cause any serious problem. On the other hand, when the rock has a number of discontinuities and low tensile strength, the concrete must be reinforced to behave as a monolithic body in order to reduce high tensile stresses in the rock, which may cause the failure of surrounding rock in the shape of a cone rather than as a cylindrical failure form.

2.6   Masonry structures

2.6.1   Shuri Castle arch gate

A series of discrete finite element method (DFEM) analyses were performed on the masonry arch gates of Shuri Castle in Okinawa island, Japan. Figures 2.37 and 2.38 show a computational example on the arch gate under a horizontal seismic load and gravity. As noted, the failure occurs due to the sliding of the sidewall and subsequent rotation and fall of the arch blocks.

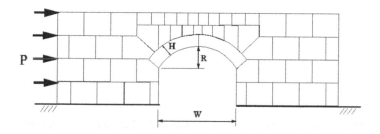

Figure 2.37 Mesh used in the DFEM analysis

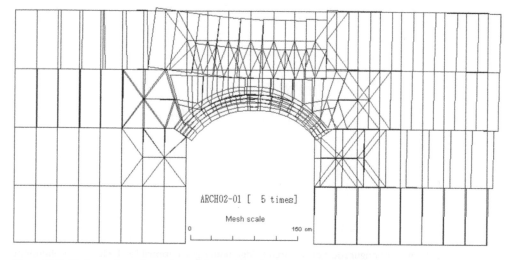

Figure 2.38 Deformed configuration of the arch under horizontal seismic load

2.6.2   ledonchi masonry arch bridge

Some numerical analyses by the discrete finite element method have been performed to investigate the effect of horizontal shaking using the seismic coefficient method. Figure 2.39 shows an example of the computation for a horizontal seismic coefficient of 0.2.

Figure 2.40 shows the deformed configuration of the masonry arch bridge under gravitational load and the removal of one of the blocks on the left abutment. This computation was carried out to investigate the effect of configurations of the abutment of the bridge. As noted from the computed results, the removal of the rock block reduces the shear resistance of the abutment, the bridge starts to exhibit large deformation and it fails.

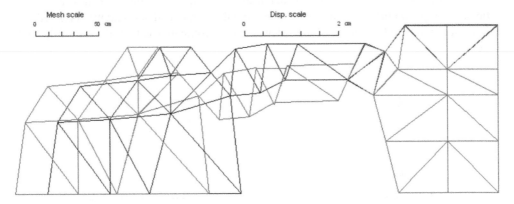

Figure 2.39 Deformed configuration of the arch bridge under horizontal seismic load ($k = 0.2$; $\phi = 10$ degree)

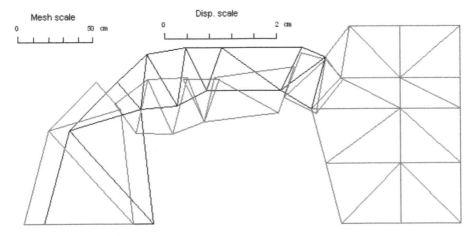

Figure 2.40 Deformed configuration of the arch bridge under gravitational load with lesser abutment block ($\phi = 10$ degree)

2.7   Reinforcement of dam foundations

The failure of the foundation of the Malpasset Arch Dam in France in 1959 is one of the most catastrophic events in rock engineering. This event had a profound effect on the design of dams on rock foundation with a strong emphasis on the effect of major structural weaknesses existing in the rock mass under high water pressures. Figure 2.41 shows some illustrations of possible modes of failure of gravity dams. The analyses of the foundations of dams can be done through some limiting equilibrium approaches for simple conditions. However, the use of numerical analyses with the consideration of discontinuities in rock mass would be necessary for complex conditions.

2.7.1   Limit equilibrium method for foundation design

A dam subjected to base-shearing and reinforced by rock anchors shown in Figure 2.42 is considered. The safety factors of the dam against base shearing and overturning about point O may be easily obtained, respectively, as given here:

Safety factor against base shearing

$$SF_s = \frac{cL_b + N \tan \varphi}{U_s - T_a \cos \alpha} \tag{2.42}$$

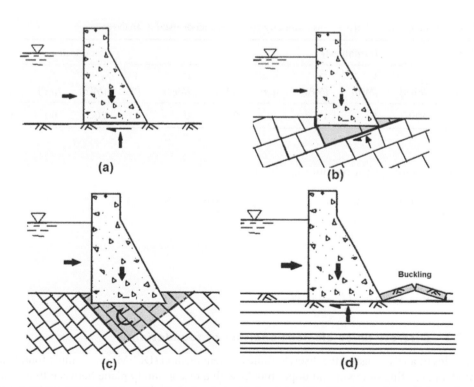

Figure 2.41 Possible failure modes of gravity dams: (a) base shearing, (b) planar sliding along a thoroughgoing discontinuity plane, (c) flexural or block toppling failure, (d) buckling failure

Source: Modified and redrawn from U.S. Army Corps of Engineers (1994)

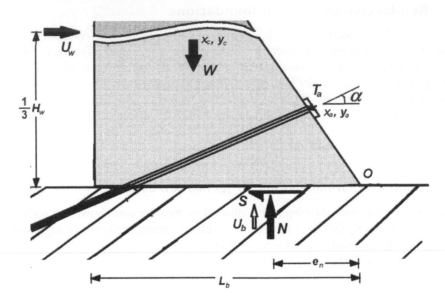

Figure 2.42 Possible force conditions acting on gravity dam foundations

Table 2.9 Properties of blocks, discontinuity and rock anchor used in analyses

Blocks		Discontinuity						Rock Anchor		
λ (MPa)	μ (MPa)	E_n (MPa)	G_s (MPa)	h mm	σ_t (MPa)	C (MPa)	ϕ (°)	σ_t (MPa)	A_b (m²)	G_g (GPa)
100	100	5	1	10	0	0	25	200	0.1	3

Safety factor against overturning about point O

$$SF_m = \frac{W(L_b - x_c) + T_a(y_a \cos\alpha + (L_b - x_a)\sin\alpha)}{\dfrac{H_w}{3}U_s + \dfrac{2}{3}L_b U_b + Ne_n} \tag{2.43}$$

where

$$N = W + T_a \sin\alpha - U_b$$

2.7.2   Discrete finite element analyses of foundations

A series of analysis using the discrete finite element method (DFEM) on a simple foundation model consisting of lower and upper blocks with a discontinuity plane between two blocks was carried out by changing the inclination of the rock anchor with respect to the normal of discontinuity plane and its yielding strength. Material properties used in the analyses are given in Table 2.9. The upper block was assumed to be subjected to a uniform 10 MN compressive normal load, while a 7 MN uniformly distributed shear load was applied over the discontinuity plane.

Five different cases were analyzed (Figures 2.43–2.47)

Case 1: No rock anchor (Figure 2.43)
Case 2: Rock anchor inclined at an angle of 45 degrees with respect to the normal of discontinuity (Figure 2.44)
Case 3: Rock anchor inclined at an angle of +45 degrees with respect to the normal of discontinuity (Figure 2.45)
Case 4: Rock anchor inclined at an angle of 0 degree with respect to the normal of discontinuity and rock anchor behaves elastically (Figure 2.46)
Case 5: Rock anchor inclined at an angle of 0 degree with respect to the normal of discontinuity and rock anchor yields (Figure 2.47). Tensile strength of rock anchor is reduced to 100 MPa.

When the friction angle of the discontinuity plane was set to 40 degrees, the behavior was elastic and no relative slip occurred. The limiting friction angle for elastic behavior is 35 degrees, and if the friction angle of the discontinuity plane is less than 35 degrees, the relative slip is likely.

The friction angle of the discontinuity plane was reduced from 40 degrees to 25 degrees. The computation results for this case (Case 1) is shown in Figure 2.43. The nonlinear analysis was based on the secant method. As noted from Figure 2.43, the relative slip of the upper block accelerates after each computation step. The installation of a rock anchor at different inclinations with respect to the normal of the discontinuity plane restricts the relative slip of the upper block. Particularly, the effect of the rock anchor is largest among all cases when its inclination is +45 degrees. When the relative slip of the discontinuity plane is considered, the largest relative slip occurs when the inclination is −45 degrees. On the other hand, the relative slip is smallest when the inclination of the rock anchor is +45 degrees.

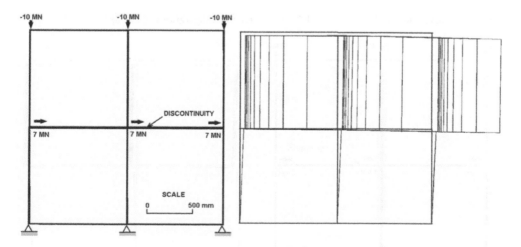

Figure 2.43 Illustration of the DFEM model and deformed configurations (Case 1)

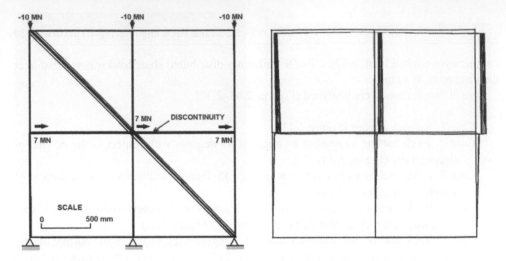

Figure 2.44 Illustration of the DFEM model and deformed configurations (Case 2)

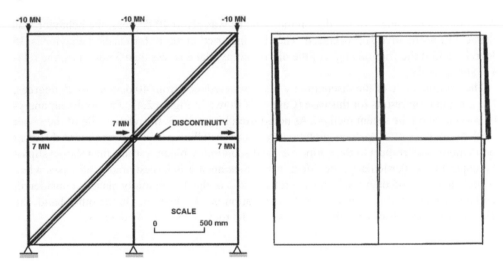

Figure 2.45 Illustration of the DFEM model and deformed configurations (Case 3)

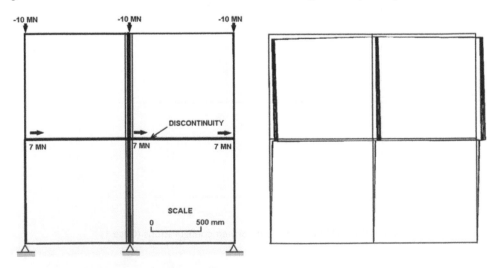

Figure 2.46 Illustration of the DFEM model and deformed configurations (Case 4)

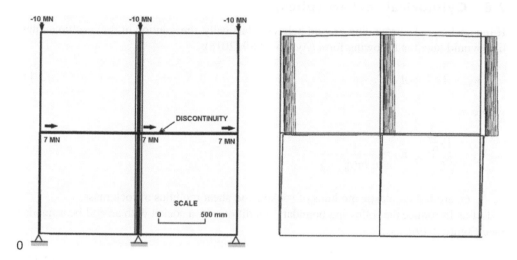

Figure 2.47 Illustration of the DFEM model and deformed configurations (Case 5)

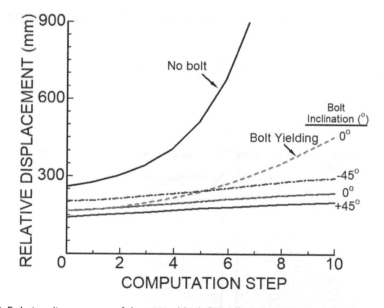

Figure 2.48 Relative slip response of the upper block for different inclinations of rock anchor

When the inclination of the rock anchor is 0 degree, the relative slip is just between those for Case 2 and Case 3. However, if the rock anchor yields, the relative slip gradually increase after its yielding. The computational results are compared in Figure 2.48 as a function of computation step. It should be noted that the analysis is a pseudo dynamic type, and if the behavior of the analyzed domain remains elastic, no further deformation takes place, and it remains constant with respect to the increase of computation step number. This series of analyses clearly illustrates that rock anchors/rock bolts can be quite effective in reinforcing the foundations.

2.8 Cylindrical sockets (piles)

If the socket and surrounding ground behaves elastically, the solution of the resulting equation would take the following form (Aydan, 1989, 2018):

$$w_b = A_1 e^{\alpha z} + A_2 e^{-\alpha z} \tag{2.44}$$

where

$$\alpha = \sqrt{\frac{2K_r}{E_b r_b}}., \quad K_r = \frac{G_r}{r_b \ln(r_0 / r_b)}$$

r_b, E_b, G_r are radius, elastic modulus of socket, and shear modulus of rock mass.

Let us introduce the following boundary conditions for a socket with an end-bearing also (see Figure 2.49):

$$\sigma_z = \sigma_0 \quad \text{at} \quad z = 0 \tag{2.45a}$$
$$w_b = w_e \quad \text{at} \quad z = L \tag{2.45b}$$

Thus, integration constants A_1, A_2 of Equation (2.44) can be obtained as given here:

$$A_1 = \frac{1}{e^{\alpha L} + e^{-\alpha L}}\left(w_e - \frac{\sigma_0}{E_b \alpha}e^{-\alpha L}\right), \quad A_2 = \frac{1}{e^{\alpha L} + e^{-\alpha L}}\left(w_e + \frac{\sigma_0}{E_b \alpha}e^{\alpha L}\right) \tag{2.46}$$

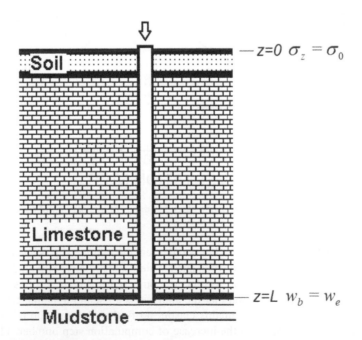

Figure 2.49 Illustration of boundary conditions for a socket with an end bearing

The axial displacement and axial stress of the socket and shear stress at the interface between the socket and surrounding rock can be expressed using the integration constants as follows:

Axial displacement

$$w_b = \frac{1}{e^{\alpha L} + e^{-\alpha L}} \left[w_e \left(e^{\alpha z} + e^{-\alpha z} \right) + \frac{\sigma_0}{E_b \alpha} \left(e^{\alpha(L-z)} - e^{-\alpha(L-z)} \right) \right] \tag{2.47a}$$

Axial stress

$$\sigma_b = \frac{1}{e^{\alpha L} + e^{-\alpha L}} \left(-E_b \alpha w_e \left(e^{\alpha z} - e^{-\alpha z} \right) + \sigma_0 \left(e^{\alpha(L-z)} + e^{-\alpha(L-z)} \right) \right) \tag{2.47b}$$

Interface shear stress

$$\tau_b = \frac{K_r}{e^{\alpha L} + e^{-\alpha L}} \left[w_e \left(e^{\alpha z} + e^{-\alpha z} \right) + \frac{\sigma_0}{E_b \alpha} \left(e^{\alpha(L-z)} - e^{-\alpha(L-z)} \right) \right] \tag{2.47c}$$

If the end of the socket is rigidly supported so that:

$$w_e = 0 \tag{2.48}$$

For this particular case, the axial displacement and axial stress of the socket and shear stress at the interface between the socket and surrounding rock can be expressed using the integration constants as follows:

Axial displacement

$$w_b = \frac{\sigma_0}{E_b \alpha} \frac{e^{\alpha(L-z)} - e^{-\alpha(L-z)}}{e^{\alpha L} + e^{-\alpha L}} \tag{2.49a}$$

Axial stress

$$\sigma_b = \sigma_0 \frac{e^{\alpha(L-z)} + e^{-\alpha(L-z)}}{e^{\alpha L} + e^{-\alpha L}} \tag{2.49b}$$

Interface shear stress

$$\tau_b = \frac{\sigma_0 K_r}{E_b \alpha} \frac{e^{\alpha(L-z)} - e^{-\alpha(L-z)}}{e^{\alpha L} + e^{-\alpha L}} \tag{2.49c}$$

However, it is very unlikely that the ends of the sockets would be rigidly supported. Therefore, the displacement of the surrounding ground beneath the socket tip may be approximated using the following equations proposed by Timoshenko and Goodier (1951, 1970), Aydan et al. (2008) and Aydan (2016):

Timoshenko and Goodier (1951/1970)

$$w_e = \frac{\pi(1 - \nu_r^2)}{2E_r} r_b p_e \tag{2.50a}$$

Aydan *et al.* (2008), Aydan (2016)

$$w_e = \frac{(1+\nu_r)}{E_r} r_b p_e \tag{2.50b}$$

where E_r, ν_r are elastic modulus and Poisson's ratio of rock mass, and p_e is the pressure acting at the tip of the socket. p_e is fundamentally unknown. One needs to obtain it, by requiring the continuity of displacement of the socket tip and rock mass beneath at z = L using an iterative technique such as Runga-Kutta or Newton-Raphson method.

This method was applied to a reinforced concrete socket for a foundation of a bridge. The socket was 20 m long and had a diameter of 1.2 m. Table 2.10 gives the material properties.

Table 2.10 Properties of blocks, discontinuity and rock anchor used in analyses

Rock Burst		Socket			
E_r (MPa)	V_r	E_s (GPa)	r_s (m)	L m	σ_0 (MPa)
720	0.25	20	0.6	20	2.476

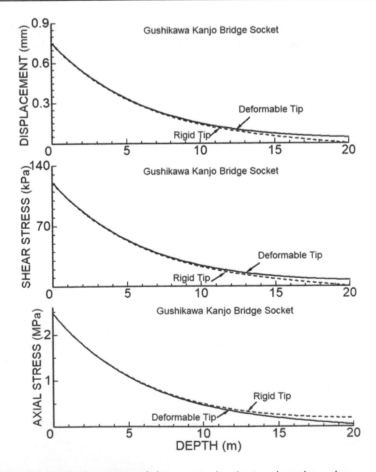

Figure 2.50 Axial stress, displacement and shear stress distribution along the socket

Figure 2.50 shows the distributions of axial stress, displacement and shear stress for two cases: For CASE 1, the end was rigidly supported, and for CASE 2, the socket displacement obeys Equation (2.47). The dotted lines correspond to CASE 1 while continuous lines correspond to CASE 2. As the socket is relatively long, the stress at the socket tip is only 7.53% of that at the socket top.

References

Aydan, Ö. (1989) *The Stabilisation of Rock Engineering Structures by Rockbolts*. Doctorate Thesis, Nagoya University.

Aydan, Ö. (2016) *Time Dependency in Rock Mechanics and Rock Engineering*, London, CRC Press, ISRM Book Series, No. 2, 246 pages, 9781138028630.

Aydan, Ö. (2017) *Rock Dynamics*. CRC Press, Taylor and Francis Group, 462p, ISRM Book Series No. 3, ISBN 9781138032286.

Aydan, Ö. (2018) *Rock Reinforcement and Rock Support*. CRC Press, Taylor and Francis Group, 486p, ISRM Book Series, No. 6, ISBN 9781138095830.

Aydan, Ö., Daido, M., Ito, T., Tano, H. & Kawamoto, T. (2006) Prediction of post-failure motions of rock slopes induced by earthquakes. *4th Asian Rock Mechanics Symposium, Singapore*, Paper No. A0356 (on CD).

Aydan, Ö. & Kawamoto, T. (1992) The stability of slopes and underground openings against flexural toppling and their stabilisation. *Rock Mechanics and Rock Engineering*, 25(3), 143–165.

Aydan, Ö. & Kumsar, H. (2010) An experimental and theoretical approach on the modeling of sliding response of rock wedges under dynamic loading. *Rock Mechanics and Rock Engineering*, 43(6), 821–830.

Aydan, Ö., Mamaghani, I.H.P. & Kawamoto, T. (1996a) Application of discrete finite element method (DFEM) to rock engineering structures. *NARMS'96*. pp. 2039–2046.

Aydan, Ö, Shimizu, Y. & Kawamoto, T. (1992) The stability of rock slopes against combined shearing and sliding failures and their stabilisation. *Int. Symp. on Rock Slopes, New Delhi*. pp. 203–210.

Aydan, Ö. & Ulusay, R. (2002) Back analysis of a seismically induced highway embankment during the 1999 Düzce earthquake. *Environmental Geology*, 42, 621–631.

Aydan, Ö., Ulusay, R. & Atak, V.O. (2008) Evaluation of ground deformations induced by the 1999 Kocaeli earthquake (Turkey) at selected sites on shorelines. *Environmental Geology*, Springer Verlag, 54, 165–182.

Aydan, Ö., Ulusay, R., Kumsar, H. & Ersen, A. (1996b) Buckling failure at an open-pit coal mine. *EUROCK'96*. pp. 641–648.

Aydan, Ö., Takahashi, Y., Iwata, N., Kiyota, R. & Adachi, K. (2018) Dynamic response and stability of un-reinforced and reinforced rock slopes against planar sliding subjected ground shaking. *Journal of Earthquake and Tsunami*, 12(4), 1841001.

Ikeda, K. (1970). *Classification of rock conditions for tunnelling. 1st Int. Congress on Engineering Geology*, IAEG, Paris, 1258-1265.

Kawamoto, T., Aydan, Ö., Shimizu, Y. & Kiyama, H. (1992) An investigation into the failure of a natural rock slope. *The 6th Int. Symp. Landslides, ISL 92*, 1, 465–470, Christchurch.

Kovari, K. & Fritz, P. (1975). *Stability analysis of rock slopes for plane and wedge failure with the aid of a programmable pocket calculator*. 16th US Rock Mech. Symp., Minneapolis, USA, 25-33.

Kumsar, H., Aydan, Ö. & Ulusay, R. (2000) Dynamic and static stability of rock slopes against wedge failures. *Rock Mechanics and Rock Engineering*, 33(1), 31–51.

Mamaghani, I.H.P, Aydan, Ö. & Kajikawa, Y. (1999) Analysis of masonry structures under static and dynamic loading by discrete finite element method. *JSCE Geotechnical Journal* (626), 1–12.

Timoshenko, S. & Goodier, J.N. (1951/1970) *Theory of Elasticity*. McGraw Hill Book Company, New York, 519 pages.

Tokashiki, N. & Aydan, Ö. (2010) Kita-Uebaru natural rock slope failure and its back analysis. *Environmental Earth Sciences*, 62(1), 25–31.

Tokashiki, N. & Aydan, Ö. (2011) A comparative study on the analytical and numerical stability assessment methods for rock cliffs in Ryukyu islands. *The 13th International Conference of the*

International Association for Computer Methods and Advances in Geomechanics, Melbourne, Australia. pp. 663–668.

Ulusay, R., Ersen, A. & Aydan, Ö. (1995) Buckling failure at an open-pit coal mine and its back analysis. *The 8th Int. Congress on Rock Mechanics, ISRM*, Tokyo. pp. 451–454.

Ulusay, R., Aydan, Ö. & Ersen, A. (2019) Assessment of a complex large slope failure at Kışlaköy open pit mine, Turkey. *Proceedings of 2019 Rock Dynamics Summit in Okinawa.* pp. 45–52.

United States Army Corps of Engineers (1994) *Engineering and Design: Rock Foundations.* EM1110-1-2908, Washington, 121 pages.

Chapter 3

Applications to underground structures

3.1 Stress concentrations around underground openings

The stability of underground openings during and after excavation is always of great concern to engineers. The size and location of possible yielding or failure zones are always necessary for providing support and reinforcement for safety (Aydan, 1989, 2018). For a quick assessment of approximate size and locations, elastic solutions are used together with some yield criteria. Common yield criteria follow.

3.1.1 Criteria for stability assessment

(a) Energy methods

Energy methods have been used in mining for a long time, and it is based on the linear behavior of materials. One energy method is called strain energy, and it is expressed in terms of principal strain and stress components as follows (e.g. Jaeger and Cook, 1979; Aydan and Kawamoto, 2001):

$$W_s = \frac{1}{2}\left[\sigma_1\varepsilon_1 + \sigma_2\varepsilon_2 + \sigma_3\varepsilon_3\right] \tag{3.1}$$

If Hooke's law is introduced, Equation (3.1) takes the following form:

$$W_s = \frac{1}{2E}\left[\sigma_1^2 + \sigma_1^2 + \sigma_1^2 - 2\nu(\sigma_1\sigma_2 + \sigma_2\sigma_3 + \sigma_3\sigma_1)\right] \tag{3.2}$$

For the uniaxial condition, that is, $\sigma_1 = \sigma_c$, $\sigma_2 = 0$, $\sigma_3 = 0$, strain energy at the time of yielding is reduced to the following form:

$$W_{su} = \frac{1}{2E}\left[\sigma_c^2\right] \tag{3.3}$$

In addition, the distortion energy concept or shear strain concept is introduced instead of the energy concept. The distortion energy is given in the following form:

$$W_d = W_s - W_v \tag{3.4}$$

The volumetric strain energy may be written as:

$$W_v = \frac{1}{2}\left[\sigma_v\varepsilon_v\right] \tag{3.5}$$

where

$$\sigma_v = \frac{1}{3}(\sigma_1 + \sigma_2 + \sigma_3), \ \varepsilon_v = \frac{1}{3}(\varepsilon_1 + \varepsilon_2 + \varepsilon_3) \tag{3.6}$$

Accordingly, the distortion energy may be rewritten as follows:

$$W_d = \frac{1}{4G}\left[(\sigma_1 - \sigma_2)^2 + (\sigma_2 - \sigma_3)^2 + (\sigma_3 - \sigma_1)^2\right] \tag{3.7}$$

Under the uniaxial condition, that is, Equation (3.6) reduces to the following form:

$$W_{du} = \frac{1}{2G}\left[\sigma_c^2\right] \tag{3.8}$$

However, it should be noted that it becomes difficult to define the energy when the material behavior becomes nonlinear.

(b)   Extensional strain method

Stacey (1981) proposed the extensional strain method for assessing the stability of underground openings in hard rocks. He stated that it was possible to estimate the spalling of underground cavities in hard rocks through the use of his extensional strain criterion. The extensional strain is defined as the deviation of the least principal strain from linear behavior (Figure 3.1). This definition actually corresponds to the definition of initial yielding in the theory of plasticity. This initial yielding is generally observed, at 40–60% of the deviatoric strength of materials.

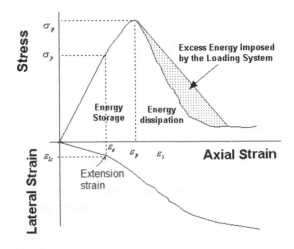

Figure 3.1 Illustration of extensional strain concept

Source: Aydan et al. (2001, 2004)

(c) Elasto-plastic method

In the elasto-plastic method, there are several models to model the strength of rock using the criteria of Mohr-Coulomb, Drucker-Prager, Hoek-Brown, Aydan. The simplest one is the brittle plastic model, in which the strength is abruptly reduced from the peak strength to its residual value. Aydan *et al.* (2001) recently combined both squeezing and rockbursting phenomena, and a more general strength reduction model is proposed as a function of strain level. This model, at least, treats both phenomena in a unified manner.

3.1.2 Stress distributions around underground openings in biaxial stress state

The stress state around underground openings is evaluated using the semi-numerical technique proposed by Gerçek (1996). Figure 3.2 show principal stress distribution of an underground openings with different shapes subjected to biaxial far field. The largest far-field stress is inclined at an angle of 20 degrees. As expected, stress concentrations occur at

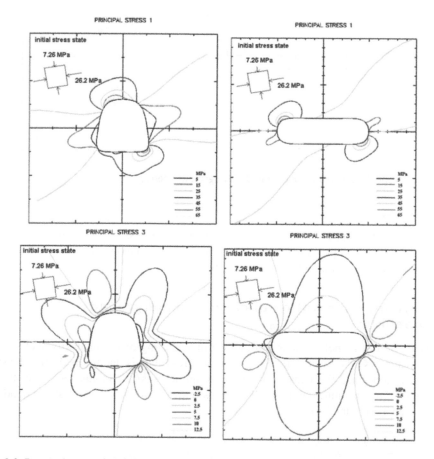

Figure 3.2 Principal stress distribution around underground openings

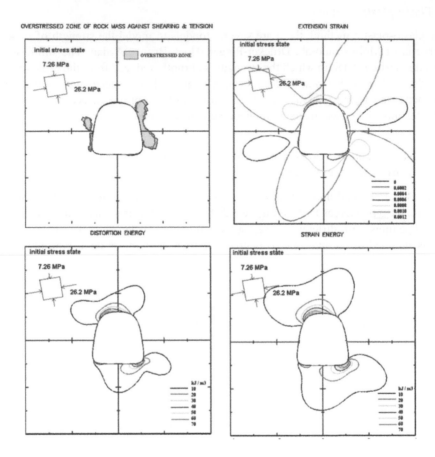

Figure 3.3 Potential yield/failure zone around underground openings

the lower-right and upper-left corners of the openings. Assessment of potential yield/failure zones around the openings are shown in Figure 3.3. Uniaxial compressive and tensile strength of intact rock was 218 and 8.0 MPa, respectively.

3.2   Dynamic excavation of circular underground openings

The excavation of tunnels is done through drilling-blasting or mechanically such as TBM and/or excavators. The most critical situation on stress state is due to the drilling-blasting–type excavation since the excavation force is applied almost impulsively. The dynamic response of circular tunnels during excavations under hydrostatic *in-situ* stress conditions can be given as (Figure 3.4):

$$\frac{\partial \sigma_r}{\partial r} + \frac{\sigma_\theta - \sigma_r}{r} = p\frac{\partial^2 u}{\partial t^2} \tag{3.9}$$

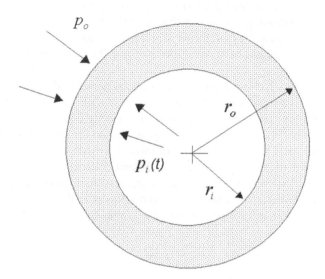

Figure 3.4 Illustration of dynamic excavation of a circular opening under hydrostatic *in-situ* stress condition

where σ_r, σ_θ, u, ρ and r are radial, tangential stresses, radial displacement, density and distance from the center of the circular cavity, respectively. Let us assume that the surrounding rock behaves in a visco-elastic manner of Kelvin-Voigt type, which is specifically given as:

$$\begin{Bmatrix} \sigma_r \\ \sigma_\theta \end{Bmatrix} = \begin{bmatrix} D_1 & D_2 \\ D_2 & D_1 \end{bmatrix} \begin{Bmatrix} \varepsilon_r \\ \varepsilon_\theta \end{Bmatrix} + \begin{bmatrix} C_1 & C_2 \\ C_2 & C_1 \end{bmatrix} \begin{Bmatrix} \dot{\varepsilon}_r \\ \dot{\varepsilon}_\theta \end{Bmatrix} \tag{3.10}$$

where ε_r, ε_θ and $\dot{\varepsilon}_r$, $\dot{\varepsilon}_\theta$ are radial and tangential strain and strain rates, respectively. The strain and strain rates are related to the radial displacement in the following form:

$$\varepsilon_r = \frac{\partial u}{\partial r}; \varepsilon_\theta = \frac{u}{r} \text{ and } \dot{\varepsilon}_r = \frac{\partial \varepsilon_r}{\partial_t}; \dot{\varepsilon}_\theta = \frac{\partial \varepsilon_\theta}{\partial t} \tag{3.11}$$

Eringen (1961) developed closed-form solution for Equation (3.9) under blasting loads. In order to deal with more complex boundary and initial conditions and material behavior, Equation (3.9) is preferred for solving using a dynamic finite element code. The discretized finite element form of Equation (3.9) together with the constitutive law given by Equation (3.10) takes the following form:

$$\mathbf{M\ddot{U} + C\dot{U} + KU = F} \tag{3.12}$$

Equation (3.12) has to be discretized in time-domain, and the resulting equation would take the following form:

$$\mathbf{K}^* \mathbf{U}_{n+1} = \mathbf{F}^*_{n+1} \tag{3.13}$$

The specific form of matrices in Equation (3.13) may change depending upon the method adopted in the discretization procedure in time-domain. For example, if the central difference technique is employed, the final forms would be the same as those given in Subsection 10.2. A finite element code has been developed by the author and used in the examples presented in this subsection.

In this subsection, the dynamic response of a circular tunnel under the impulsive application of excavation force is presented. The results were initially reported in Aydan (2011).

Figure 3.5 shows the responses of displacement, velocity and acceleration of the tunnel surface with a radius of 5 m. As noted from the figure, the sudden application of the excavation force, in other words, and the sudden release of ground pressure result in 1.6 times the static ground displacement at the tunnel perimeter, and shaking disappears almost within 2 s. As time progresses, it becomes asymptotic to the static value, and velocity and acceleration disappear.

The resulting tangential and radial stress components near the tunnel perimeter (25 cm from the opening surface) are plotted in Figure 3.6 as a function of time. It is of great interest

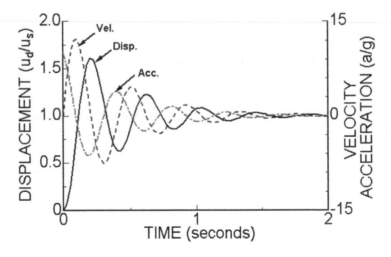

Figure 3.5 Responses of displacement, velocity and acceleration of the tunnel surface

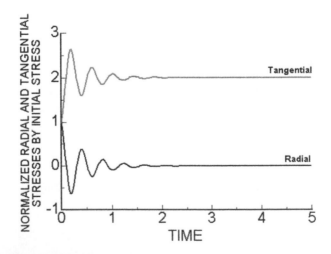

Figure 3.6 Responses of radial and tangential stress components nearby the tunnel surface (25 cm away from the perimeter)

that the tangential stress is greater than that under static condition. Furthermore, very high radial stress of tensile character occurs near the tunnel perimeter. This implies that the tunnel may be subjected to transient stress state, which is quite different from that under static conditions. However, if the surrounding rock behaves elastically, they will become asymptotic to their static equivalents. In other words, the surrounding rock may become plastic even though the static condition may imply otherwise.

3.3   Evaluation of tunnel face effect

Advancing tunnels utilizing support systems consisting of rock bolts, shotcrete, steel ribs and concrete lining are three-dimensional complex structures and is a dynamic process. However, tunnels are often modeled as one-dimensional axisymmetric structures subjected to hydrostatic initial stress state as a static problem. The effect of tunnel face advance on the response and design of support systems is often replaced through an excavation stress release factor determined from pseudo three-dimensional (axisymmetric) or pure three-dimensional analyses as given here:

$$f = \frac{e^{-bx/D}}{1/B + e^{-bx/D}} \tag{3.14}$$

where x is distance from tunnel face, and the values for coefficients B and b suggested by Aydan (2011) are 2.33 and 1.7, respectively.

Figure 3.7a illustrates an unsupported circular tunnel subjected to an axisymmetric initial stress state. The variation of displacement and stresses along the tunnel axis were computed using the elastic finite element method. The radial displacement at the tunnel wall is normalized by the largest displacement and is shown in Figure 3.7b. As seen from the figure, the radial displacement takes place in front of the tunnel face. The displacement is about 28–30% of the final displacement. Its variation terminates when the face advance is about +2-D. Almost 80% of the total displacement takes place when the tunnel face is about +1-D. The effect of the initial axial stress on the radial displacement is almost negligible.

Figure 3.7c shows the variation of radial, tangential and axial stress around the tunnel at a depth of 0.125R. As noted from the figure, the tangential stress gradually increases as the distance increases from the tunnel face. The effect of the initial axial stress on the tangential stress is almost negligible. The radial stress rapidly decreases in the close vicinity of the tunnel face, and the effect of the initial axial stress on the radial stress is also negligible. The most interesting variation is associated with the axial stress distribution. The axial stress increases as the face approaches, and then it gradually decreases to its initial value as the face effect disappears. This variation is limited to a length of 1R(0.5 D) from the tunnel face. It is also interesting to note that if the initial axial stress is nil, even some tensile axial stresses may occur in the vicinity of tunnel face.

Figure 3.7d shows the stress distributions along the r-axis of the tunnel at various distances from the face when the initial axial stress is equal to initial radial and tangential stresses. As noted from the figure, the maximum tangential stress is 1.5 times the initial hydrostatic stress, and it becomes twice the distance from the tunnel face is +5R, which is almost equal to theoretical estimations for tunnels subjected to the hydrostatic initial stress state. The stress state near the tunnel face is also close to that of the spherical opening subjected to the hydrostatic stress state. The stress state seems to change from spherical state to the cylindrical state (Aydan, 2011). It should be noted that it would be almost impossible to simulate exactly the same displacement and stress changes of 3-D analyses in the vicinity of

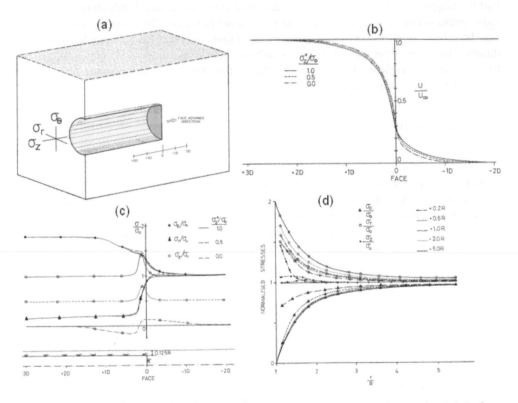

Figure 3.7 (a) Computational model for elastic finite element analysis, (b) normalized radial displacement of the tunnel surface, (c) normalized stress components along the tunnel axis at a
distance of 0.125R, (d) the variation of stresses along the r-direction at various distances
from the tunnel face

tunnels by 2-D simulations using the stress-release approach irrespective of the constitutive
law of surrounding rock as a function of distance from the tunnel (Aydan *et al.*, 1988; Aydan
and Geniş, 2010).

3.4   Abandoned room and pillar lignite mines

When abandoned lignite mines and quarries are of room and pillar type, their short-term
and long-term stability may be evaluated using some simple analytical techniques. Roof
stability is generally evaluated using beam theory and/or arching theory under gravitational, earthquake and point loading (i.e. Coates, 1981; Obert and Duvall, 1967; Aydan,
1989; Aydan and Tokashiki, 2011). The tributary area method is quite widely used in
mining engineering for assessing the pillar stability. Aydan and his coworkers (Aydan and
Geniş, 2007; Aydan and Tokashiki, 2011; Geniş and Aydan, 2013) extended the method
to cover the effects of earthquake and point loading, in addition to gravitational loading,
creep and degradation of geomaterials, in order to evaluate the stability of roof and pillars
(Figure 3.8).

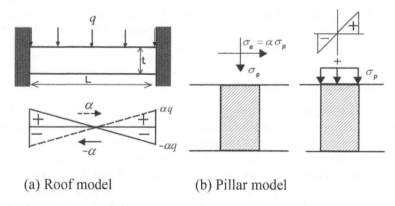

(a) Roof model   (b) Pillar model

Figure 3.8 Models for roof and pillars

Source: From Aydan and Geniş (2007)

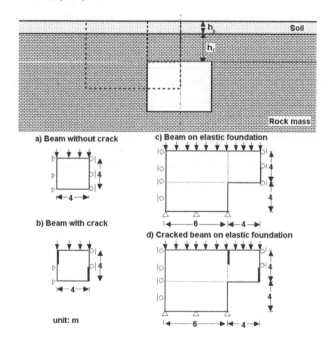

Figure 3.9 Illustration of finite element models with boundary conditions for roof of natural underground openings

Aydan and Tokashiki (2011) performed a series of finite element analyses for assessing the stability of roofs of shallow natural underground openings. Basically, two different situations were considered (Figure 3.9):

1   Beam with or without cracks
2   Beam on soft and rigid elastic foundations with or without cracks at abutments and center of the beam

All models are assumed to be subjected to body forces together with a distributed load due to the dead weight of 5 m thick topsoil with a unit weight of 22 kN m⁻³. The material properties used in the analyses are given in Table 3.1. As the problem is symmetric, half of the region is modeled in finite element analyses (Figure 3.9).

3.4.1   Beams with and without cracks

Figure 3.10 shows the principal stresses in roof beam with and without cracks. Cracks were assumed to propagate to the half thickness of the roof beam at abutments and at the center. As seen in Figure 3.10a, the principal stress distribution is almost the same as what could be obtained from the bending theory of beams. The tensile stress is highest at the uppermost fiber of the roof beam at abutments. When the crack occurs, tensile stresses are drastically reduced in the roof beam. Nevertheless, high tensile stresses occur near the tips of cracks. The compressive stresses become higher at abutments and the top center of the cracked beam, as expected. Definitely, the arch action is induced in the cracked beam.

Table 3.1 Material properties used in analyses

Layer	Elastic Modulus (GPa)	Poisson's ratio	Unit weight (kN m⁻³)
Roof	0.6	0.25	23
Foundation	6	0.25	23
Abutment	60	0.25	23

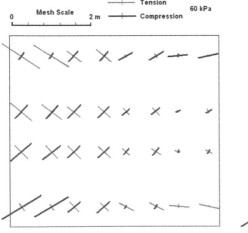

(a) Beam without crack

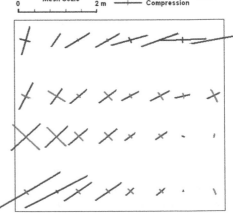

(b) Beam with crack (crack length 1m)

Figure 3.10 Principal stress distributions in beams with and without cracks

3.4.2   Beams on elastic foundations with overburden soil

Figure 3.11 compares the principal stress distributions of beams with or without topsoil on soft elastic foundations, whose elastic modulus is 10 times that of the beam. As noted from the figure, stresses are lower than those shown in Figure 3.10. This is due to the stresses being distributed over a large area and the consideration of Poisson's ratio in finite element analyses. Furthermore, the highest tensile stress occurs at the center of the lowermost fiber of the beam. The highest tensile stress occurs near the ground surface slightly farther away from the abutment. These results are somewhat different from those of the beam rigidly supported at abutments. The existence of the topsoil increases the amplitude of principal stresses. Nevertheless, the overall response remains basically the same.

3.4.3   Roof beams supported by rigid abutments with or without cracks

The elastic modulus of the ground and beam beyond abutments was increased to 100 times that of the roof beam. Furthermore, the existence of cracks was also considered in the analyses. Figures 3.12 and 3.13 show the principal stress distributions and deformed configurations of the models with or without cracks. If the rigidity of the abutments increases, the stress state approaches those of the single beam. Similarly, the presence of cracks causes a stress state close to those of the cracked beam shown in Figure 3.10(b).

As shown, simple analytical models and computations from the two-dimensional elastic finite element method yield very similar results. Nevertheless, stresses computed from the FEM in the roof are less than those computed from the beam theory with built conditions. On the other hand, stresses computed from the FEM in pillars are slightly higher than those computed from the tributary area method. However, the stress state in the roof would be quite different if the opening depth increases. In such cases, the effects of gravitational load in the stress state of the roof should be also taken into account. Nevertheless, the stability of

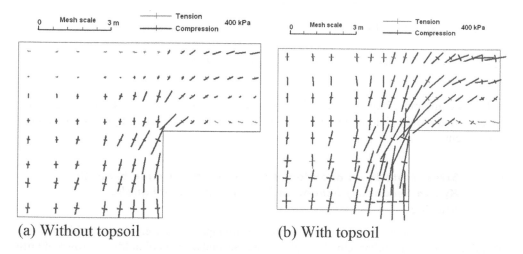

(a) Without topsoil   (b) With topsoil

Figure 3.11 Principal stress distribution of beams on elastic foundations

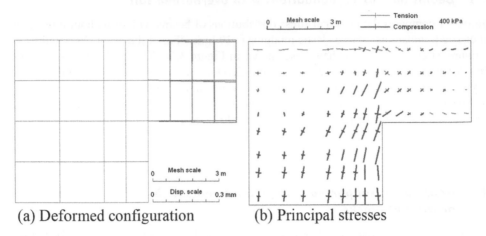

(a) Deformed configuration   (b) Principal stresses

Figure 3.12 Deformed configurations and principal stress distributions (without cracks)

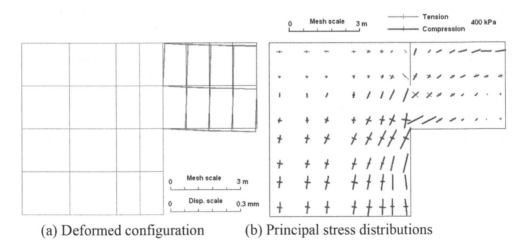

(a) Deformed configuration   (b) Principal stress distributions

Figure 3.13 Deformed configurations and principal stress distributions (with cracks)

pillars become more important than the roof itself under such conditions, and the tributary area method would yield quite reasonable values for the stress state in pillars for stability assessment.

3.4.4   Stress state in an abandoned room and pillar mine beneath Kyowa Secondary School under static and earthquake loading

Figure 3.14 shows the distribution of minimum principal stress (tension is positive) for an abandoned room and pillar mine beneath Kyowa Secondary School in Mitake Town of Gifu Prefecture, Japan, under the static loading condition. Although the maximum pillar stresses

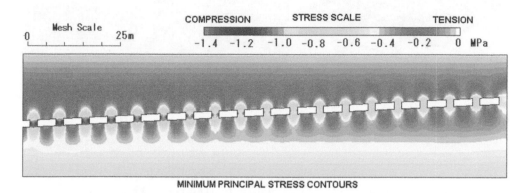

Figure 3.14 Contours of minimum principal stress beneath the Kyowa Secondary School

Table 3.2 Material properties of layers

Layer	γ kN m^{-3}	E MPa	v	c MPa	ϕ (°)
Topsoil	19	270	0.35	0.0	38
Upper Mst-Sst	19	750	0.3	0.7	25
Lignite	14	400	0.3	0.66	45
Lower Mst-Sst	19	1073	0.3	1.00	45
Chert	19	3647	0.3	3.00	45

are slightly higher than those computed from the tributary area method, the quick stability assessment using the tributary area method should be quite acceptable.

This area would be subjected to the anticipated Nankai-Tonankai-Tokai earthquake in the future, and there is a great concern about it. That authors have been involved with the stability assessment of the abandoned lignite mine beneath Kyowa Secondary School during the anticipated Nankai-Tonankai-Tokai earthquake (Aydan *et al.*, 2012; Geniş and Aydan, 2013). Material properties of investigated ground are given in Table 3.2. The authors carried out 1-D, 2-D and 3-D dynamic simulations for an estimated base ground motion data at Mitake Town obtained from the method of Sugito *et al.* (2001). Figure 3.15 illustrates the numerical model of the ground and abandoned lignite mine beneath the Kyowa Secondary School. Figure 3.16 shows the computed responses from 1-D and 3-D numerical analyses. It is interesting to note that responses from 1-D and 3-D analyses are quite similar to each other.

A three-dimensional elasto-plastic numerical analysis of abandoned lignite mine beneath the Kyowa Secondary school (Figure 3.17) uses the estimated ground motion record (Figure 3.18), based on the methods developed by Sugito *et al.* (2000) and Aydan (2012) for the anticipated Nankai-Tonankai-Tokai mega earthquake. The Nankai earthquake terminates at 43 s, and the Tonankai earthquake starts and terminates at 75 s. The last earthquake is Tokai earthquake, and it terminates at about 125 s.

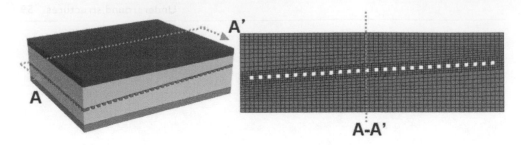

Figure 3.15 Illustration of models used in numerical analyses and selected section

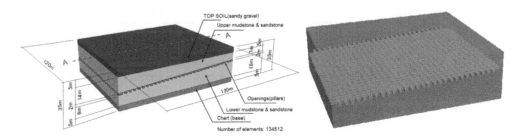

(a) 1D Analysis   (b) 3D Analysis

Figure 3.16 Acceleration responses at selected section from 1-D and 3-D numerical analyses

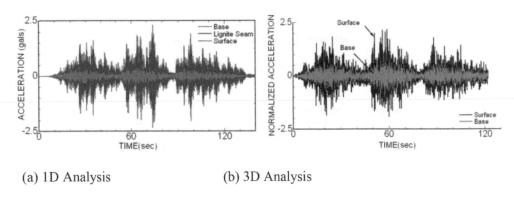

Figure 3.17 3-D views of the numerical model and abandoned room and pillar mine

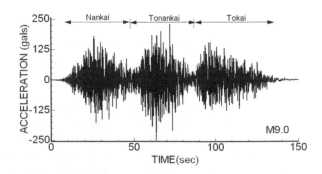

Figure 3.18 Input base acceleration record

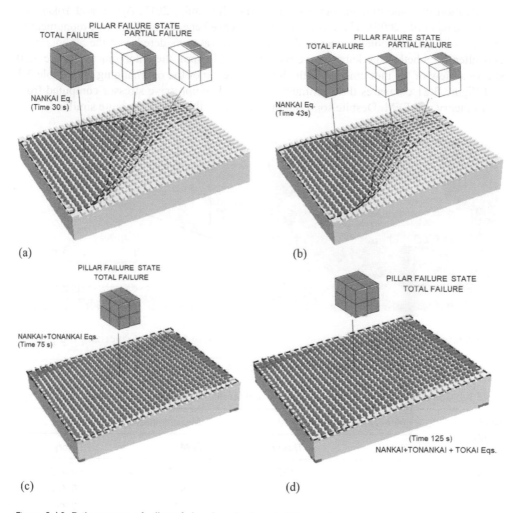

Figure 3.19 Failure state of pillar of abandoned mine at different time steps

The failure of the pillars starts at the deepest site and propagates towards shallower parts as estimated from 2-D numerical analyses (Figure 3.19). The failure state of pillars can be broadly classified as total failure and partial failure, as illustrated. When the Nankai earthquake terminates, about 60% of the pillars were already totally or partially yielded. The Tonankai earthquake is the nearest one to the Mitake town, and all the pillars are in total failure state. The Tokai earthquake has no further effect on the failure state.

3.5 Karstic caves

Karstic caves are quite common worldwide whenever limestone and evaporate deposits exist. In the coral limestone formation in the Ryukyu islands of Japan, there are many karstic caves, which present many geoengineering problems. The authors are involved with the stability assessment of some karstic caves in relation to some engineering projects or preservation of some monumental structures (i.e. Tokashiki, 2011; Aydan and Tokashiki, 2011; Geniş *et al.*, 2009). There is a huge karstic cave beneath the Himeyuri monument in Okinawa island. The enlargement of the monument was considered, and the authors were consulted on whether the karstic cave would be stable upon the enlargement. Figure 3.20 shows a view of the monument and the beam modeling of the overhanging part. Table 3.3 and Figure 3.21 compares the maximum tensile and compressive stresses computed from beam theory and FEM. Despite some slight differences, the results are quite similar.

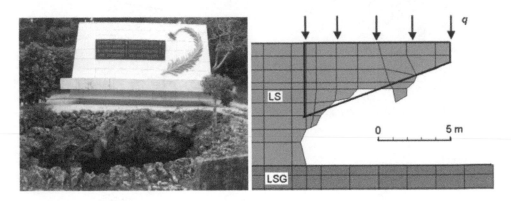

(a) View of the monument   (b) Beam modeling of overhanging part

Figure 3.20 View and beam modeling of overhanging part

Table 3.3 Comparison of maximum compressive and tensile stresses from FEM and bending theory

Loading Condition	Max. Tensile Stress (MPa)		Max. Compressive Stress (MPa)	
	FEM	Theory	FEM	Theory
Natural	0.557	0.677	−1.363	−0.677
Present	0.631	0.713	−1.402	−0.713
Planned-2	0.770	0.991	−1.478	−0.991

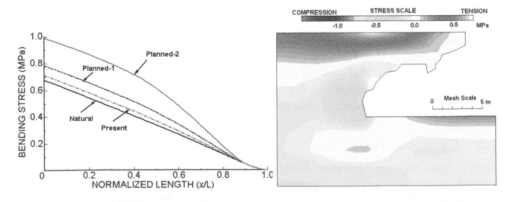

(a) Stress distribution from bending theory (b) Maximum principal stress from FEM

Figure 3.21 Comparison of stresses obtained from the bending theory and FEM

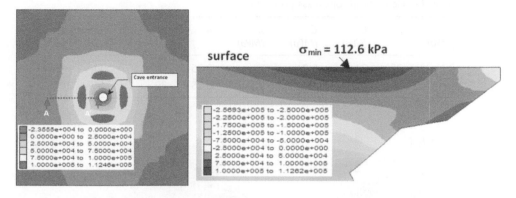

(a) Surface stresses   (b) Maximum principal stress contours along A-A'

Figure 3.22 Computed maximum principal stress distributions from 3-D numerical analysis

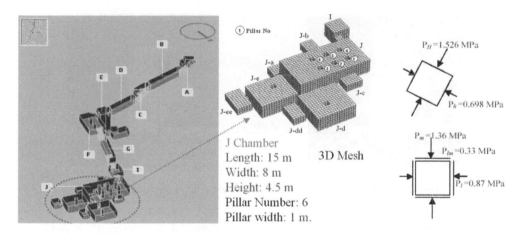

Figure 3.23 Numerical model of the tomb and assumed *in-situ* stress state

A 3-D analysis of the vicinity of Himeyuri monument and the cave beneath it was carried out with the consideration of surface loading due to the deadweight of the monument structure (Aydan *et al.*, 2011). The cave was considered to be circular in plain view. The maximum tensile stress was much smaller than that computed from the bending theory and 2-D FEM analysis. An additional axisymmetric FEM analysis was also performed, and it yielded similar results. However, the cave has an ovaloid shape in plan, and the actual stress state is expected to be closer to that of 2-D analyses. Furthermore, there are some cross-joints in the rock mass so that the actual stress state should be quite close to that of 2-D-FEM analyses (Figure 3.22).

3.6   Stability analyses of tomb of Pharaoh Amenophis III

Three-dimensional elasto-plastic numerical analyses of the tomb were carried out for 12 different situations using the FLAC code developed by ITASCA (1997) (Figure 3.23). Only two cases, in which the properties of rock mass are assumed to be equivalent to those of

Table 3.4 Considered rock mass properties.

S_c (MPa)	f	E (GPa)	C (MPa)	S_t (MPa)	n	g (kN m^{-3})	Remarks
20	35	2	5.2	2	0.25	20	Dry
5	30	0.8	1.44	0.5	0.25	20	Wet

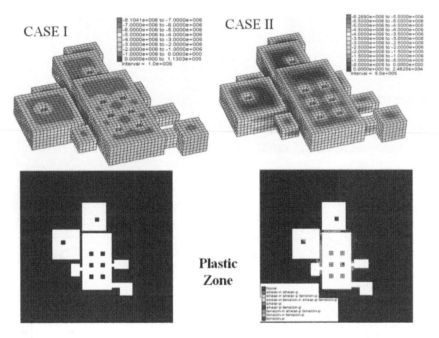

Figure 3.24 Stress distribution and plastic zone development around the tomb

intact rock under dry and wet conditions, as given in Table 3.4, are presented (Aydan *et al.*, 2008, Egypt). The computational results indicated that if rock mass is assumed to be dry, there should be no plastic zone development in rock mass around the tomb (Figure 3.24). However, if rock becomes saturated, the computational results indicated that the damage in the walls between J-room and Jd-room has a great influence on the overall stability of the J-room and adjacent rooms. the saturation of rock mass from time to time due to floods may also have a negative effect on the stability of the tomb.

3.7   Retrofitting of unlined tunnels

The Unten tunnel in Nakijin region of Okinawa island was an unlined single-lane roadway tunnel. Following the collapse of Toyohama tunnel in Hokkaido island in 1996, the authorities were ordered to check the safety of all roadway tunnels in Japan. The Unten tunnel was designated as unsafe after the checking procedure, and it was decided to close it to traffic. However, the strong demand by local residents to keep the tunnel open to traffic resulted in the reassessment of the stability of the tunnel and its retrofitting.

The site investigations revealed several thoroughgoing discontinuities as shown in Figure 3.25. Model experiments using the base friction apparatus indicated that the tunnel might be unstable if the frictional properties of discontinuities decrease with time. The reduction of friction angle of discontinuities was achieved by introducing the double-layer Teflon sheets along discontinuities in the model tests.

The discrete finite element method (DFEM) was used to assess the stability of the tunnel (see Tokashiki *et al.*, 1997, for the details of numerical analyses). The DFEM analyses also indicated that the tunnel might become unstable if the frictional properties of discontinuities were drastically reduced. For retrofitting the tunnel, glass-fiber rock bolts were installed, and reinforced concrete lining was constructed. Figure 3.26 shows the deformed configurations

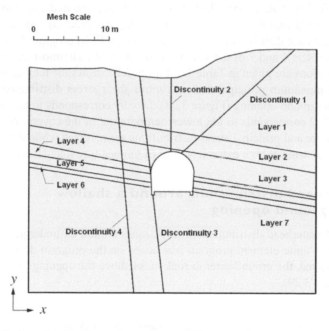

Figure 3.25 Distribution of major discontinuities around Unten tunnel

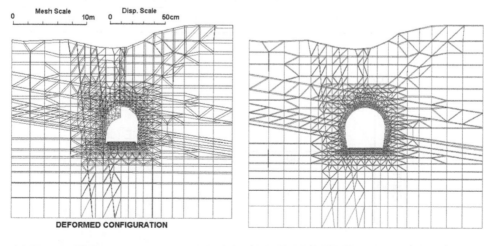

(a) No retrofitting   (b) With retrofitting

Figure 3.26 Deformation of surrounding rock mass with/without the measures of retrofitting

of the tunnel without and with retrofitting, respectively. As noted from the figures, the tunnel should be stable if the selected measures of retrofitting are employed.

3.8   Temperature and stress distributions around an underground opening

The example given here is concerned with temperature fluctuation in rock mass in an underground opening subjected to temperature variation applied on the surface of the opening (see details by Aydan *et al.*, 2008). The temperature fluctuation in the opening was based on actual measurements, and it was subjected to ±10 degrees yearlong sinusoidal temperature variation. Temperature distribution is computed from an FEM program based on the theory presented in Chapters 4 and 7 of the Volume 1 of this book. Thermo-mechanical properties used in computations are given in Table 3.5. Figure 3.27 shows the temperature distribution and associated maximum principal and maximum shear stress distributions in rock mass around the underground opening (Figure 3.28). Day 99 corresponds to the highest temperature, and Day 272 corresponds to the lowest temperature in the cavern. As noted from the figure, temperature and principal stress distributions are reversed while the maximum shear stress remains the same at both extreme values of temperature fluctuations.

3.9   Waterhead distributions around a shallow underground opening

An example of waterhead distributions in rock mass around an underground opening was investigated. The finite element program was based on the program developed by Verruijt (1982). As expected, the groundwater in rock mass above the opening was close to drained condition (Figure 3.29).

Table 3.5 Thermo-hydro-mechanical properties of surrounding rock mass

Unit weight (kN m⁻³)	Elastic Modulus (GPa)	Poisson's ratio	Cohesion (MPa)	Friction Angle (°)	Thermal Diffusivity (m² day⁻¹)	Thermal Expansion Coefficient (1 °C⁻¹)
26	5–10	0.25	3	40	0.1	1.0×10^{-5}

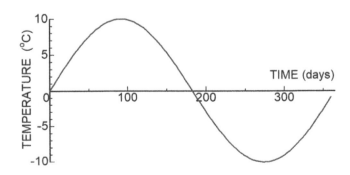

Figure 3.27 Applied temperature at the surface of the cavern

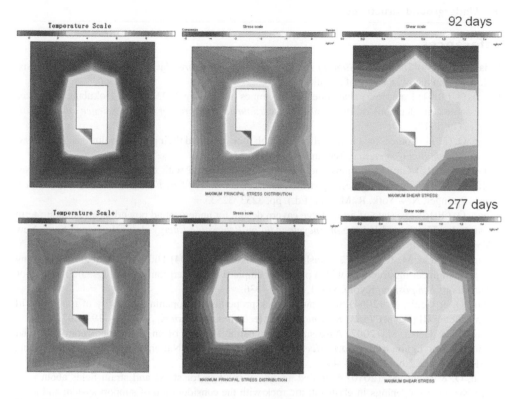

Figure 3.28 Computed temperature stress distributions

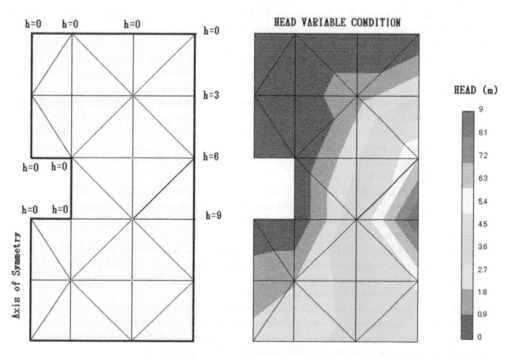

Figure 3.29 FEM mesh and computed water head distribution in rock mass around underground opening

References

Aydan, Ö. (1989) *The Stabilisation of Rock Engineering Structures by Rockbolts*. Doctorate Thesis, Nagoya University, Faculty of Engineering.

Aydan, Ö. (2004) Damage to abandoned lignite mines induced by 2003 Miyagi-Hokubu earthquakes and some considerations on its possible causes. *Journal of School of Marine Science and Technology*, 2(1), 1–17.

Aydan, Ö. (2011) Some issues in tunnelling through rock mass and their possible solutions. *Proc. First Asian Tunnelling Conference,* Tehran, ATS-15. pp. 33–44.

Aydan, Ö. (2012) Ground motions and deformations associated with earthquake faulting and their effects on the safety of engineering structures. *Encyclopedia of Sustainability Science and Technology*, Springer, New York. R. Meyers (Ed.), pp. 3233–3253.

Aydan, Ö. (2016) The state of art on large cavern design for underground powerhouses and long-term issues. In: *The Second Volume of Encyclopedia on Renewable Energy*. John Wiley and Sons. New York. pp. 467–487.

Aydan, Ö., Daido, M., Owada, Y., Tokashiki, T. & Ohkubo, K. (2004) The assessment of rock bursting in rock engineering structures with a particular emphasis on underground openings. *3rd Asian Rock Mechanics Symposium*, Kyoto, Vol. 1, pp. 531–536.

Aydan, Ö. & Geniş, M. (2004) Surrounding rock properties and openings stability of rock tomb of Amenhotep III (Egypt). *ISRM Regional Rock Mechanics Symposium, Sivas*. Pp. 191–202.

Aydan, Ö. & Geniş, M. (2007) Assessment of dynamic stability of an abandoned room and pillar underground lignite mine. *Rock Mechanics Bulletin*. Turkish National Rock Mechanics Group, ISRM, Ankara, No. 16. pp. 23–44.

Aydan, Ö. & Geniş, M. (2010) A unified analytical solution for stress and strain fields about radially symmetric openings in elasto-plastic rock with the consideration of support system and long-term properties of surrounding. *International Journal of Mining and Mineral Processing (IJMMP)*, 1(1–32).

Aydan, Ö., Geniş, M., Akagi, T. & Kawamoto, T. (2001) Assessment of susceptibility of rockbursting in tunneling in hard rocks. *International Symposium on Modern Tunnelling Science and Technology*, IS-KYOTO, 1, 391–396.

Aydan, Ö., Geniş, M., Sugiura, K. & Sakamoto, A. (2012) Characteristics and amplification of ground motions above abandoned mines. *First International Symposium on Earthquake Engineering, JAEE*, Tokyo, Vol. 1, pp. 75–84.

Aydan, Ö. & Kawamoto, T. (2001) The stability assessment of a large underground opening at great depth. *17th International Mining Congress and Exhibition of Turkey, Ankara*. pp. 277–278.

Aydan, Ö., Kyoya, T., Ichikawa, Y., Kawamoto, T., Ito, T. & Shimizu, Y. (1988). Three-dimensional simulation of an advancing tunnel supported with forepoles, shotcrete, steel ribs and rockbolts. *The 6th International Conference on Numerical and Analytical Methods in Geomechanics,* Innsbruck, Austria. Vol. 2, pp. 1481–1486.

Aydan, Ö., Ohta, Y., Daido, M., Kumsar, H. Genis, M., Tokashiki, N., Ito, T. & Amini, M. (2011) Chapter 15: Earthquakes as a rock dynamic problem and their effects on rock engineering structures. In *Advances in Rock Dynamics and Applications*, Editors Y. Zhou and J. Zhao, CRC Press, Taylor and Francis Group, London, pp. 341–422.

Aydan, Ö. & Tokashiki, N. (2011) A comparative study on the applicability of analytical stability assessment methods with numerical methods for shallow natural underground openings. *The 13th International Conference of the International Association for Computer Methods and Advances in Geomechanics,* Melbourne, Australia. pp. 964–969.

Aydan, Ö., Tsuchiyama, S., Kinbara, T., Uehara, F., Tokashiki, N. & Kawamoto, T. (2008) A numerical analysis of non-destructive tests for the maintenance and assessment of corrosion of rockbolts and rock anchors. *The 12th International Conference of International Association for Computer Methods and Advances in Geomechanics (IACMAG)*, Goa, India. pp. 40–45.

Coates, D.F. (1981) *Rock Mechanics Principles*. Canadian Government Pub Centre, Ottawa, 410 pages.

Eringen, A.C. (1961) Propagations of elastic waves generated by dynamical loads on a circular cavity. *Journal of Applied Mechanics, ASME*, 28, 218–222.

Geniş, M. & Aydan, Ö. (2013) A numerical study on the ground amplifications in areas above abandoned room and pillar mines and longwall old mines. *The 2013 ISRM EUROCK International Symposium*, Wroclaw, pp. 733–738.

Geniş, M., Tokashiki, N. & Aydan, Ö. (2009) The stability assessment of karstic caves beneath Gushikawa Castle remains (Japan). EUROCK 2010, pp. 449–454.

Gerçek, H. (1996) Special elastic solutions for underground openings. *Milestones in Rock Engineering: The Bieniawski Jubilee Collection,* Balkema, Rotterdam. pp. 275–290.

ITASCA. 1997 *FLAC3D-Fast Lagrangian Analysis of Continua (Version 2.0)*. 5 Vols. Minneapolis: Itasca Consulting Group, Inc.

Jaeger, J.G. & Cook, N.G.W. (1979) *Fundamentals of Rock Mechanics*. 3rd Ed. Chapman and Hall, London.

Obert, L. & Duvall, W.I. (1967) *Rock Mechanics and the Design of Structures in Rock*. John Wiley & Sons, New York.

Sugito, M., Furumoto, Y. & Sugiyama, T. (2000) Strong motion prediction on rock surface by superposed evolutionary spectra, *12th World Conference on Earthquake Engineering*, 2111/4/A, CD-ROM.

Stacey, T.R. (1981) A simple extension strain criterion for fracture of brittle rock. *International Journal Rock Mechanics and Mining Science & Geomechanics Abstracts*, Oxford, 18, 469–474.

Tokashiki, N. (2011) Study on the Engineering Properties of Ryukyu Limestone and the Evaluation of the Stability of its Rock Mass and Masonry Structures. PhD Thesis, 221p., Engineering and Science Graduate School, Waseda University.

Tokashiki, N., Aydan, Ö., Shimizu, Y. & Kawamoto, T. (1997) The assessment of the stability of a very old tunnel by discrete finite element method (DFEM), *Numerical Methods in Geomechanics, NUMOG VI,* Montreal. pp. 495–500.

Verruijt, A. (1982) *Groundwater Flow*. 2nd Ed. Macmillan Press Ltd., London, 144 pages.

Chapter 4

Rock mass classifications and their engineering utilization

4.1 Introduction

Rock mass classifications are used for various engineering design and stability analyses, and they are initially proposed for the design of a given rock structure. However, this trend has been now changing, and the main objectives of rock mass classifications have become to identify the most significant parameters influencing the behavior of rock masses, to divide a particular rock mass formulation into groups of similar behavior, to provide the characterizations of each rock mass class, to derive quantitative data and guidelines for engineering design, and to provide a common basis for engineers and geologists. These are based on empirical relations between rock mass parameters and engineering applications like tunnels and other underground caverns.

In the empirical methods, rock mass classification systems are extensively used for feasibility and predesign studies and often also for the design. Although the history of rock classifications for a given specific structure is old, the rock mass classification system proposed by K. Terzaghi (1946) for tunnels with steel set support has become the basis for the follow-up quantitative rock mass classifications. Currently, there are many classification systems in rock engineering, particularly in tunneling, such as Rock Mass Rating (RMR) (Bieniawski, 1974, 1989), Q system (Barton *et al.*, 1974), RSR (Wikham *et al.*, 1974), Rock Mass Quality Rating (RMQR) proposed by Aydan *et al.*, 2014. In addition, rock mass classifications of NEXCO (known as DOROKODAN) and JR (KYU-KOKUTETSU) are commonly used to design tunnels in Japan. Nevertheless, the utilization of these systems for characterization of complex rock mass conditions is a challenge for engineers and is not always possible in some cases.

In this chapter, several classification systems have been briefly explained, and computations have been done based on RMQR.

4.2 Rock Mass Rating (RMR)

Bieniawski (1974) published the details of a rock mass classification called the Geomechanics Classification or the Rock Mass Rating (RMR) system. Over the years, this system has been refined as more case records have been examined, and the reader should be aware that Bieniawski (1989) has made significant changes in the ratings assigned to different parameters and that the 1989 version of the classification is suggested by Bieniawski (1989). The following six parameters are used to classify a rock mass using the RMR system (Table 4.1):

1   Uniaxial compressive strength (UCS) of rock material
2   Rock Quality Designation (RQD)
3   Spacing of discontinuities
4   Condition of discontinuities
5   Groundwater conditions
6   Orientation of discontinuities

Table 4.1 Ratings of RMR classification system

A. Classification Parameters and Their Ratings

Parameter		Range of Values						
1	Strength of intact rock material	Point-load strength index	>10 MPa	4–10 MPa	2–4 MPa	1–2 MPa	For this low range-uniaxial compressive test is preferred	
		Uniaxial comp. strength	>250 MPa	100–250 MPa	50–100 MPa	25–50 MPa	5–25 MPa	1–5 MPa	< 1 MPa
	Rating		15	12	7	4	2	1	0
2	Drill core quality RQD		90%–100%	75%–90%	50%–75%	25%–50%	<25%	
	Rating		20	17	13	8	3	
3	Spacing of discontinuities		> 2 m	0.6–2 m	200–600 mm	60–200 mm	<60 mm	
	Rating		20	15	10	8	5	
4	Condition of discontinuities (See E)		Very rough surfaces Not continuous No separation Unweathered wall rock	Slightly rough surfaces Separation <1 mm Slightly weathered walls	Slightly rough surfaces Separation <1 mm Highly weathered walls	Slickensided surfaces or Gouge <5 mm thick or Separation 1–5 mm Continuous	Soft gouge >5 mm thick or Separation >5 mm Continuous	
	Rating		30	25	20	10	0	
5	Groundwater	Inflow per 10 m tunnel length (l/m)	None	<10	10–25	25–125	>125	
		(Joint water press)/ (Major principal σ)	0	<0.1	0.1,–0.2	0.2–0.5	>0.5	
		General conditions	Completely dry	Damp	Wet	Dripping	Flowing	
	Rating		15	10	7	4	0	

B. Rating Adjustment for Discontinuity Orientations (See F)

Strike and Dip Orientations		Very Favorable	Favorable	Fair	Unfavorable	Very Unfavorable
Rating	Tunnels and mines	0	-2	-5	-10	-12
	Foundations	0	-2	-7	-15	-25
	Slopes	0	-5	-25	-50	-25

C. Rock Mass Classes Determined from Total Ratings

	Very Favorable	Favorable	Fair	Unfavorable	Very Unfavorable
Rating	100 — 81	80 — 61	60 — 41	40 — 21	<21
Class number	I	II	III	IV	V
Description	Very good rock	Good rock	Fair rock	Poor rock	Very poor rock

D. Meaning of Rock Classes

Class number	I	II	III	IV	V
Average stand-up time	20 a for 15 m span	1 a for 10 m span	1 wk for 5 m span	10 h for 2.5 m span	30 m. for 1 m span
Cohesion of rock mass (kPa)	>400	300–400	200–300	100–200	<100
Friction angle of rock mass (deg)	>45	35–45	25–35	15–25	<15

E. Guidelines for Classification of Discontinuity conditions

Discontinuity length (persistence)	<1 m	1–3 m	3–10 m	10–20 m	>20 m
Rating	6	4	2	1	0
Separation (aperture)	None	<0.1 mm	0.1–1.0 mm	1–5 mm	>5 mm
Rating	6	5	4	1	0
Roughness	Very rough	Rough	Slightly rough	Smooth	Slickensided
Rating	6	5	3	1	0
Infilling (gouge)	None	Hard filling <5 mm	Hard filling >5 mm	Soft filling <5 mm	Soft filling >5 mm
Rating	6	4	2	2	0
Weathering	Unweathered	Slightly weathered	Moderately weathered	Highly weathered	Decomposed
Ratings	6	5	3	1	0

(Continued)

Table 4.1 (Continued)

F. Effect of Discontinuity Strike and DIP Orientation in Tunneling**

Strike Perpendicular to Tunnel Axis		Strike Parallel to Tunnel Axis	
Drive with dip – Dip 45–90 degrees	Drive with dip – Dip 20–45 degrees	Dip 45–90 degrees	Dip 20–45 degrees
Very favorable	Favorable	Very unfavorable	Fair
Drive against dip – Dip 45–90 degrees	Drive against dip – Dip 20–45 degrees	Dip 0–20 – Irrespective of strike degrees	
Fair	Unfavorable	Fair	

* Some conditions are mutually exclusive. For example, if infilling is present, the roughness of the surface will be overshadowed by the influence of the gouge. In such cases use A.4 directly.

** Modified after Wickham et al. (1974).

In applying this classification system, the rock mass is divided into a number of structural regions, and each region is classified separately. The boundaries of the structural regions usually coincide with a major structural feature such as a fault or with a change in rock type. In some cases, significant changes in discontinuity spacing or characteristics, within the same rock type, may necessitate the division of the rock mass into a number of small structural regions. It should be noted that RQD and the spacing of discontinuities are actually double-counted, as pointed out by Aydan *et al.* (2014). Furthermore, the use of UCS is also another problem with this rock classification system as the rock mass rating value decreases as the UCS rock mass decreases, even though it has no discontinuities.

Bieniawski (1989) published a set of guidelines for the selection of support in tunnels in rock for which the value of RMR has been determined. These guidelines are reproduced in Table 4.2. Note that these guidelines have been published for a 10 m span horseshoe-shaped

Table 4.2 Guidelines for excavation and support of 10 m span rock tunnels in accordance to RMR system

Rock mass class	Excavation	Rock bolts (20 mm diameter, fully grouted)	Shotcrete	Steel sets
I – Very good rock RMR: 81–100	Full face, 3 m advance	Generally no support required except spot boiling.		
II – Good rock RMR: 61–80	Full face, 1–1.5 m advance. Complete support 20 m from face	Locally, bolts in crown 3 m long, spaced 2.5 m with occasional wire mesh	50 mm in crown where required	None
III – Fair rock RMR: 41–60	Top heading and bench 1.5–3 m advance in lop Heading Commence support after each blast Complete support 10 m from face	Systematic bolts 4 m long, spaced 1.5–2 m in crown and walls with wire mesh in crown	50–100 mm in crown and 30 mm in sides	None
IV – Poor rock RMR: 21–40	Top heading and bench 1.0–1.5 m advance in top heading. Install support concurrently with excavation, 10 m from face	Systematic bolts 4–5 m long, spaced 1–1.5 m in crown and walls with wire mesh	100–150 mm in crown and 100 mm in sides	Light to medium ribs spaced 1.5 m where required
V – Very poor rock RMR: < 20	Multiple drifts 0.5–1.5 m advance in lop heading Install support concurrently with excavation Shotcrete as soon as possible after blasting	Systematic bolts 5–6 m long, spaced 1–1.5 m in crown and walls with wire mesh Bolt invert	150–200 mm in crown, 150 mm in sides, and 50 mm on face	Medium low heavy ribs spaced 0.75 m with steel lagging and forepoling if required Close invert

Source: After Bieniawski (1989).

tunnel, constructed using drill and blast methods, in a rock mass subjected to a vertical stress < 25 MPa (equivalent to a depth below surface of < 900 m).

It should be noted that Table 1.2 has not had a major revision since 1973. In many mining and civil engineering applications, steel fiber–reinforced shotcrete may be considered in place of wire mesh and shotcrete.

4.3   Q-system (rock Tunneling Quality Index)

On the basis of an evaluation of a large number of case histories of underground excavations, Barton *et al.* (1974) of the Norwegian Geotechnical Institute proposed a Tunneling Quality Index (Q) for the determination of rock mass characteristics and tunnel support requirements.

The Q-system also gives a description of the rock mass stability of an underground opening in jointed rock masses. High Q-value indicates good stability, and low values mean poor stability. Based on six parameters, the Q-value is calculated using the following equation: The numerical value of the index Q varies on a logarithmic scale from 0.001 to a maximum of 1000 and is defined by:

$$Q = \frac{RQD}{J_n} x \frac{J_r}{J_a} x \frac{J_w}{SRF} \tag{4.1}$$

where the six parameters are RQD = degree of jointing (Rock Quality Designation), J_n = joint set number, J_r = joint roughness number, J_a = joint alteration number, J_w = joint water reduction factor, and SRF = stress reduction factor.

In explaining the meaning of the parameters used to determine the value of Q, Barton *et al.* (1974) offer the following comments:

The first quotient (RQD/Jn), representing the structure of the rock mass, is a crude measure of the block or particle size, with the two extreme values (100/0.5 and 10/20) differing by a factor of 400. If the quotient is interpreted in units of centimeters, the extreme "particle sizes" of 200 to 0.5 cm are seen to be crude but fairly realistic approximations. Probably the largest blocks should be several times this size and the smallest fragments less than half the size. (Clay particles are of course excluded). The second quotient (J_r/J_a) represents the roughness and frictional characteristics of the joint walls or filling materials. This quotient is weighted in favor of rough, unaltered joints in direct contact. It is to be expected that such surfaces will be close to peak strength, that they will dilate strongly when sheared, and that they will therefore be especially favorable to tunnel stability. When rock joints have thin clay mineral coatings and fillings, the strength is reduced significantly. Nevertheless, rock wall contact after small shear displacements have occurred may be a very important factor for preserving the excavation from ultimate failure.

The individual parameters are determined during geological mapping using tables that give numerical values to be assigned to a described situation, Paired, the six parameters express the three main factors that describe the stability in underground openings.

Where no rock wall contact exists, the conditions are extremely unfavorable to tunnel stability. The "friction angles" (Table 4.3) are a little below the residual strength values for most clays and are possibly downgraded by the fact that these clay bands or fillings may tend to consolidate during shear, at least if normal consolidation or if softening and swelling have occurred. The swelling pressure of montmorillonite may also be a factor here.

Table 4.3 Classification of individual parameters used in the Tunneling Quality Index Q

Description	Value	Notes
1. Rock quality designation	**RQD**	
A. Very poor	0–25	1. Where RQD is reported or measured as ≤10 (including 0), a nominal value of 10 is used to evaluate Q.
B. Poor	25–50	
C. Fair	50–75	
D. Good	75–90	2. RQD intervals of 5, i.e. 100, 95, 90 etc. are sufficiently accurate.
E. Excellent	90–100	
2. Joint set number	J_n	
A. Massive, no or few joints	0.5–1.0	
B. One joint set	2	
C. One joint set plus random	3	
D. Two joint sets	4	
E. Two joint sets plus random	6	
F. Three joint sets	9	1. For intersections use $(3.0 \times J_n)$
G. Three joint sets plus random	12	
H. Four or more joint sets, random, heavily jointed, "sugar cube," etc.	15	2. For portals use $(2.0 \times J_n)$
J. Crushed rock, earthlike	20	
3. Joint roughness number	J_r	
a. Rock wall contact		
b. Rock wall contact before 10 cm shear		
A. Discontinuous joints	4	
B. Rough and irregular, undulating	3	
C. Smooth undulating	2	
D. Slickensided undulating	1.5	1. Add 1.0 if the mean spacing of the relevant joint set is greater than 3 m.
E. Rough or irregular, planar	1.5	
F. Smooth, planar	1.0	
G. Slickensided, planar	0.5	2. $J_r = 0.5$ can be used for planar, slickensided joints having lineations, provided that the lineations are oriented for minimum strength.
c. No rock wall contact when sheared		
H. Zones containing clay minerals thick enough to prevent rock wall contact	1.0 (nominal)	

(Continued)

Table 4.3 (Continued)

Description	Value	Notes
J. Sandy, gravely or crushed zone thick enough to prevent rock wall contact	1.0 (nominal)	
4. Joint alteration number	J_a	ϕr degrees (approx.)
a. Rock wall contact		
A. Tightly healed, hard, nonsoftening, Impermeable filling	0.75	1. Values of ϕr, the residual friction angle, are intended as an approximate guide to the mineralogical properties of the alteration products, if present.
B. Unaltered joint walls, surface staining only	1.0	25–35
C. Slightly altered joint walls, nonsoftening mineral coatings, sandy particles, clay-free disintegrated rock, etc.	2.0	25–30
D. Silty-, or sandy-clay coatings, small day-fraction (nonsoftening)	3.0	20–25
E. Softening or low-friction clay mineral coatings, i.e. kaolinite, mica. Also chlorite, talc, gypsum and graphite etc., and small quantities of swelling clays. (Discontinuous coatings. 1–2 mm or less)	4.0	8–16

Source: After Barton et al. (1974)

The third quotient (J_w/SRF) consists of two stress parameters. SRF is a measure of (1) loosening load in the case of an excavation through shear zones and clay bearing rock, (2) rock stress in competent rock, and (3) squeezing loads in plastic incompetent rocks. It can be regarded as a total stress parameter. The parameter J_w is a measure of water pressure, which has an adverse effect on the shear strength of joints due to a reduction in effective normal stress. Water may, in addition, cause softening and possible outwash in the case of clay-filled joints. It has proved impossible to combine these two parameters in terms of interblock effective stress, because paradoxically a high value of effective normal stress may sometimes signify less stable conditions than a low value, despite the higher shear strength. The quotient (J_w/SRF) is a complicated empirical factor describing the "active stress." It appears that the rock-tunneling quality Q can now be considered to be a function of only three parameters, which are crude measures of:

$$\frac{RQD}{J_n} = \text{degree of jointing (block size)}$$

$$\frac{J_r}{J_a} = \text{joint friction (interblock shear strength)}$$

$$\frac{J_w}{SRF} = \text{active stress}$$

Undoubtedly, several other parameters could be added to improve the accuracy of the classification system. One of these would be the joint orientation. Although many case records include the necessary information on structural orientation in relation to excavation axis, it was not found to be the important general parameter that might be expected. Part of the reason for this may be that the orientations of many types of excavations can be and normally are adjusted to avoid the maximum effect of unfavorably oriented major joints. However, this choice is not available in the case of tunnels, and more than half the case records were in this category. The parameters J_n, J_r and J_a appear to play a more important role than orientation because the number of joint sets determines the degree of freedom for block movement (if any), and the frictional and dilatational characteristics can vary more than the down-dip gravitational component of unfavorably oriented joints. If joint orientations had been included, the classification would have been less general, and its essential simplicity lost.

Table 4.3 gives the classification of individual parameters used to obtain the Tunneling Quality Index Q for a rock mass.

(a) Application of Q-system to tunnel support system

In relating the value of the index Q to the stability and support requirements of underground excavations, Barton *et al.* (1974) defined an additional parameter that they called the Equivalent Dimension, D_e, of the excavation. This dimension is obtained by dividing the span, diameter or wall height of the excavation by a quantity called the Excavation Support Ratio (ESR). Hence:

$$D_e = \frac{\text{Excavation span diameter or height (m)}}{\text{Excavation Support Ratio ESR}} \tag{4.2}$$

The value of ESR is related to the intended use of the excavation and to the degree of security that is demanded of the support system installed to maintain the stability of the excavation. Barton *et al.* (1974) suggest the following values shown in Table 4.4.

The equivalent dimension, D_e, plotted against the value of Q, is used to define a number of support categories in a chart published in the original paper by Barton *et al.* (1974). This chart has recently been updated by Grimstad and Barton (1993) to reflect the increasing use

Table 4.4 Values of ESR for various structures

Excavation category		ESR
A	Temporary mine openings	3–5
B	Permanent mine openings, water tunnels for hydropower (excluding high-pressure penstocks), pilot tunnels, drifts and headings for large excavations	1.6
C	Storage, access tunnels	1.3
D	Power stations, major road and railway tunnels, civil defiance chambers, portal intersections	1.0
E	Underground nuclear power stations, railway stations, sports and public facilities, factories	0.8

of steel fiber–reinforced shotcrete in underground excavation support. Figure 4.1 is reproduced from this updated chart.

From Figure 4.1, a value of D_e of 9.4 and a value of Q of 4.5 place this crusher excavation
in category (4) which requires a pattern of rock bolts (spaced at 2.3 m) and 40 to 50 mm of
unreinforced shotcrete.

Because of the anticipated mild to heavy rockburst conditions, it may be prudent to
destress the rock in the walls of this crusher chamber. This is achieved by using relatively
heavy production blasting to excavate the chamber and omitting the smooth blasting usually used to trim the final walls of an excavation such as an underground powerhouse at
shallower depth. Caution is recommended in the use of destress blasting, and, for critical
applications, it may be advisable to seek the advice of a blasting specialist before embarking
on this course of action.

Løset (1992) suggests that, for rocks with $4 < Q < 30$, blasting damage will result in the
creation of new "joints" with a consequent local reduction in the value of Q for the rock surrounding the excavation. He suggests that this can be accounted for by reducing the RQD
value for the blast-damaged zone.

Assuming that the RQD value for the destressed rock around the crusher chamber drops
to 50%, the resulting value of $Q = 2.9$. From Figure 4.3, this value of Q, for an equivalent

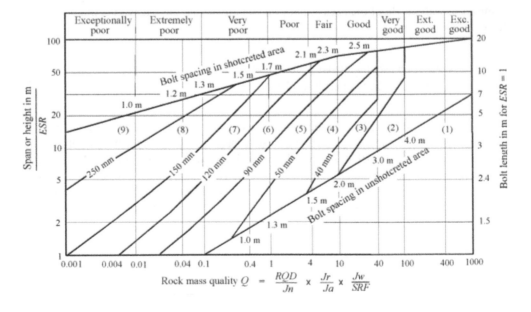

REINFORCEMENT CATEGORIES
1) Unsupported
2) Spot bolting
3) Systematic bolting
4) Systematic bolting with 40-100 mm
   unreinforced shotcrete

5) Fibre reinforced shotcrete, 50 - 90 mm, and bolting
6) Fibre reinforced shotcrete, 90 - 120 mm, and bolting
7) Fibre reinforced shotcrete, 120 - 150 mm, and bolting
8) Fibre reinforced shotcrete, > 150 mm, with reinforced
   ribs of shotcrete and bolting
9) Cast concrete lining

Figure 4.1 Estimated support categories based on the tunneling quality index Q

Source: After Grimstad and Barton (1993), reproduced from Palmstrom and Broch (2006)

dimension, D_e of 9.4, places the excavation just inside category (5), which requires rock bolts, at approximately 2 m spacing, and a 50 mm thick layer of steel fiber–reinforced shotcrete.

Barton *et al.* (1980) provide additional information on rock bolt length, maximum unsupported spans and roof support pressures to supplement the support recommendations published in the original 1974 paper.

The length L of rock bolts can be estimated from the excavation width B and the Excavation Support Ratio (ESR):

$$L = 2 + \frac{0.15B}{ESR} \tag{4.3}$$

The maximum unsupported span can be estimated from

$$\text{Maximum span (unsupported)} = 2ESRQ^{0.4} \tag{4.4}$$

Based on analyses of case records, Grimstad and Barton (1993) suggest that the relationship between the value of Q and the permanent roof support pressure is estimated from:

$$P_{roof} = \frac{2\sqrt{J_n}Q^{-\frac{1}{3}}}{3J_r} \tag{4.5}$$

4.4   Rock Mass Quality Rating (RMQR)

The Rock Mass Quality Rating (RMQR) is a new rock classification, developed by Aydan *et al.*, 2014. This new rock classification quantifies the state of rock mass and helps to estimate geomechanical properties (UCS, cohesion, friction angle, deformation modulus, Poisson's ratio and tensile strength) of rock masses using a unified formula considering RMQR together with intrinsic geomechanical properties of intact rock. The comparisons of the empirical formula, together with the values of constants, were found to be quite consistent with the experimental results for data compiled from various rock engineering projects in Japan. Based on the databases of the authors, the application of the system was also extended to rock support selection for underground caverns and tunnels by considering the type of instability mode in relation to RMQR value. Some empirical relations, established between RMQR and dimensions of the elements of support systems, seem sufficient for many engineering applications and act as guidelines. The rating of parameters of this system is given in Table 4.5.

The most commonly used factors in engineering descriptions of rock masses are the condition and geometrical characteristics of discontinuities such as the discontinuity set number (DSN), discontinuity spacing (DS) and discontinuity condition (DC). The weathering of rocks causes the weakening of bonds and decomposition of constituting minerals into clayey materials. The alteration process is due to percolating hydrothermal fluids in rock mass, and it may act on rock mass in a positive or negative way. The positive action of the alteration may heal existing rock discontinuities by rewelding through the deposition of ferrous oxides, calcite or siliceous filling material. On the other hand, the negative action of the alteration would cause the weakening of bonding of particles of rocks and producing clayey materials. As the intact rock is one of the most important parameters influencing the mechanical

Table 4.5 Rating of parameters of RMQR rock classification system

Degradation degree (DD)	Fresh	Stained	Slight degradation	Moderate degradation	Heavy degradation	Decomposed
Rating (R_{DD})	15	12	9	6	3	1–0
Discontinuity set number (DSN)	None (solid or massive)	One set plus random	Two sets plus random	Three sets plus random	Four sets plus random	Crushed or shattered
Rating (R_{DSN})	20	16	12	8	4	1–0
Discontinuity spacing (DS) or RQD	None or DS ≥ 24 m	24 > DS ≥ 6 m	6m > DS ≥ 1.2 m	1.2m > DS ≥ 0.3 m	0.3 > DS ≥ 0.07 m	0.07 m > DS
	RQD= 100			100 > RQD2 ≥ 75	75 > RQD ≥ 35	35 > RQD
Rating (R_{DS})	20	16	12	8	4	1–0
Discontinuity condition (DC)	None	Healed or intermittent	Rough	Relatively smooth and light	Slicken sided with thin infill or separation (t < 5 mm)	Thick fill or separation (t > 10 mm)
Rating (R_{DC})	30	26	22	15	7	1

Or, alternatively, excluding "None" and "Healed or intermittent" classes

Discontinuity condition (DC)
$R_{DC} = R_{DCA} + R_{DCI} + R_{DCR}$

	Fresh	Stained	Slight degradation	Moderate degradation	Heavy degradation	Decomposed
Aperture or separation	None or Very tight, < 0.1 mm	0.1–0.25 mm	0.25–0.5 mm	0.5–2.5 mm	2.5–10 mm	> 10 mm
Rating (R_{DCA})	6	5	4	3	2	1
Infilling	None	Surface staining only	Thin coating < 1 mm	Thin filling 1 < t 10 mm	Thick filling 60 > t > 10 mm	Very thick filling or shear zone t > 60 mm
Rating (R_{DCI})	6	5	4	3	2	1
Rating (R_{DCR}) Roughness Descriptive	Very rough	Rough	Smooth undulating	Smooth planar	Slicken-sided	Shear band/ zone
Profile No. in ISRM (2007)	10 9	8 7	6 5	4 3	2	1–0

Groundwater seepage condition (GWSC)	Dry	Rating (R_{DCR}) Damp	10 Wet	9	8	7 Dripping	6	5 Rowing	4	3	2 Gushing	1–0
Rating (R_{GWSC})	9	7	5			3		1			0	
Groundwater absorption condition (GWAC)	Non absorptive	Capillarity or electrically absorptive	Slightly absorptive			Moderately absorptive		Highly absorptive			Extremely absorptive	
Rating R_{GWAC}	6	5	4			3		2			1–0	

*RQMR = $R_{DD} + R_{DSN} + R_{DS} + R_{DC} + R_{GWSC} + R_{GWAC}$

response of rock masses, weathering and/or the negative action of hydrothermal alteration may be accounted for as the degradation degree (DD) of intact rock. Groundwater is also an important parameter affecting the mechanical response of rock masses. There are also cases, where some rocks may absorb groundwater electrically or chemically, resulting in the drastic reduction of material properties and/or swelling. RMQR system incorporates important parameters of the available quantitative modern rock classifications, and Table 4.5 provides the descriptions of each parameter and their ratings. In the following subsections, first, the basic concepts involving each parameter and their ratings on the basis of knowledge gained in rock mechanics and rock engineering are briefly explained.

(a)   Degradation Degree (DD)

The degradation processes generally cause weakening of the bonds between particles or grains constituting rocks, and, physically, they cause the reduction of the strength and deformation modulus of intact rock and also influence the joint spacing and discontinuity filling material in the form of clay. Therefore, in RMQR, degradation degree, which is considered as one of the elements of the joint condition parameter in some previously developed classifications, is taken as one of the input parameters.

(b)   Discontinuity set number (DSN)

The rock mass structurally would have at least one discontinuity set associated with the surface shape of erosion. There may be some cases where rock mass is completely shattered and crushed. Therefore, the discontinuous nature of rock masses may be described through some adjectives, such as none, one set plus random, two sets plus random, three sets plus random, four sets or more, and crushed/shattered. It should be noted that, if the discontinuity set number is four or more, it would definitely imply that it was subjected to tectonic events in the past.

(c)   Discontinuity spacing (DS)

The modern rock mass classifications consider that the rock mass is massive when the discontinuity spacing is greater than 2–3 m. This definition may not be so important when the underground openings have a smaller size, say, less than 8–6 m in diameter or span. However, when one considers the present common size of major underground powerhouses and storage caverns for crude oil and gas, the rock mass around the underground opening would look very blocky. Therefore, the present discontinuity spacing definitions are not compatible with actual circumstances, and it needs some improvements with consideration of the actual size of underground structures. To describe the representative discontinuity spacing, RMQR includes six categories of discontinuity spacing, as given in Table 4.5 with their ratings. As understood from Table 4.3, RQD is not sensitive to the variation of discontinuity spacing greater than 1 m, and RQD should not be used to determine the rating for discontinuity spacing if the discontinuity spacing is greater than 1 m. However, by considering that RQD is a commonly used parameter, particularly in borehole cores, it is also included in Table 4.5 as an alternative parameter to discontinuity spacing, provided that it is inferred to be less than 1 m and free of drilling-induced disturbance.

(d) Discontinuity condition (DC)

The causes of the formation of discontinuities in rock masses are various, and the condition of discontinuities is closely related to their genesis. The condition of discontinuities not involving tectonic events are generally favorable unless they are filled with clayey material or discontinuity walls are subjected to weathering. However, the tectonically induced fractures may be associated with relative shear displacement, and they may produce slickensided discontinuities with a certain thickness of clayey gouge. Such conditions would considerably reduce the shear strength of discontinuities, and they may be squeezed out under redistributed *in-situ* stress or washed away under high groundwater pressure. Table 4.6 describes the possible discontinuity conditions and the ratings suggested for visual observations.

(e) Groundwater condition

The effects of groundwater on rock mass are described through adjectives such as dry, damp, wet, dripping, flowing and gushing. It is known that the strength and deformation modulus of weak rocks such as water-absorbing rocks decrease drastically with water content. It is also reported that even such properties of hard rocks may decrease with saturation. Rocks containing water-absorbing minerals have this feature, and the geomechanical properties of the surrounding rock mass may be drastically reduced. Furthermore, it may also show large volumetric changes (swelling, contraction) during excavation and cyclic groundwater changes and/or disintegration. Therefore, in addition to the seepage condition of groundwater (GWSC), the groundwater absorption characteristics of rocks (GWAC) are taken into account. The descriptions and their ratings for these two characteristics are determined from Table 4.5.

The value of RMQR ranges between 0 and 100. Rock mass is divided into six classes and their rating ranges are given in Table 4.7.

Table 4.6 The possible discontinuity conditions and the ratings suggested for visual observations

Aperture or separation		None or very tight, <0.1 mm	0.1–0.25 mm	0.25–0.5 mm		0.5–2.5 mm	2.5–10 mm	>10 mm
Rating (R_{DCA})*		6	5	4		3	2	1–0
Infilling		None	Surface staining only	Thin coating <1 mm		Thin filling 1 < t < 10 mm	Thick filling 60 > t > 10 mm	Very thick filling or shear zones t > 60 mm
Rating (R_{DCI})*		6	5	4		3	2	1–0
Roughness	Descriptive	Very rough	Rough	Smooth undulating		Smooth planar	Slicken-sided	Shear band/zone
	ISRM Profile No.	10	9 8	7	6	5 4	3 2	1–0
Rating (R_{DCR})*		10	9 8	7	6	5 4	3 2	1–0

* $R_{DC} = R_{DCA} + R_{DCI} + R_{DCR}$

(f)   Application of RMQR to rock support design

The design of support systems of tunnels in rock engineering is of great importance, as these structures are required to be stable during their service lifetime. Provided that the elements of support systems are resistant against chemical actions due to environmental conditions and their long-term behavior is satisfactory, the support systems must be designed against anticipated load conditions. As rock masses have many geological discontinuities and weakness zones, the load acting on support systems may be due to the deadweight of potential unstable blocks formed by rock discontinuities, which may be designated as structurally controlled or local instability modes and independent of the *in-situ* stress state or inward displacement of rock mass due to elasto-plastic or elasto-visco plastic behavior induced by *in-situ* stresses (Figures. 4.2 and 4.3). Therefore, the main purpose of the design of support systems must be well established with due considerations of these situations.

Aydan and Kawamoto (1999) developed a database system for large underground openings, and it was named CAVERN. This database system was modified recently to include the RMQR classification system. It was renamed as UGCAVERN database system and converted to the MS-ACCESS environment from the previous development environment dBasePLUS III. The system includes parameters related to the geometry, support system,

Table 4.7 Rock classes of RMQR classification system

Rock Class	I	II	III	IV	V	VI
Description of rock mass	Solid or Rock material	Very good	Good	Medium	Poor or Weak	Very poor/ very weak
RMQR	$100 \geq RMQR > 95$	$95 \geq RMQR > 80$	$80 > RMQR \geq 60$	$60 > RMQR \geq 40$	$40 > RMQR \geq 20$	$20 > RMQR$
DENKEN	A	B	CH	CM	CL	D or F

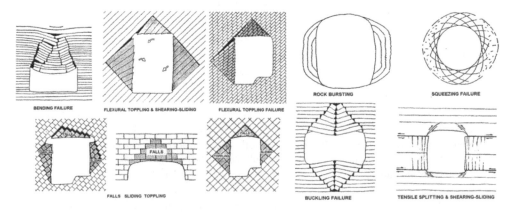

Figure 4.2 Instability modes of underground openings

Source: Arranged from Aydan (1989)

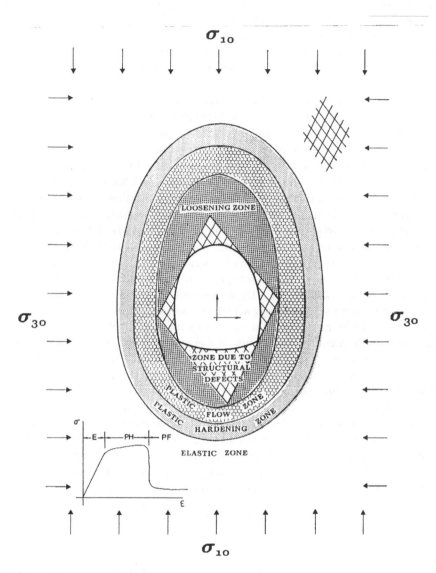

σ_{10}

σ_{30}

σ_{30}

LOOSENING ZONE

ZONE DUE TO
STRUCTURAL
DEFECTS

PLASTIC FLOW ZONE

PLASTIC

FLOW

HARDENING ZONE

ELASTIC ZONE

σ_{10}

σ

E — PH — PF

ε

Figure 4.3 Load conditions in a tunnel

rock classifications, *in-situ* stress and geomechanical properties of intact rock and rock mass
and measured displacements of the large underground openings. It has about 110 entries of
worldwide large underground openings. Aydan (Aydan *et al.*, 1993, 1996) also developed
two databases for tunnels through squeezing rocks as well as for their geomechanical properties named as SQTUN (103 entries) and SQUZROCK (171 entries), respectively, and they
were developed originally on dBasePLUS III environment. The both databases have been
converted to their equivalents on MS-ACCESS environment.

Recently, Aydan and Genis (2010) expanded SQTUN to case histories of rockbursting and
renamed as SQROCKBURST (146 entries). These databases originally include RMR and

Q system as two rock classification systems, namely Japan Railway Classification (JRAC) and Japan Roadway Classification (JROC). The interrelations among several parameters can be explored using a code developed in True-BASIC programming language.

The competency implies that the intrinsic rock material does not yield under an induced stress state, and the ratio of the UCS of the intact rock over the major *in-situ* stress is generally more than 4.

Tunnels, which are also becoming large in recent years (width is up to 14 m), are relatively smaller in size (10–11 m wide, 7–9 m high) and long linear structures. There is rich worldwide experience in tunneling under diverse rock conditions. Tunnels may be excavated in various rock masses, which may be subjected to squeezing, rockbursting and structural failure. Even all these failure modes may be experienced in a single tunnel. Except new large tunnels, the support system of tunnels generally consists of rock bolts, shotcrete, steel ribs as primary support members and concrete lining to smooth the airflow, to prevent direct seepage of groundwater into the tunnel and auxiliary extra safety measure against rock loads after the introduction of the New Austrian Tunneling Method (NATM). When rock mass is not competent against stress-induced yielding, they may be lined with the invert concrete liner. When tunnels are excavated by TBMs, rock bolts and shotcrete may be totally disappeared.

Using these databases and adopting the approach of Aydan and Kawamoto (1999), several interrelations have been established for the dimensions of support members and related size parameters of the underground openings with the consideration of structurally controlled and stress-induced instability modes using the databases mentioned before. However, the interrelations could not be presented in this article due to lack of space. The design of support systems is relatively simple once the modes of structural instability, which may be also categorized as local instability modes by Aydan (1989), are defined. The procedures described by Aydan (1989, 1994) and Kawamoto *et al.* (1991) can be easily adopted for such a purpose.

As for the design of support members against stress-induced instability modes, the use of past experiences, analytical and numerical methods (i.e. Bieniawski, 1989; Barton *et al.*, 1974; Barton and Grimstad, 1993 Wickham *et al.*, 1974; Aydan, 1989, 1994; Aydan *et al.*, 1992, 1993, 1996) using the geomechanical properties of rock mass, which may be obtained with the use of RMQR and intrinsic properties of intact rock, is necessary, together with *in-situ* stress state and geometry of underground openings. Aydan *et al.* (2014) suggested Tables 4.8 and 4.9 for the empirical design of support systems for tunnels in competent rock, which may be subjected to even stress-induced failure modes such as squeezing and

Table 4.8 Support system for cavern in competent discontinuous rock mass (S: 20–25 m, H: 40–50 m)

RMQR range	Bolts				Anchors				Shotcrete	
	Roof		Sidewall		Roof		Sidewall		Roof	Sidewall
	L (m)	e (m)	L (m)	e (m)	L (m)	e (m)	L (m)	e (m)	t (mm)	t (mm)
100 ≥ RMQR > 95	–	–	–	–	–	–	–	–	–	–
95 ≥ RMQR > 80	3	2.5	5	3.0	8	4.0	10	5.0	100	80
80 ≥ RMQR > 60	4	2.2	6	2.7	10	3.7	12	4.3	150	120
60 ≥ RMQR > 40	5	1.9	7	2.3	12	3.3	15	3.6	200	150

L is length; e is spacing; t is thickness; bolt is 200 kN; anchor is 400 kN; UCS of shotcrete: 10 MPa; S is span (width); H is Height)

Table 4.9 Support system for tunnels (D and B; 10 m span)

RMQR Range	Rock bolts		Shotcrete	Steel Ribs	Wire Mesh	Lining (mm)	Concrete Invert	
	L_b	e_b	t_s					
	(m)	(m)	(mm)				t_i (mm)	Bolt
100 ≥ RMQR > 95	None	None	None	None	None	None	None	–
95 ≥ RMQR > 80	2–3	2.5	50	None	None	None	None	–
80 ≥ RMQR > 60	3–4	2.0	100	Light	Yes	200	None	–
60 ≥ RMQR > 40	4–5	1.5	150	Medium	Yes	300	300	–
40 ≥ RMQR > 20	5–6	1.0	200	Heavy	Yes	500	500	5–6
20 ≥ RMQR	6–7	0.5	250	Very heavy	Yes	800	800	6–7

rockbursting, respectively. The numbers in Tables 4.8 and 4.9 are based on the databases previously mentioned, together with the considerations of past experiences as well as some empirical, analytical and numerical methods (i.e. Aydan and Kawamoto, 1999; Aydan and Ulusay, 2014 Aydan et al., 1993, 1996, 2000; Aydan, 2011; Kawamoto et al., 1991). Nevertheless, Tables 4.8 and 4.9 should actually be sufficient for the design of the support system of many tunnels.

Engineers generally employ empirical methods to estimate the properties of rock mass for the stability assessment of structures and feasibility studies since in-situ tests are usually expensive to perform. For such purposes, rock mass classification systems such as RMR and Q system are often used. Besides the utilization of rock classification for preliminary structural design, some empirical relations among RMR, Q-value, GSI and the like and the rock mass properties such as unit weight, deformation modulus, uniaxial compressive strength, friction angle, elastic wave velocity are proposed and used in practice. In the following section, a brief history of rock classifications and modern classifications and some recent trends are presented.

4.5   Geological Strength Index classification

The Geological Strength Index (GSI) was introduced by Hoek (1994). The rock mass characterization is straightforward, and it is based upon the visual impression of the rock structure, in terms of blockiness, and the surface condition of the discontinuities indicated by joint roughness and alteration. The combination of these two parameters provides a practical basis for describing a wide range of rock mass types, with diversified rock structure ranging from very tightly interlocked strong rock fragments to heavily crushed rock masses. Based on the rock mass description the value of GSI is estimated from the contours given in his table. Hoek also proposed establishing some empirical relations between his GSI numbers and RMR.

4.6   Denken's classification and modified Denken's classification

An example of rock classification, which is called Denken's classification in Japan, is briefly described. This classification system is widely used in Japan, and it constitutes the basis for other rock classifications proposed for some specific structures. The evaluation of rock mass based on visual impressions of intact rock, jointing, weathering, the sound of geologic hammer and rock classes is denoted as A, B, C_H, C_M, C_L, D and F in descending order of rock quality.

4.7   Estimations of engineering properties

There are presently four engineering approaches to assess the strength of rock masses:

1   *In-situ* testing
2   Empirical relations based on the elastic wave velocity of rock masses
3   Empirical relations based on indices of rock mass classifications
4   Reduction factor using the elastic wave velocity of intact rock and of rock mass and properties

Although several rock classifications are used in many countries, it seems that RMR, Q-system and recently RMQR are the most widely known rock classifications (Bieniawski, 1974; Barton *et al.*, 1974; Aydan *et al.*, 2014). Hoek and Brown (1980, 1988) tried to establish some relations between the parameters of their empirical yield criterion and RMR. Bieniawski and his coworkers (Bieniawski, 1974; Kalamaras and Kalamaras, 1995; Jasarevic and Kovacevic, 1996; Serafim and Pereira, 1983; Aydan *et al.* (1997; Aydan and Dalgıç, 1998) also tried to establish some empirical relations between RMR-value and deformation modulus and the compressive uniaxial strength values of rock masses (Table 4.10).

4.7.1   Elastic modulus

Bieniawski (1978) proposed the following function between RMR-value and elastic modulus of rock masses:

$$E_m = 2 * RMR - 100 \tag{4.6}$$

The unit of E_m is GPa. However, this function could not be applied to rock masses having a RMR value less than 50. Serafim and Pereira (1983) put forward the following function in view of experimental data from dam sites:

$$E_m = 10^{((RMR-10)/40)} \tag{4.7}$$

The unit of E_m is GPa in the preceding equation. Jasarevic and Kovacevic (1996) have come up with the following empirical relation between RMR value and elastic modulus of rock masses by taking into account experiments performed on Adriatic limestone:

$$E_m = e^{(4.407+0.081*RMR)} \tag{4.8}$$

Recently, Aydan *et al.* (1997) developed the following function by using experimental data from sites in Japan:

$$E_m = 0.0097 RMR^{3.54} \tag{4.9}$$

The unit of E_m is MPa in Equations (4.5) and (4.6).

4.7.2   Uniaxial compressive strength

The following relation between RMR value and the uniaxial strength of rock masses was proposed by Aydan *et al*. (1997):

$$\sigma_{cm} = 0.0016 RMR^{2.5} \tag{4.10}$$

The unit is MPa. Although this function is applicable to rock masses, it could not cover all experimental data. The scattering is likely to be associated with the strength of intact rock. It is more desirable to establish a reduction coefficient as a function of RMR value between the mass strength and intact rock strength, which may handle the scattering due to the intrinsic strength of rocks. Hoek and Brown (1980) proposed such a relation as given here:

$$\sigma_{cm} = \sigma_{ci} \sqrt{e^{(RMR-100)/B}} \tag{4.11}$$

Hoek and Brown (1980) initially suggested the value of 6 for constant B, and they remodified the value of constant B as 9 later. On the other hand, Kalamaras and Bieniawski (1995) suggested the following formula between the uniaxial strength of intact rock and that of rock mass as a function of RMR value:

$$\sigma_{cm} = 0.5 \frac{RMR - 15}{85} \sigma_{ci} \tag{4.12}$$

Aydan and Dalgıç (1998) put forward the following function for estimating the mass strength of squeezing rocks:

$$\sigma_{cm} = \frac{RMR}{RMR + 6(100 - RMR)} \sigma_{ci} \tag{4.13}$$

4.7.3   Friction angle

In the literature, it is difficult to find any relation between friction angle of rock mass and RMR value. Aydan and Dalgıç (1998) suggested the use of an empirical relation proposed by Aydan *et al*. (1993) after obtaining the mass strength. Aydan and Kawamoto (1999) recently proposed the following function:

$$\phi_m = 20 + 0.05 * RMR \tag{4.14}$$

Although no direct relation between RMR value and the cohesion of rock mass was proposed, the cohesion of rock masses may be obtained from the following theoretical relation among cohesion, uniaxial compressive strength and friction angle once the uniaxial compressive strength and friction angle are known.

$$c_m = \frac{\sigma_{cm}}{2} \frac{1 - \sin\phi_m}{\cos\phi_m} \tag{4.15}$$

4.7.4   Relation between rock mass properties and Rock Mass Quality Rating (RMQR)

The design of many geoengineering structures is based on the equivalent properties of rock masses. For this purpose, *in-situ* tests on the strength properties of rock masses are carried out using uniaxial and triaxial compression, direct shear and plate loading tests (e.g. Nose, 1992; Lama, 1974, Archambault and Ladanyi, 1970, 1972; Brown and Trollope, 1970; Cording et al. 1972; Einstein et al. 1969; Goldstein et al. 1966; Kawamoto 1970; Protodyakonov and Koifman, 1964, Van Heerden, 1975; Vardar 1977; Ulusay et al. 1993, Walker 1971). However, it is very rare to carry out *in-situ* triaxial compression experiments due to their cost. Using the available experimental data, some empirical direct relations among different mechanical properties and some rock mass classification parameters are proposed by various researchers. Most of these relations are concerned only with elastic modulus and rock mass strength, except those by the authors. As discussed by Aydan *et al.* (1997), the scattering of experimental data and rock classification indexes is very large, and such approaches generally fail when intact rock itself is a soft rock. Therefore, the property of intact rock and rock mass classification indexes must be involved in such evaluations.

The recent tendency is to obtain mass properties from the utilization of properties of intact rock and rock mass classification indexes (i.e. Hoek and Brown, 1980, 1988; Hoek, 1994; Aydan and Kawamoto, 2000). There are several proposed relations between the normalized properties of rock mass by those of intact rock and rock mass classification indexes as listed in Table 4.10.

Aydan and Dalgıç (1998) proposed an empirical relation between RMR and rock mass strength in terms of strength of intact rock. This relation was extended to other geomechanical properties of rock mass by Aydan and Kawamoto (2000). Recently Aydan and Ulusay (2014 provided relations for six different mechanical properties of rock mass using the proposed relation by Aydan and Kawamoto (2000). In this study, RMR is replaced by RMQR, and it is given in the following form for any mechanical properties of rock mass in terms of those of intact rock.

$$\alpha = \alpha_0 - (\alpha_0 - \alpha_{100}) \frac{RMQR}{RMQR + \beta(100 - RMQR)} \tag{4.16}$$

where $\acute{a}_0$ and $\acute{a}_{100}$ are the values of the function at RMQR = 0 and RMQR = 100 of property $\acute{a}$, and $\hat{a}$ is a constant to be determined by using a minimization procedure for experimental values of given physical or mechanical properties. The authors proposed some values for these empirical constants with the consideration of *in-situ* experiments carried out in Japan as given in Table 4.11. When a representative value of RMQR is determined for a given site, the geomechanical properties of rock mass can be obtained using Equation (4.16) together with the values of constants given in Table 4.10 and the values of intact rock for a desired property.

The empirical relations for normalized properties presented in the previous section are compared with the experimental results from *in-situ* tests carried out at various large projects (underground power houses, dams, nuclear power plants and underground crude oil and gas storage caverns) in Japan. Figure 4.4 compares the experimental results for elastic modulus and Poisson's ratio of rock mass. The experimental results on normalized elastic modulus of

Table 4.10 Empirical relations between rock mass classification and normalized properties of rock mass

Property	Relation	Proposed by
Deformation modulus, E_M	$\dfrac{E_m}{E_i} = 0.009\,e^{RMR/22.82} + 0.000028\,RMR^2$	Nicholson and Bieniawski 1990
	$\dfrac{E_m}{E_i} = \dfrac{RMR}{RMR + \beta(100 - RMR)}$	Aydan and Kawamoto (2000)
	$\dfrac{E_m}{E_i} = 0.02 + \dfrac{(1 - 0.5D)}{1 + e^{((60+15D-GSI)/11)}}$	Hoek and Diederichs (2006)
	$\dfrac{E_m}{E_i} = \dfrac{1}{2}\left(1 - \cos\left(\pi\,\dfrac{RMR}{100}\right)\right)$	Mitri et al. (1994)
	$\dfrac{E_m}{E_i} = 10^{0.0186RQD-1.91}$	Zhang and Einstein (2004)
	$\dfrac{E_m}{E_i} = e^{(RMR-100)/36}$	Galera et al. (2005)
Uniaxial compressive strength, σ_{cm}	$\dfrac{\sigma_{cm}}{\sigma_{ci}} = \sqrt{s}$ $\quad(s = e^{(RMR-100)/9})$	Hoek and Brown (1980)
	$\dfrac{\sigma_{cm}}{\sigma_{ci}} = e^{(RMR-100)/24}$	Kalamaras and Bieniawski (1995)
	$\sigma_{cm} = \dfrac{RMR}{RMR + 6(100 - RMR)}\,\sigma_{ci}$	Aydan and Dalgıç (1998)
Cohesion, c_m	$c_m = \dfrac{RMR}{RMR + 6(100 - RMR)}\,c_i$	Aydan et al. (2012)
Friction angle, ϕ_m	$\dfrac{\varphi_m}{\varphi_i} = 0.3 + 0.7\,\dfrac{RMR}{RMR + \beta(100 - RMR)}$	Aydan and Kawamoto (2000)
Poisson's ratio, v_m	$\dfrac{v_m}{v_i} = 2.5 - 1.5\,\dfrac{RMR}{RMR + (100 - RMR)}$	Aydan et al. (2012)
Tensile strength, σ_m	$\dfrac{\sigma_{tm}}{\sigma_{ti}} = \dfrac{RMR}{RMR + 6(100 - RMR)}$	Tokashiki (2011)

s, a are rock mass constants, c_i is cohesion of intact rock, f is the friction angle of intact rock, and n_i is Poisson's ratio of intact rock.

Table 4.11 Values of α_0, α_{100} and β for various properties of rock mass

Property (α)	α_0	α_{100}	β
Deformation modulus	0.0	1.0	6
Poisson's ratio	2.5	1.0	0.3
Uniaxial compressive strength	0.0	1.0	6
Tensile strength	0.0	1.0	6
Cohesion	0.0	1.0	6
Friction angle	0.3	1.0	1.0

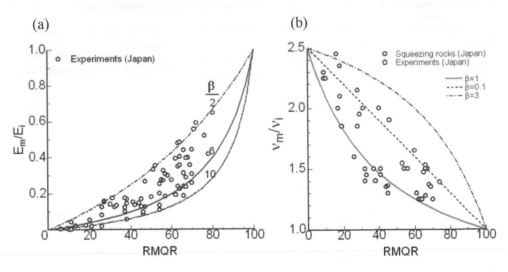

Figure 4.4 Comparison of experimental data for (a) deformation modulus, (b) Poisson's ratio of rock mass with empirical relation (Equation 4.16) together with values of parameters given in Table 4.11

rock mass are closely represented by the empirical relation (4.16), together with the values given in Table 4.11, and they are clustered around the curve with the value of coefficient β as 6.

It should be noted that experiments on the Poisson's ratio of rock masses are quite rare. In this particular comparison, Poisson's ratio of rock mass in tunnels through squeezing rocks correlated with RMQR using the approach proposed by Aydan and Dalgıç (1998) and Aydan et al. (2000) is also included. The data for RMQR value less than 50 is mainly from those of rock masses exhibiting squeezing behavior (Aydan et al., 1993, 1996). The measured data is well enveloped by the empirical relation with the values of coefficient β ranging between 0.1 and 3. The authors suggest that the values of α_0, α_{100} and β should be 2.5, 1.0 and 1, respectively as given in Table 4.11.

Figure 4.5 compares experimental results with empirical relations for normalized uniaxial compressive strength and tensile strength of rock masses by those of intact rock. The uniaxial compressive strengths of rock masses plotted in this figure are mostly obtained using rock shear test together with the Mohr-Coulomb failure criterion. The experimental results generally confirm the empirical relation given in Equation (4.16) in analogy to that proposed by Aydan and Dalgıç (1998).

In literature, there is almost no *in-situ* experimental procedure or experimental results for the tensile strength of rock mass to the knowledge of the authors. However, there is a possibility of utilizing plate loading tests, large-scale water chamber experiments and borehole jacking tests for indirect inference of tensile strength of rock mass from the measured responses.

The authors investigated the Ryukyu limestone cliffs along the shores of Okinawa, Miyako, Kurima, Ikema, Ishigaki, Ikejima, Heianza and Miyagi and Iriomote islands of Japan for inferring the tensile strength of rock masses. The authors also back-analyzed the stable and unstable (failed) cliffs using a theory based on the cantilever theory (Tokashiki

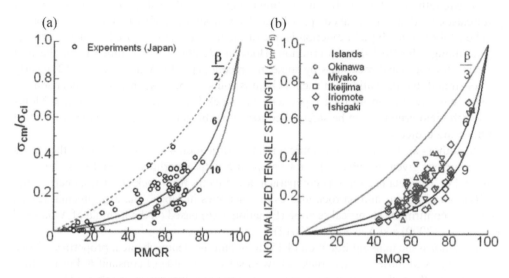

Figure 4.5 Comparison of experimental data for (a) uniaxial compression and (b) tensile strengths of rock masses with empirical relations (Equation 4.16), together with values of parameters given in Table 4.11

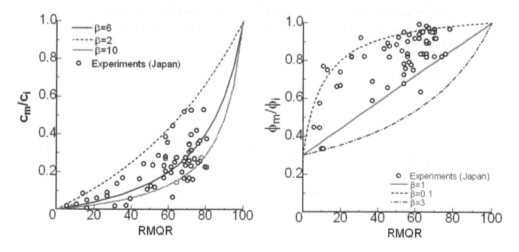

Figure 4.6 Comparison of (a) cohesion and (b) friction angle of rock mass with empirical relation (Equation 4.16), together with values of parameters given in Table 4.11

and Aydan, 2010). Tokashiki and Aydan (2011a, 2011b) fitted the inferred tensile strength of the rock mass normalized by that of intact rock using the empirical relation of Aydan and Kawamoto (2000). In this study, such evaluations were revisited, and RMQR values were recalculated. The results are plotted in Figure 4.6 by varying the value of the empirical constant β between 5 and 7. As the ratio of the uniaxial compressive strength of rock to its tensile strength is within the range of 10–20 and remains constant for the same rock type, it is found that the value of empirical constant β could be designated as 6 in view of the inferred tensile strength of rock mass. It is interesting to notice that the values of empirical constant β for elastic modulus, uniaxial compressive and tensile strength of rock masses are the same.

The Mohr-Coulomb yield criterion is one of the most commonly used in rock engineering. Although the Hoek-Brown criterion (Hoek *et al.*, 2002) was claimed to be the best criterion for rocks and rock mass by some, the recent paper by Aydan *et al.* (2012) clearly demonstrated that the validity of such a claim is found to be false through comparisons of experimental results on all rock types with the Hoek-Brown criterion. The linear Mohr-Coulomb yield criterion can be safely used for a possible stress state encountered in actual engineering projects.

The authors again utilize the empirical relation (Equation 4.16), together with the values of parameters given in Table 4.11, for comparing with experimental results as shown in Figure 4.6. The data used in this comparison are directly from rock shear tests carried out on Ryukyu limestone and on rock masses in other sites in Japan. The experimental results generally confirm the empirical relation (Equation 4.16) based on the formula of Aydan and Kawamoto (2000) and Aydan *et al.* (2012).

The major issue in using Equation (4.15) to obtain the geomechanical properties of rock mass in terms of those of intact rock is how to select the value of constant β. For practical applications, the authors strongly suggest the use of the values given in Table 4.11.

Aydan *et al.* (2012) recently proposed a procedure to evaluate the direct shear tests on rock masses with the use of both Mohr-Coulomb and Aydan yield/failure criteria (Aydan, 1995), together with the use of the unified formula of Aydan and Kawamoto (2000). The same approach can be adopted herein, replacing RMR with RMQR in the respective equations. The specific form of the Mohr-Coulomb yield criterion in shear stress and normal stress space may be written as:

$$\tau = c_m + \sigma_n \tan \phi_m \tag{4.17}$$

where

$$c_m = \frac{RMQR}{RMQR + 6(100 - RMQR)} c_i, \quad \phi_m = \left(0.3 + 0.7 \frac{RMQR}{100}\right) \phi_i \tag{4.18}$$

Similarly, Aydan's criterion is written in terms of shear stress and normal stress by neglecting the effect of temperature as:

$$\tau = c_{m\infty}\left[1 - \left(1 - \frac{c_{mo}}{c_{m\infty}}\right)\right] e^{-b_m \sigma_n} + \sigma_n \tan \phi_{m\infty} \tag{4.19}$$

where

$$c_{m\infty} = \frac{RMQR}{RMQR + 6\,(100 - RMQR)} c_\infty, \quad c_{mo} = \frac{RMQR}{RMQR + 6\,(100 - RMQR)} c_o,$$

$$b_m = \left(0.3 + 0.7 \frac{RMQR}{RMQR + (100 - RMQR)}\right) b_i \tag{4.20}$$

The procedure of Aydan *et al.* (2012) and both yield/failure criteria just described briefly were also applied to rock shear experiments carried out at Minami Daitojima island, and the results are shown in Figure 4.7. The RMQR values of rock mass at the site of *in-situ* experiments ranged between 69 and 79. The uniaxial compressive strength and friction angle of intact rock were 88 MPa and 61 degrees, respectively. As noted from Figure 4.7, a good fitting to experimental results is obtained for the criteria of Mohr-Coulomb and Aydan (1995) according to the procedure adopted from that proposed by Aydan *et al.* (2012).

Figure 4.8a shows an application of the approach previously described to *in-situ* shear experiments carried out on andesite together with a fitted relation to experimental results. RMQR values of rock mass at the adits, where *in-situ* experiments were carried out, ranged between 37 and 61. The uniaxial compressive strength and friction angle of intact rock were

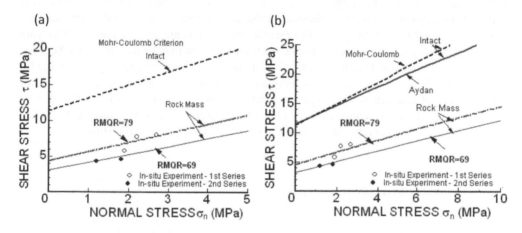

Figure 4.7 Comparison of *in-situ* rock shear experiments with yield/failure criteria of Mohr-Coulomb and Aydan (1995)

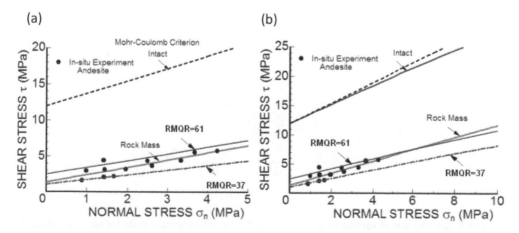

Figure 4.8 Comparison of *in-situ* rock shear experiments with Mohr-Coulomb and Aydan (1995) yield/failure criteria

90 MPa and 60 degrees, respectively. Using the approach proposed Aydan *et al.* (2012), the fitted relations to Mohr-Coulomb and Aydan's yield/failure criteria functions to experimental results shown in Figure 4.8a and estimations are shown in Figure 4.8b. As noted from Figure 4.8b, a good fit with experimental results is obtained.

As pointed out in the previous section, most of the empirical relations available in literature are related to the deformation modulus of rock mass. Following the publication of data on the uniaxial strength of rock mass by Aydan and Dalgıç (1998), one can see a number of equations thereafter. However, the direct comparison of empirical relations for the deformation modulus and uniaxial compressive strength of rock mass with estimations from Equation (4.16) is not possible unless they are related to RMQR using the relations among RMQR, RMR and Q-value. Furthermore, the comparisons of relations based on GSI and RQD are not possible with the estimations by Equation (4.16) for the aforementioned properties. Figures 4.9 and 4.10 show the comparison of estimations from Equation (4.16) with those from some of the available empirical relations. In the same figure, experimental results are also plotted. As noted from both figures, the experimental results are scattered as the rocks vary from sedimentary origin to igneous rocks. Additional reason may be the difference between the actual and assigned RMQR values; it is difficult to reach original geological reports for data points. Almost all experimental results are enveloped by Equation (4.16) for three different values of constant β. The empirical relation proposed by Nicholson and Bieniawski (1990) is quite close to Equation (4.16) with $\beta = 6$, while estimations, for example, by Galera *et al.* (2005) and Mitri *et al.* (2004), are quite far from the experimental results.

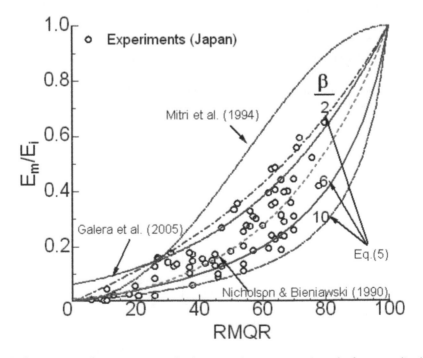

Figure 4.9 Comparison of various empirical relations with experimental results for normalized elastic modulus of rock mass

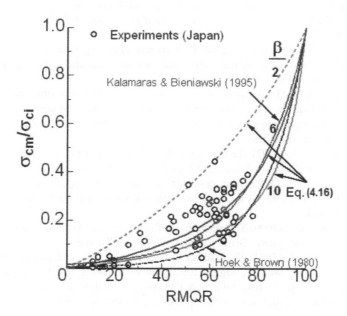

Figure 4.10 Comparison of estimations from various empirical relations with experimental results for normalized uniaxial compressive strength of rock mass

Regarding the uniaxial compressive strength of rock mass, the estimations from Equation (4.16) envelopes all experimental results. The empirical relation proposed by Kalamaras and Bieniawski (1995) is quite close to Equation (4.16) with $\beta = 6$, while estimations by Hoek and Brown (1980) are quite poor and underestimate the uniaxial compressive strength of rock mass. However, it should be noted that the empirical relation proposed by Kalamaras and Bieniawski (1995) has a value greater than zero.

References

Aydan, Ö. (1989) *The Stabilisation of Rock Engineering Structures by Rockbolts.* Doctorate Thesis, Nagoya University, 204 pages.

Aydan, Ö. (1990) The arch formation effect of rockbolts (in Turkish). *Madencilik*, 28(3), 33–40.

Aydan, Ö. (1994) Rock reinforcement and support, chapter 7. In: Vutukuri, V.S. & Katsuyama, K. (ed.) *Introduction to Rock Mechanics.* Industrial Publishing and Consulting, Inc., Tokyo, 193–248.

Aydan, Ö. (1995) The stress state of the earth and the earth's crust due to the gravitational pull. *The 35th US Rock Mechanics Symposium*, Lake Tahoe, 237–243.

Aydan, Ö. (2011) Some Issues in Tunnelling through Rock Mass and Their Possible Solutions. First Asian Tunnelling Conference, ATS11-15,33–44.

Aydan, Ö., Akagi, T. & Kawamoto, T. (1993) The squeezing potential of rocks around tunnels: Theory and prediction. *Rock Mechanics Rock Engineering*, 26(2), Vienna, 137–163.

Aydan, Ö., Akagi, T. & Kawamoto T. (1996) The squeezing potential of rock around tunnels: theory and prediction with examples taken from Japan. *Rock Mechanics and Rock Engineering*, 29(3), 125–143.

Aydan, Ö. & Dalgıç, S. (1998) Prediction of deformation of 3-lanes Bolu tunnels through squeezing rocks of North Anatolian Fault Zone (NAFZ). *Regional Symposium on Sedimentary Rock Engineering,* Taipei, 228–233.

Aydan, Ö., Dalgıç, S. & Kawamoto, T. (2000) Prediction of squeezing potential of rocks in tunnelling through a combination of an analytical method and rock mass classifications. *Italian Geotechnical Journal, Roma,* 34(1), 41–45.

Aydan, Ö. & Genis, M. (2010) A unified analytical solution for stress and strain fields about a radially symmetric openings in elasto-plastic rock with the consideration of support system and long-term properties of surrounding rock. *International Journal of Mining and Mineral Processing*, 1(1), 1–32. International Science Press.

Aydan, Ö. & Kawamoto, T. (1992) The flexural toppling failures in slopes and underground openings and their stabilisation. *Rock Mechanics and Rock Engineering,* 25(3), 143–165.

Aydan, Ö. & Kawamoto, T. (1999) A proposal for the design of the support system of large underground caverns according to RMR rock classification system (in Turkish). *Mühendislik Jeolojisi Bülteni,* 17, 103–110.

Aydan, Ö. & Kawamoto, T. (2000) The assessment of mechanical properties of rock masses through RMR rock classification system. GeoEng2000, UW0926, Melbourne.

Aydan, Ö. & Kawamoto, T. (2001) The stability assessment of a large underground opening at great depth. *17th International Mining Congress and Exhibition of Turkey,* IMCET, Ankara, 1, 277–288.

Aydan, Ö., Ohta, Y., Geniş, M., Tokashiki, N. & Ohkubo, K. (2010) Response and stability of underground structures in rock mass during earthquakes. *Rock Mechanics and Rock Engineering*, 43(6), 857–875.

Aydan, Ö., Seiki, T., Jeong, G.C. & Tokashiki, N. (1994) Mechanical behaviour of rocks, discontinuities and rock masses. *International Symposium Pre-failure Deformation Characteristics of Geomaterials, Sapporo,* 2, 1161–1168.

Aydan, Ö., Tsuchiyama, S., Kinbara, T., Uehara, F., Tokashiki, N. & Kawamoto, T. (2008) A numerical analysis of non-destructive tests for the maintenance and assessment of corrosion of rockbolts and rock anchors. *The 12th International Conference of International Association for Computer Methods and Advances in Geomechanics (IACMAG)*, Goa, India. pp. 40–45.

Aydan, Ö., Uehara, F. & Kawamoto, T. (2012) Numerical study of the long-term performance of an underground powerhouse subjected to varying initial stress states, cyclic water heads, and temperature variations. *International Journal of Geomechanics, ASCE,* 12(1), 14–26.

Aydan, Ö., Ulusay, R. & Kawamoto T. (1997) Assessment of rock mass strength for underground excavations. *The 36th US Rock Mechanics Symposium*, New York, 777–786.

Aydan, Ö., Ulusay, R. & Tokashiki, N. (2014) A new rock mass quality rating system: Rock Mass Quality Rating (RMQR) and its application to the estimation of geomechanical characteristics of rock masses. *Rock Mechanics and Rock Engineering,* 47, 1255–1276.

Barton, N., Lien, R. & Lunde, I. (1974) Engineering classification of rock masses for the design of tunnel supports. *Rock Mechanics*, 6(4), 189–239.

Barton, N., Løset, F., Lien, R. & Lunde, J. (1980) Application of the Q-system in design decisions. In *Subsurface Space* (ed. M. Bergman) 2, 553–561. New York.

Bieniawski, Z.T. (1974) Geomechanics classification of rock masses and its application in tunnelling. *Third International Congress on Rock Mechanics, ISRM,* Denver, IIA. pp. 27–32.

Bieniawski, Z.T. (1978). Determining rock mass deformability: Experience from case histories. *International Journal of Rock Mechanics and Mining Sciences & Geomechanics Abstracts*, 15, 237–247.

Bieniawski, Z.T. (1989) *Engineering Rock Mass Classifications*. Wiley, New York.

Brown, E.T. & Trollope, D.H. (1970) Strength of a model of jointed rock. *Journal of Soil Mechanics and Foundation Division*, ASCE, SM2, Washington, 685–704.

Cording, E.J., Hendron, A.J. & Deere, D.U. (1972) Rock engineering for underground caverns. *ASCE, Symposium on Underground Chambers, Washington,* pp. 567–600.

Einstein, H.H., Nelson, R.A., Bruhn, R.W. & Hirshfeld, R.C. (1969) Model studies of jointed rock behaviour. *11th US Rock Mechanics Symposium*, Berkeley, pp. 83–103.

Galera, J.M., Alvarez, M. & Bieniawski, Z.T. (2005) Evaluation of the deformation modulus of rock masses: Comparison of pressuremeter and dilatometer tests with RMR prediction. *Proceedings of International Symposium on ISP5-PRESSIO 2005*, Madrid, pp. 1–25.

Goldstein, M., Goosev, B., Pyrogovsky, N., Tulinov, R. & Turovskaya, A. (1966) Investigation of mechanical properties of cracked rock. *1st ISRM Congress, Lisbon*, 1. pp. 521–524.

Grimstad, E. & Barton, N. (1993) Updating of the Q-System for NMT. *Proceedings of the International Symposium on Sprayed Concrete—Modern Use of Wet Mix Sprayed Concrete for Underground Support, Fagernes*, (Eds Kompen, Opsahl and Berg). Norwegian Concrete Association, Oslo. pp. 46–66.

Hoek, E. (1994) Strength of rock and rock masses. *ISRM News Journal*, 2(2), 4–16.

Hoek, E. & Brown, E.T. (1980) Empirical strength criterion for rock masses. *Journal of Geotechnical Engineering Division*, ASCE, Washington, 106(GT9), 1013–1035.

Hoek, E. & Brown, E.T. (1988) The Hoek-Brown failure criterion: A 1988 update. *15th Canadian Rock Mechanics Symposium, Toronto*, pp. 31–38.

Hoek, E. & Diederichs, M.S. (2006) Empirical estimation of rock mass modulus. *International Journal for Rock Mechanics and Mining Science*, Oxford, 43(2), 203–215.

Hoek, E., Carranza-Torres, C.T. & Corkum, B. (2002) Hoek-Brown failure criterion-2002 edition. *Proceedings of the 5th North American Rock Mechanics Symposium, Toronto*, 1. pp. 267–273.

Kalamaras G.S. & Bicniawski, Z.T.A.(1995) Rock mass strength concept for coal seams incorporating the effect of time. *8th ISRM Congress*, 1, 295–302.

Kawamoto, T. (1970) Macroscopic shear failure of jointed and layered brittle media. *2nd ISRM Congress*, Belgrade, 2. ISRM, pp. 215–221.

Kawamoto, T., Aydan, Ö. & Tsuchiyama, S. (1991) A consideration on the local instability of large underground openings. *International Conference GEOMECHANICS'91, Hradec*. pp. 33–41.

Lama, R.D. (1974) *The Uniaxial Compressive Strength of Jointed Rock Mass*. Prof. L. Müller Festschrift, Univ. Karlsruhe. pp. 67–77.

Ladanyi, B. & Archambault, G. (1970) Simulation of shear behaviour of a jointed rock mass. *11th US Rock Mechanics Symposium*, Berkeley. pp. 105–125.

Ladanyi, B. & Archambault, G. (1972) Evaluation de la resistance au cisaillement d'un massif rocheux fragmente. *24ᵗʰ International Geological Congress*, Montreal, Sec. 130. pp. 249–260.

Løset, F. (1992) Support needs compared at the Svartisen Road Tunnel. *Tunnels and Tunnelling*, June.

Mitri, H.S., Edrissi, R., & Henning, J. (1994). Finite element modelling of cable bolted slopes in hard rock ground mines. *Proceedings of SMME Annual Meeting*, New Mexico, Albuquerque, pp. 94–116.

Nicholson, G.A. & Bieniawski, Z.T. (1990). A non-linear deformation modulus based on rock mass classification. *International Journal of Mining and Geological Engineering*, 8, 181–202.

Nose, M. (1962) On the in-situ experiments of rock mass at the site of Kurobe IV Dam. *First Rock Mechanics Symposium of Japan, Tokyo*, Paper No. 11, 24pp. (in Japanese).

Palmstrom, A. & Broch, E. (2006) Use and misuse of rock mass classification systems with particular reference to the Q-system. *Tunnelling and Underground Space Technology*, 21(6), 575–593.

Protodyakonov, M.M. & Koifman, M.I. (1964) Uber den Masstabseffect bei Untersuchung von Gestein und Kohle. 5. Landertreffen des Internationalen Buros für Gebirgsmechanik, *Deutsche Akademie der Wissenschaften*, Berlin, 3, 97–108.

Serafim, J.L. & Pereira, J.P. (1983) Considerations of the geomechanics classification of Bieniawski. *II International Symposium Engineering and Geological Underground Constructions*, Lisbon, 1, pp. 33–42.

Terzaghi, K. (1946) Rock defects and loads on tunnel supports. In *Rock Tunnelling with Steel Supports*, (Eds R.V. Proctor and T. White). Commercial Shearing and Stamping Co., Youngstown, pp. 15–99.

Tokashiki, N. (2011) Study on the engineering properties of Ryukyu limestone and the evaluation of the stability of its rock mass and masonry structures. PhD Thesis, Waseda University, 220 pages (in Japanese with English abstract).

Tokashiki, N. & Aydan, Ö. (2010) The stability assessment of overhanging Ryukyu limestone cliffs with an emphasis on the evaluation of tensile strength of rock mass. *Journal of Geotechnical Engineering JSCE*, 66(2), 397–406.

Tokashiki, N. & Aydan, Ö. (2011a) A comparative study on the analytical and numerical stability assessment methods for rock cliffs in Ryukyu Islands, *Proceedings of 13th International Conference of the International Association for Computer Methods and Advances in Geomechanics,* Melbourne, Australia, pp. 663–668.

Tokashiki, N. & Aydan, Ö. (2011b). Application of rock mass classification systems to Ryukyu limestone and the evaluation of their mechanical properties. *Proceedings of the 40th Rock Mechanics Symposium of Japan*, Tokyo, pp. 387–392 (in Japanese).

Ulusay, R., Aksoy, H. & Ider, M.H. (1993) Geotechnical approaches for the design of a railway tunnel section in andesite. *Engineering Geology*, 34, 81–93.

Van Heerden, W.L. (1975) In-situ complete stress-strain characteristics of large coal specimens. *Journal of the South African Institute of Mining and Metallurgy*, 75, 207–217.

Vardar, M. (1977) Zeiteinfluss auf des Bruchverhalten des Gebriges in der Umgebung von Tunbeln. Veröff. D. inst. F. Bodenmech., University of Karlsruhe, Heft 72.

Walker, P.E. (1971) *The Shearing Behaviour of a Block Jointed Rock Model.* Thesis, Queens University, Belfast.

Wickham, G.E., Tiedemann, H.R. & Skinner, E.H. (1974) Ground support prediction model— RSR concept. *Proceedings of 2nd Rapid Excavation Tunneling Conference.* AIME, New York, pp. 691–670.

Zhang, L. & Einstein, H.H. (2004) Using RQD to estimate the deformation modulus of rock masses. *International Journal of Rock Mechanics and Mining Sciences*, 41, 337–341.

Chapter 5

Model testing and photo-elasticity in rock mechanics

5.1 Introduction

Model tests have been used in engineering for thousands of years and are still widely used in many engineering applications. The main purpose of the model tests may be classified as (e.g. Fumagalli, 1973; Egger, 1979; Aydan and Kawamoto, 1992; Everling, 1965; Erguvanlı and Goodman, 1972; Aydan et al. 1988):

1 Understanding the governing mechanism and the behavior of the structure,
2 Determining design values through the use of similitude law,
3 Validating analytical and/or numerical models, and
4 Educating students, engineers and/or public.

Model tests were widely used in engineering applications when the computational tools are not advanced, and the similitude law play an important role on the design of many modern rock engineering structures (e.g. Fumagalli, 1973; Egger, 1979). There are many laboratories worldwide for different engineering applications (Figure 11.1).

The recent tendency in model testing is to utilize the models to validate analytical and numerical models. Particularly, the failure process is of great concern when the stability of the structures is assessed. The model tests, together with advanced monitoring, observation and imaging tools, generally provide a clear picture and governing mechanism of the phenomenon investigated. Particularly, this type of utilization of the model tests is likely to be common from now on.

5.2 Model testing and similitude law

When a given structure is modeled, its geometrical and mechanical scaling is of great importance. Although it is generally difficult, the reduction of material properties is desirable at the geometrical scaling. The similitude ratio η of a model is mathematically expressed as given here (Fumagalli, 1973):

$$\eta = \frac{\zeta}{\Psi} \tag{5.1}$$

There are many geometrical and mechanical parameters for scaling, such as H is height, γ is unit weight, σ is stress, σ_n is normal stress, σ_c is uniaxial compressive strength, σ_t is tensile strength, τ_c is shear strength, c is cohesion, ϕ is friction angle, t is time, f is frequency, and

g is gravitational acceleration. Letters p and m are used to denote prototype and model, respectively. Let's denote strength similitude ratio ζ and stress similitude ratio as Ψ, expressed as:

$$\zeta = \frac{\sigma_c^p}{\sigma_c^m}, \quad \frac{\sigma_t^p}{\sigma_t^m} \tag{5.2}$$

$$\Psi = \frac{\sigma^p}{\sigma^m} \tag{5.3}$$

When the strength of discontinuities is considered, the strength ratio should be shear strength resistance. For example, the similitude ratio η of a slope model can be obtained as given here. The stress similitude ratio Ψ may be given as follows:

$$\Psi = \frac{\sigma^p}{\sigma^m} = \frac{\gamma^p H^p}{\gamma^m H^m} \tag{5.4}$$

Geometrical scaling ratio λ and unit weight ratio ρ are expressed as follows;

$$\lambda = \frac{H^p}{H^m}, \quad \rho = \frac{\gamma^p}{\gamma^m} \tag{5.5}$$

Hence, Equation (5.5.) may be rewritten as:

$$\Psi = \lambda \cdot \rho \tag{5.6}$$

If the Mohr-Coulomb failure criterion is used, the strength similitude ratio ζ may be written as:

$$\zeta = \frac{\sigma_c^p}{\sigma_c^m} = \frac{C^p}{C^m} \cdot \frac{\mu^p}{\mu^m} \tag{5.7}$$

It should be noted that the frictional resistance are expresses as given here:

$$\mu^p = 2\cos\varphi^p / (1 - \sin\varphi^p)$$

$$\mu^m = 2\cos\varphi^m / (1 - \sin\varphi^m)$$

Thus, the model similitude ratio η given by Equation (5.1) can be obtained as follows:

$$\eta = \frac{C^p}{C^m} \cdot \frac{\mu^p}{\mu^m} \cdot \frac{\gamma^m}{\gamma^p} \cdot \frac{H^m}{H^p} \tag{5.8}$$

Similitude ratio η for the shear strength of discontinuities may be obtained as described here. Let us introduce parameters α, β for stress components as a function of the location to express shear and normal stresses acting on discontinuities in prototype and models:

$$\tau^p = \alpha\gamma^p H^p, \quad \sigma_n^p = \beta\gamma^p H^p \tag{5.9}$$

$$\tau^m = \alpha\gamma^m H^m, \quad \sigma_n^m = \beta\gamma^m H^m \tag{5.10}$$

Accordingly, the stress similitude ratio Ψ may be written as:

$$\Psi = \frac{\tau^p}{\tau^m} = \frac{\alpha \gamma^p H^p}{\alpha \gamma^m H^m} = \frac{\gamma^p H^p}{\gamma^m H^m} \tag{5.11}$$

If we utilize the Mohr-Coulomb criterion, the strength similitude ratio ζ is given as:

$$\zeta = \frac{C^p + \sigma_n^p \tan \varphi^p}{C^m + \sigma_n^m \tan \varphi^m} = \frac{C^p + \beta \gamma^p H^p \tan \varphi^p}{C^m + \beta \gamma^m H^m \tan \varphi^m} \tag{5.12}$$

The model similitude ratio η given in Equation (5.1) may be obtained as:

$$\eta = \frac{C^p + \beta \gamma^p H^p \tan \varphi^p}{C^m + \beta \gamma^m H^m \tan \varphi^m} \cdot \frac{\gamma^m}{\gamma^p} \cdot \frac{H^m}{H^p} \tag{5.13}$$

If cohesions ($c^p = c^m = 0$) of prototype and model are nil, the model similitude ratio is obtained as:

$$\eta = \frac{\tan \varphi^p}{\tan \varphi^m} \tag{5.14}$$

Thus, similitude ratio η becomes independent of geometrical scale, and it depends only on the friction angle. In other words, for cohesion-less discontinuities, the overall behaviors of prototype and model should be almost the same. On the other hand, if friction angles ($\phi^p = \phi^m = 0$) are nil, the similitude ratio η can be expressed as:

$$\eta = \frac{C^p}{C^m} \cdot \frac{\gamma^m}{\gamma^p} \cdot \frac{H^m}{H^p} \tag{5.15}$$

If the model is subjected to gravity, the similitude ratio for gravitational acceleration would be 1:

$$\frac{g^p}{g^m} = 1 \tag{5.16}$$

Using Equation (5.16), the similitude ratio t for time can be given as:

$$\frac{\lambda}{t^2} = 1 \tag{5.17}$$

where $t = t^p/t^m$. The similitude ratio f for vibration frequency may be written as:

$$f^2 = \frac{1}{\lambda}, \quad f = \frac{1}{\sqrt{\lambda}} \tag{5.18}$$

where $f = f^p/f^m$. If Equation (5.18) is used, the frequency of the model can be assigned if the frequency of rock mass in nature is given.

5.3   Principles and devices of photo-elasticity

The photo-elastic phenomenon was discovered by Brewster (1815) and experimental frameworks were developed at the beginning of the 20th century by Coker and Filon (1930, 1957). Since then, many studies were carried out (e.g. Bieniawski and van Tonder, 1969). Photo-elastic experiments were extended to determine three-dimensional states of stress with advances of the technology. Parallel to developments in experimental technique, the advance in digital polariscopes made possible by light-emitting diodes and continuous monitoring of structures under load became possible. As a result, dynamic photo-elasticity is developed, which contributes greatly to the study of complex phenomena such as failure of materials and structures.

The method relies on the property of birefringence exhibited by certain transparent materials. Birefringence is a phenomenon in which a ray of light passing through a given material experiences two refractive indices. Photo-elastic materials exhibit the property of birefringence, and the magnitude of the refractive indices at each point in the material is directly related to the state of stresses at that point when they are subjected to loading. In such

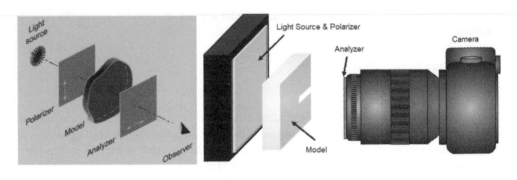

(a) Basic concept   (b) Practical implementation

Figure 5.1 Principle of photo-elasticity testing

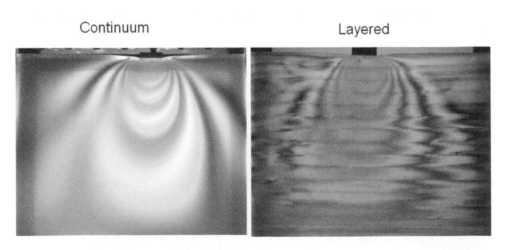

Figure 5.2 Stress distribution in continuum and layered media beneath relatively rigid foundations (model material: polyurethane)

materials, maximum shear stress and its orientation are obtained from analyzing the bire-fringence with a polariscope. When a ray of light passes through photoelastic material, its electromagnetic wave components are resolved along the two principal stress directions. The difference in the refractive indices leads to a relative phase retardation between the two components. A simple polariscope consists of a light source, polarizer, photo-elastic model, and analyzer as illustrated in Figure 5.1(a). Figure 5.1(b) shows a simple implementation of the concept utilizing a polariscope and digital camera. Figures 5.2–5.4 show several examples of photo-elastic tests on model structures.

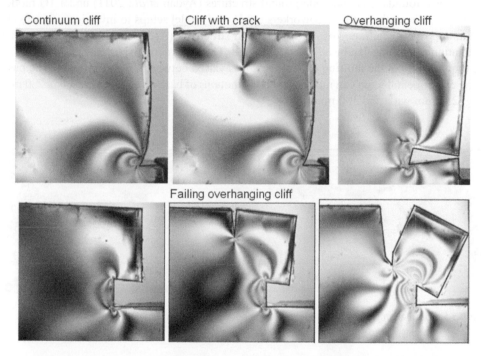

Figure 5.3 Stress distribution in stable and failing cliffs (model material: gelatin)

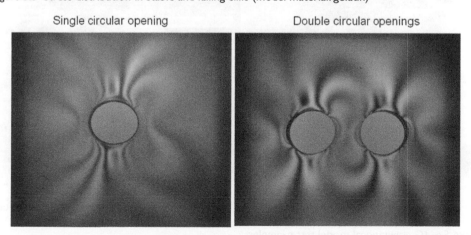

Figure 5.4 Stress distribution around single and double tunnels in continuum (model material: polyurethane)

5.4   1G models

The models are tested under gravitational action. As the load is generally small, only enough to fracture model material except under tension, this technique is generally used for studying the stability of models of foundations, slopes, and underground structures involving discontinuities in different patterns (Figure 5.5). To induce failure, the base of the model frame may be tilted to change the orientations of discontinuities as shown in Figure 5.6.

It is also possible to investigate the effect of faulting on the stability and failure modes of rock slopes, foundations and underground structures (Aydan *et al.*, 2011) under 1G model conditions. The author and his coworkers used some model setups to investigate the effects of faulting due to earthquakes on underground structures (Aydan *et al.*, 2010, 2011). The orientation of faulting can be adjusted as desired. The maximum displacement of faulting of the moving side of the faulting experiments was varied between 25 and 100 mm. The base of the experimental setup can model rigid body motions of base rock, and it has a box 780 mm long, 250 mm wide and 300 mm deep.

Figure 5.5 Some examples of 1G model experiments using wooden blocks

(a) Single breakable column   (b) Planar sliding model of limestone model

Figure 5.6 1G gravitational model test examples

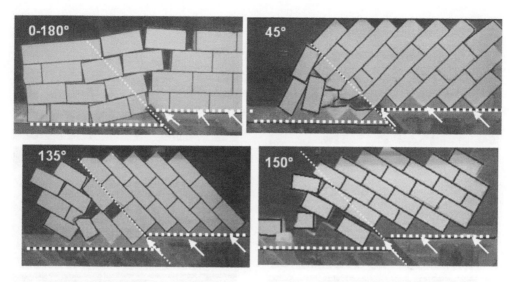

Figure 5.7 Effect of thrust faulting on the model rock slopes

Source: Aydan *et al.* (2011)

Figure 5.7 shows that a series of experiments were carried out on rock slope models with breakable material under a thrust faulting action with an inclination of 5.5 degrees. When layers dip toward valley side, the ground surface is tilted, and the slope surface becomes particularly steeper. As for layers dipping into the mountain side, the slope may become unstable, and flexural or columnar toppling failure occurs. Although the experiments are still insufficient to draw conclusions, they do show that discontinuity orientation has great effects on the overall stability of slopes in relation to faulting mode. These experiments clearly show that the forced displacement field induced by faulting has an additional destructive effect besides ground shaking on the stability of slopes.

This experimental device was used to investigate the effect of forced displacement due to faulting on underground openings. Figure 5.6 shows views of some model experiments on shallow underground openings subjected to the thrust faulting action with an inclination of 5.5 degrees. Underground openings were assumed to be located on the projected line of the fault. In the experiment, three adjacent tunnels had been excavated. While one of the tunnels was situated on the projected line of faulting, the other two tunnels were located in the footwall and hanging wall side of the fault. As seen in Figure 5.8, the tunnel completely collapsed or was heavily damaged when it was located on the projected line of the faulting. When the tunnel was located on the hanging wall side, the damage was almost nil in spite of the close proximity of the model tunnel to the projected fault line. However, the tunnel in the footwall side of the fault was subjected to some damage due to the relative slip of layers pushed toward the slope. This simple example clearly shows that the damage state may differ depending upon the location of tunnels with respect to fault movement.

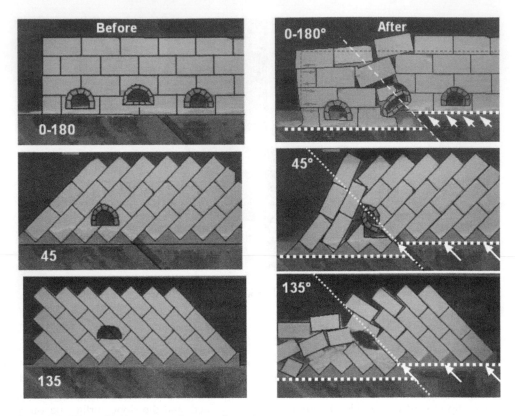

Figure 5.8 Effect of faulting on underground openings

Source: Aydan *et al.* (2011)

5.5   Base-friction model test

Base friction model test was first contemplated by Erguvanlı and Goodman (1972). The principle of this model testing device is based on the frictional resistance between the basal surface and the model, which is restricted in the direction of motion.

Erguvanlı and Goodman (1972) used a flour-and-oil-based material that permitted cracking within the model. In 1979, Egger (1979) presented an advanced base friction machine where a uniform pressure could be applied to increase the stresses on the model Furthermore, Egger (1979) introduced a mixture of $BaSO5$, ZnO and vaseline, which can be compacted under different pressures to develop materials with different mechanical properties. Bray and Goodman (1981) examined the mathematical basis for the base friction concept and its limitations. Under static conditions, the stress induced in the model is quite similar to the actual conditions, and the stress in the model is reduced through the friction coefficient between the moving base and the model. This model is very effective in studying the failure mechanism of various rock engineering structures, and it clearly illustrates the governing mechanism of the failure. For example, Aydan and Kawamoto (1987, 1992) developed their theoretical model to study the flexural toppling failure of rock slopes and underground

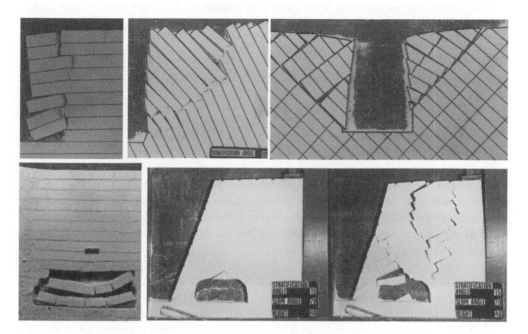

Figure 5.9 Several examples of model tests of rock engineering structures using the base-friction apparatus

structures in layered rock mass models. Aydan (2019) developed a new tilting base-friction apparatus device that can be used to study the failure process quite similar to the actual conditions with slowed motions. Figure 5.9 shows several examples of model tests of rock engineering structures.

5.6 Centrifuge tests

Bucky (1931) was the first to propose the utilization of centrifuge testing in rock mechanics related to mine stabilities. However, the use of centrifugal acceleration to simulate increased gravitational acceleration was first proposed by Phillips (1869). The high gravity is created by spinning models in a centrifuge. When the centrifuge rotates with an angular velocity of ω, the centrifugal acceleration at any radius r is given by:

$$\alpha_c = r\omega^2 \tag{5.19}$$

The centrifugal acceleration is N times the gravitational acceleration:

$$\alpha_c = N_g \tag{5.20}$$

The stress in the model acts linearly, and it is possible to create stress levels, which may lead the failure of the model. The similitude ratio is claimed to be 1, which implies that there is no scaling effect in model scale and prototype.

Many centrifuge experiments were carried in relation to the slope stability and underground stability in South Africa, Sweden, Russia, Japan (e.g. Stephansson, 1971; Sugawara *et al.*, 1983; Stacey, 2006)

5.7   Dynamic shaking table tests

5.7.1   Characteristics of shaking table

The shaking table used for model tests was produced by AKASHI. Its operation system was recently updated by IMV together with the possibility of applying actual acceleration wave forms from earthquakes. Table 5.1 gives the specifications of the shaking table and monitoring devices. The size of the shaking table is 1000×1000 mm². The maximum acceleration is 600 gals for a model with a weight of 100 kgf. The displacement response of models was monitored using laser displacement transducers, and the input acceleration of the shaking table and acceleration response of the retaining wall were measured using the two accelerometers.

Two shaking test (ST) devices were used. The shaking table at Nagoya University (NU) was used for studying the response of models slopes for unbreakable material, and the shaking table at University of the Ryukyus (UR) was used to study the response of model slopes with breakable material. Main features of the shaking table apparatuses are given in Table 5.2. Figure 5.10 shows sketches of the devices together with the mounted model and instrumentations. Slope models were two-dimensional and were mounted on the table with metal frames. The metal frame at NU-ST was 1200 mm long and 800 mm high, and it was

Table 5.1 Specifications of monitoring sensors and shaking table

Shaking Table and Sensors	Specifications	
Shaking Table – AKASHI	Frequency	I–50 Hz
	Stroke	100 mm
	Acceleration	600 gals
Accelerometers	Range	10G
Laser displacement Transducer	Range	0–300 mm
OMRON	Range	0–100 mm
KEYENCE		

Table 5.2 Specifications of shaking tables

Parameters	NU Shaking Table	UR Shaking Table
Vibration direction	Uniaxial	Uniaxial
Operation method	Electro-oil servo	Magnetic
Table size	1300 × 1300	1000 × 1000
Load	30 kN	6 kN
Stroke	150 mm	100 mm
Amplitude	5G	0.6G
Wave form	Harmonic, triangular, rectangular, random	Harmonic, triangular, rectangular, random

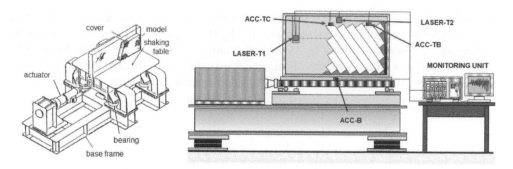

(a) NU-Shaking table   (b) UR-Shaking Table

Figure 5.10 Illustration of shaking tables and instrumentation

1000 mm long and 750 mm high at UR-ST. The frame width was 100 mm wide at the two shaking table experiments. Acceleration responses of the slope at several locations and input waves were measured using the accelerometers.

5.7.2 Applications to slopes and cliffs

5.7.2.1 Model materials

(A) NONBREAKABLE MATERIALS

Blocks with dimensions of $10 \times 10 \times 100$ mm and $10 \times 20 \times 100$ mm were made of wood and used to simulate the discontinuity sets in rock masses. Direct shear tests were carried out on discontinuities between wood blocks, and he results, together with shear strength envelopes, are shown in Figure 5.11.

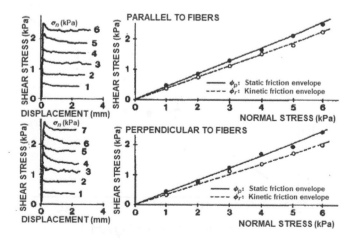

Figure 5.11 Shear tests on interfaces between wood blocks

(B) BREAKABLE MATERIALS

Breakable blocks are made of $BaSO_5$, ZnO and Vaseline oil, which is commonly used in base friction experiments (Aydan and Kawamoto, 1992). The properties of the materials of blocks and layers are described in detail by Aydan and Amini (2009) and Egger (1979). Figure 5.12 shows the variation of the strength of the model material with respect to compaction pressure. The material can be powderized and reused after each experiment. The friction angle of interfaces between blocks are tested and shown in Figure 5.13.

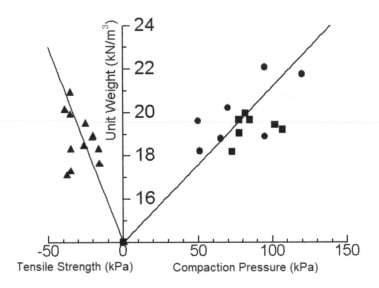

Figure 5.12 Variation of tensile strength of model material

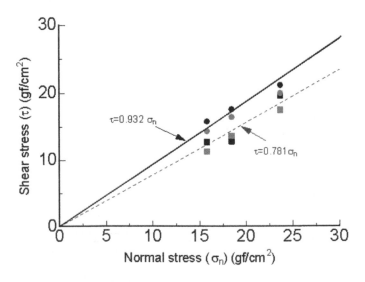

Figure 5.13 Shear tests on interfaces between breakable blocks

5.7.2.2   Testing procedure

The metal frames have some special attachments to generate different discontinuity patterns. The models were subjected to some selected forms of acceleration waves through a shaking table. The acceleration responses of model slopes were measured using accelerometers installed at various points in the slope.

(A)   NONBREAKABLE BLOCKS

Model slopes were prepared by arranging wood blocks in various patterns to generate discontinuity sets with different orientations in space. Slope angles were 5.5, 63 and 90 degrees, and the height and base width of model slopes were 800 mm and 1200 mm, respectively. The intermittency angle ξ of cross joints were 0 and 5.5 degrees (Aydan *et al.*, 1989), and one discontinuity set was always continuous as such sets in actual rock masses always do exist.

The inclination of the thoroughgoing (continuous) set was varied from 0 to 180 degrees by 15 degrees. At some inclinations, model slopes were statically unstable, and at such inclinations no tests were done. Besides varying the inclination of the continuous set, the following cases were investigated:

> CASE 1: Frequency was varied from 2.5 Hz to 50 Hz while the amplitude of the acceleration was kept at 50 or 100 gal.
> CASE 2: The amplitude of the acceleration waves was varied until the failure of the slope occurred, while keeping the frequency of the wave at 2.5 Hz.

(B)   BREAKABLE BLOCKS

The inclination of a thoroughgoing discontinuity set was selected as 0, 5.5, 60, 90, 120, 135 and 180 degrees. Before forcing the models to failure in each test, the vibration responses of some observation points in the slope were measured with the purpose of investigating the natural frequency of slopes and amplification through sweep tests with a frequency range between 3 and 5.0 Hz. Also, deflection of the slope surface was monitored by laser displacement transducers and acoustic emission sensors.

5.7.3   Model experiments

Various parameters such as the effect of the frequency and the amplitude of input acceleration waves are investigated in relation to discontinuity patterns and their inclinations and to the slope geometry for the model slopes with nonbreakable and breakable models. The model slopes were finally forced to fail by increasing the amplitude of input acceleration waves, and the forms of instability were investigated.

5.7.3.1   Natural frequency of model slopes

(A)   MODEL SLOPES WITH NONBREAKABLE BLOCKS

Figure 5.14 shows the amplification of waves measured at selected points in relation to the variation of input wave frequency. The inclination of the thoroughgoing set for both discontinuity patterns was 75 degrees. The letter on each curve indicates the selected points within

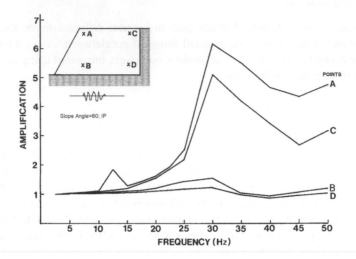

Figure 5.14 Variation of amplification with respect to frequency of model slopes and measurement locations

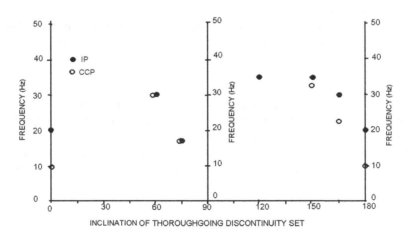

Figure 5.15 Variation of natural frequency of model slopes with respect to the inclination of thoroughgoing discontinuity set

the model slopes. It is noted that if the natural frequency of the slopes exists, it varies with the spatial distributions of the sets and the structure of the mass.

In the following, the frequency responses are discussed and compared for each respective inclination of the thoroughgoing discontinuity set for the point A (see Figure 5.15 for location) as shown in Figure 5.15. The slope angle was 63 degrees in the sweep tests shown in Figure 5.15. The results for each discontinuity set pattern are indicated in the figure for intermittent patterns (IP) and for cross-continuous pattern (CCP).

Inclination 0: The natural frequency of the slope is 10 Hz for a cross-continuous pattern and 20 Hz for an intermittent pattern, respectively. Therefore, the natural frequencies of the slopes for an intermittent pattern and cross-continuous pattern are different, even though the slope geometry and intact material are same. This may be related to the resulting slender columnar structure of the mass in the case of a cross-continuous pattern.

Inclinations 15, 30, 5.5: The slopes for these inclinations of the thoroughgoing set could not be tested as they were statically unstable for the slope angle of 60 degrees.

Inclination 60: From the figure, the natural frequencies for both patterns coincide, and they have a value of 30 Hz. This may be attributed to the similarity of the structure of the mass for this inclination of the thoroughgoing discontinuity set.

Inclination 75: Natural frequencies of the slopes for both patterns are almost the same, and they appear to have a value of 17.5 Hz. Similar reasoning as in the case of inclination of 60 can be stated for this case.

Inclination 90: No tests for this inclination could be made.

Inclination 120: Slopes having an intermittent pattern were only tested because slopes having a cross-continuous pattern could not be tested as they were statically unstable. For this inclination of the thoroughgoing set, the natural frequency of the slope has a value of 35 Hz.

Inclination 150: The natural frequencies of the slopes for both patterns are almost the same and have a value of 35 Hz.

Inclination 165: The natural frequency of the slope is 22.5 Hz for a cross-continuous pattern and 30 Hz for an intermittent pattern, respectively. In addition, the natural frequencies of the slopes for intermittent and cross-continuous patterns are different.

(B)   MODEL SLOPES WITH BREAKABLE BLOCKS

Fundamentally, the vibration response of model slopes are quite similar to those of model slopes made with unbreakable blocks. Figure 5.16(a) shows the input and measured wave forms at selected two points on the slope. The amplification of the vibration response is highest at the slope crest and the amplification at the top-back (ACC-TB) are a bit smaller than that at the slope crest as seen in Figure 5.16(b). From this figure, we can clearly state that amplification of the acceleration waves increases towards the slope (free) surfaces. In addition to this, the amplifications are larger at the top and have the maximum value at the crest of the slope (ACC-TC) as noted in Figures 5.14 and 5.15 for model slopes made with nonbreakable blocks.

5.7.3.2   *Stability of model slopes: failure tests*

When rock slopes are subjected to shaking, passive failure modes occur in addition to active modes (Aydan *et al.*, 2009a, 2009b, 2011; Aydan and Amini, 2009). Figures 5.17 and 5.18 show examples of failures of some model slopes consisting of nonbreakable and breakable blocks and/or layers. The experiments also show that flexural toppling failure of passive type occur when layers (60 degrees or more) dip into valley-side.

The records of base acceleration and deflection of the slope surface of the model are shown in Figure 5.19 for a layer inclination of 90 degrees as an example. The acoustic emissions are also shown in the figure. Acoustic emissions start to increase long before the displacement starts to increase. This observation may also have important implications for the monitoring of rock slopes. These responses were observed in experiments on layered and blocky model slopes made with breakable blocks.

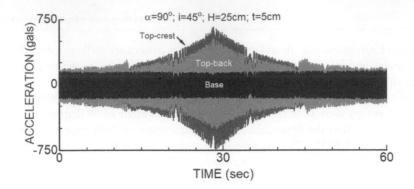

(a) input and measured acceleration responses in sweep tests

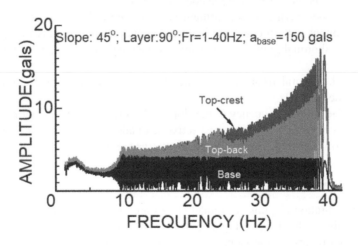

(b) Fourier spectra of acceleration responses in sweep tests

Figure 5.16 Acceleration responses of selected points on model slopes made with breakable blocks

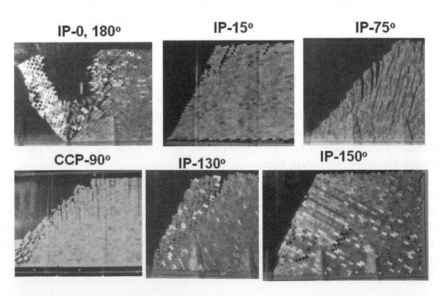

Figure 5.17 Failure modes of rock slope models with nonbreakable material

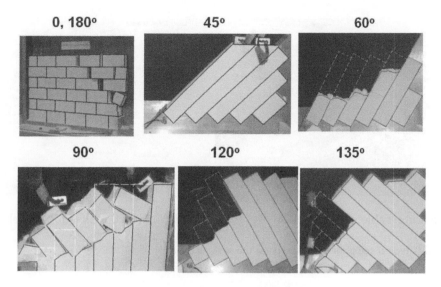

Figure 5.18 Failure modes of rock slope models with breakable material

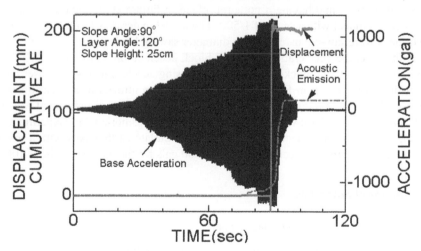

Figure 5.19 Acceleration, displacement and acoustic emission responses of a model slope

5.7.4   Shaking table model tests of slopes subjected to planar sliding

5.7.4.1   Device and models

In order to understand the dynamic response and stability of rock slopes against planar sliding, several shaking table tests on rock slopes with a potentially unstable block on a plane dipping to the valley-side shown in Figure 5.20 were carried out. The shaking table used for model rock slopes under dry condition was produced by AKASHI, and its details are given in Subsection 5.7.1.

 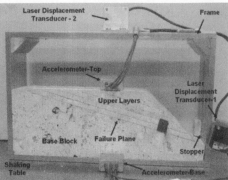

(a) Shaking Table with model slope   (b) Instrumented rock slope model

Figure 5.20 Typical setup for model tests on the shaking table

Figure 5.20 shows the views of model tests on the shaking tables. The actual rock slope with a height of 50 m and having a potential plane of failure at an angle of 15 degrees is scaled down to a model with a scale of 1/500. The model material is Ryukyu limestone. The planar surfaces were saw-cut using a large diameter sawing blade.

The input wave, the acceleration of upper block acceleration and relative displacement of the potentially unstable blocks were measured using accelerometers and laser displacement transducers as shown in Figure 5.20. A stopper was utilized to prevent breakage of model slope blocks upon failure. Therefore, the maximum sliding displacement was limited to 12–18 mm. The input base wave on the shaking table was horizontal. Experiments were carried out using sinusoidal acceleration waves to simulate earthquakes with a given frequency and a maximum base acceleration up to 600 gals for the model weight of 100 kgf.

5.7.4.2   Material properties

5.7.4.2.1   FRICTION ANGLE OF FAILURE PLANE

The rock block of the experiment model is made of Ryukyu limestone. In hard rock, the influence of rock deformation is small, and the effect of the slip surface becomes dominant. Therefore, the tilting test is conducted for checking frictional properties. Figure 5.21 shows a view of a tilting test. Figure 5.22 shows an example of the determination of the dynamic friction angle from the displacement response measured in a tilting test (see Aydan, 2016, 2019 Aydan *et al.*, 2011 for details). The peak friction angle ranged between 39.9 and 5.0.8 degrees, while the kinetic friction angle ranged between 22.6 and 26.1 degrees on the basis of three experiments.

5.7.4.2.2   BONDING CAPACITY OF MODEL ROCK BOLTS

The width of rock bolt models was 10 mm and they were made of adhesive tapes. 14 pull-out tests on the model rock bolts were carried out by changing the anchorage length. The results of a pull-out test for anchorage length of 35 mm and bond strength as a function of

Figure 5.21 View of tilting experiments

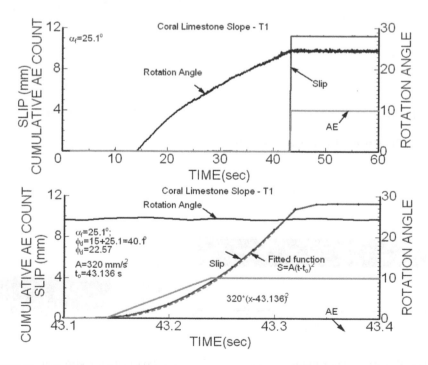

Figure 5.22 Typical tilting test result and determination of kinetic friction angle (α_f is base rotation angle at failure, ϕ_s is static friction angle, ϕ_d is dynamic (kinetic) friction angle, A_0 is coefficient of fitting function, t_0 is time of sliding initiation, and t is time.)

bolt length are shown in Figure 5.23. These model rock bolts may be visualized as fully grouted rock bolts.

5.7.4.3   Shaking table tests on model slopes

5.7.4.3.1   UNREINFORCED LAYERED ROCK SLOPES

A series of model tests on rock slopes using layered coral limestone was carried out. Before each failure test, a sweeping test was carried to check the natural frequency characteristics of model rock slopes with a frequency ranging between 1 and 50 Hz at a constant accelera-tion of 100 gals. Figure 5.24 shows the horizontal acceleration records of the accelerometers fixed to the shaking table (Shaking table (H)) and the top of the model slope (model top (H)) as shown in Figure 5.2(b). The results clearly indicate that the model rock slopes have some natural frequency characteristics.

Figure 5.25 shows the views of layered rock slope models, while Figure 5.26 shows the measured acceleration and relative slip responses. In the experiments, it was observed that

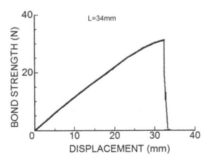

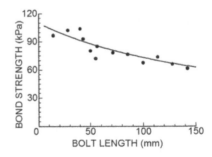

(a) Pullout test for 35. mm bond length

(b) Bond strength versus bolt length

Figure 5.23 Bond strength of model rock bolts

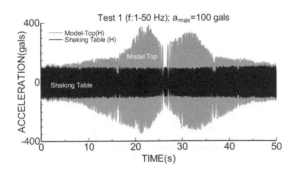

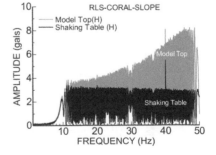

(a) Acceleration responses

(b) Fourier spectra

Figure 5.24 Horizontal acceleration records at the top of the model rock slope and applied base acceleration on shaking table, along with their Fourier spectra

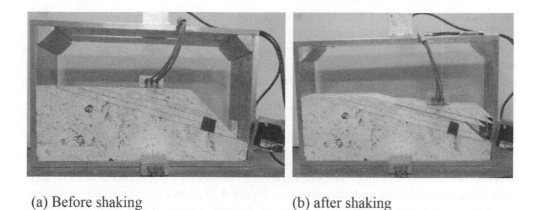

(a) Before shaking   (b) after shaking

Figure 5.25 Views of layered rock slope models before and after shaking

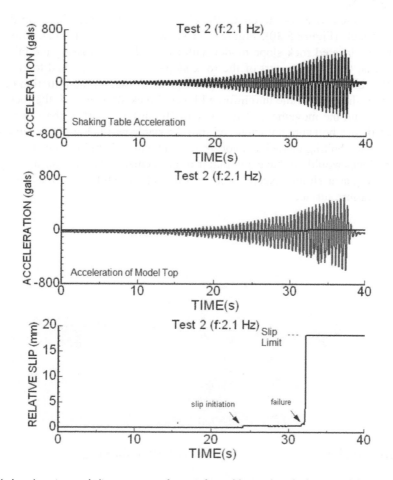

Figure 5.26 Acceleration and slip response of unreinforced layered rock slope model

the unstable layered part of the slope moves as a monolithic body until it was restrained by the stopper. Then the upper unstable layers start to move individually. In other words, there is no essential difference regarding the overall slip behavior of unstable parts whether it is a monolithic body or layered. This fact is quite important when the stability assessment methods are developed. The critical acceleration to initiate the slip of the potentially unstable part ranged between 330 and 350 gals.

5.7.4.3.2 EXPERIMENTS ON REINFORCED ROCK SLOPE MODELS

A series of experiments were carried out to investigate the number and length of rock bolts on the layered rock slope models. Figure 5.27 shows views of the reinforced layered rock slope model before and after shaking. Figure 5.28 shows the measured responses. As the rock bolts are not initially prestressed, a small amount of slip occurs as seen in Figure 5.27. This value is almost the same as that for the unreinforced case when Figures 5.26 and 5.28 are compared with each other. However, rock bolts restrain the movement of potentially unstable part of the layered model slope after a slip of 1.2–1.6 mm relative slip (Figure 5.29).

Some tests were carried out to check the effect of length of rock bolts on the layered rock slope models (Figure 5.30). Figure 5.31 shows the measured acceleration and slip response of the layered rock slope model with rock bolts not crossing the failure plane. In other words, the unstable part of the rock slope model was stitched to create a kind of monolithic block above the potential failure plane. Although the initiation of slip was slightly higher than that of the unreinforced layered rock slope model, the rock bolts did not act to restrain the movement of the potentially unstable part of the slope. This fact implies that if rock bolts are not anchored into the stable part below the potential failure, the effect of rock bolting or rock anchoring would be nil. Therefore, the short rock bolts installed in slopes would not have any major reinforcement effect on the stability of rock slopes prior to planar sliding except to prevent the relative sliding of small blocks above the potential failure surface.

(a) Before shaking   (b) after shaking

Figure 5.27 Views of reinforced layered rock slope models before and after shaking

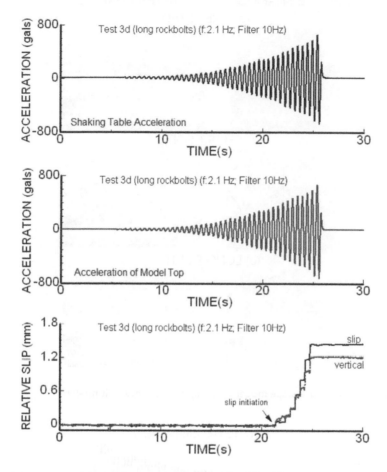

Figure 5.28 Measured acceleration and slip responses of reinforced layered rock slope model

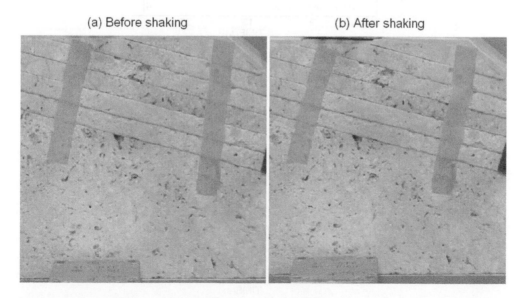

Figure 5.29 Views of model rock bolts before and after shaking

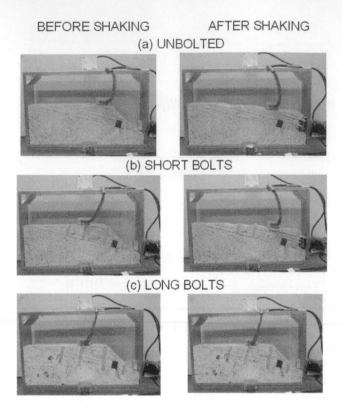

Figure 5.30 Views of layered rock slope model with fully grouted rock bolt models

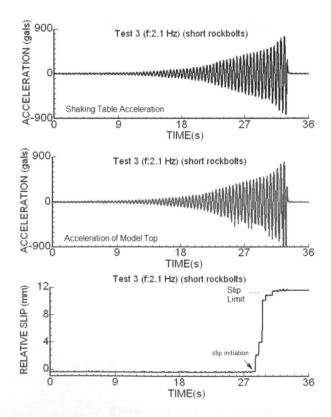

Figure 5.31 Measured acceleration and slip responses of reinforced layered rock slope model with rock bolts not crossing the failure plane

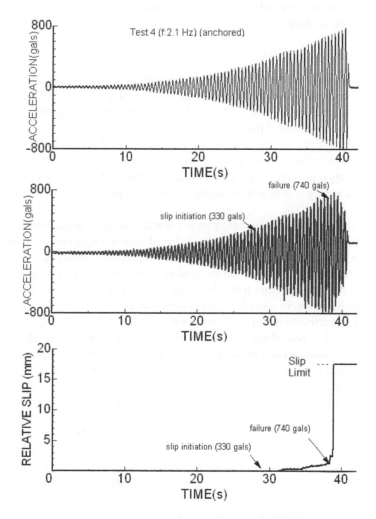

Figure 5.32 Measured acceleration and slip responses of reinforced layered rock slope model with rock bolts crossing the failure plane with insufficient anchorage length

Next, the length of rock bolts was increased, and they had anchorage length in the stable part. However, the length was not sufficient to prevent the sliding failure after a given acceleration level. Figure 5.32 shows the measured acceleration and relative slip responses. The initiation of slip was almost the same as that for the unreinforced case; the large movement of the unstable part occurred at an acceleration level of 550 gals, and total collapse was induced at the acceleration level of 740 gals. These examples clearly showed that the rock bolts must have sufficient length, anchored in the stable part of the slope, and number to prevent the failure of the slope against planar sliding.

5.7.5 Shaking table model tests of slopes subjected to wedge sliding

5.7.5.1 Preparation of models

Six special molds were prepared to cast model wedges (Kumsar *et al.*, 2000; Aydan and Kumsar, 2010). For each wedge configuration, three wedge blocks were prepared. Each base block had dimensions of $15.0 \times 100 \times 260$ mm. Base and wedge models were made of mortar and their geomechanical parameters were similar to those of rocks.

The composition of the mortar used for the preparation of the models is 1781 kgf m^{-3} of fine sand, 360 kgf m^{-3} of cement with a water-cement ratio of 0.5. The cement used in mortar was rapid hardening type, and samples were cured for about 7 days in a room with a constant temperature. The wedge angles and the initial intersection angles of wedge blocks are listed in Table 5.3.

In addition, several mortar slabs were cast to measure the friction angle of sliding planes. A number of tilting tests were performed. The inferred friction angle measured in tilting tests ranged between 30 degrees and 35 degrees with an average of 32 degrees.

Each wedge base block was fixed on the shaking table to receive same shaking with the shaking table during the dynamic test. The accelerations acting on the shaking table at the base and wedge blocks were recorded during the experiment and saved digitally in a data file (Figure 5.33). Furthermore, in the second series of dynamic tests, a laser displacement transducer was used to record the movement of the wedge block during the experiments. The reason for recording accelerations at three different locations is to determine the acceleration at the moment of failure as well as any amplification from the base to the top of the block. In fact, when the amplitude of input acceleration wave is increased, there is a sudden decrease on the wedge block acceleration records during the wedge failure, while the others are increasing. A barrier was installed at a distance of 20–30 mm away from the front of the wedge block to prevent their damage by falling off from the base block.

Dynamic testing of the wedge models was performed in the laboratory by means of a one-dimensional shaking table, which moves along horizontal plane. The applicable waveforms of the shaking table are sinusoidal, sawtooth, rectangular, trapezoidal and triangle. The shaking table has a square shape with 1 m side length. The frequency of waves to be applicable to the shaking table can range between 1 Hz and 50 Hz. The table has a maximum stroke of 100 mm and a maximum acceleration of 6 m s^{-2} for a maximum load of 980.7 N.

Table 5.3 Geometric parameters of wedges

Wedge Number	Intersection Inclination – i_a (°)	Half-wedge Angle (°)
TB1(Swedge120)	29	61.5
TB2(Swedge100)	29	51.5
TB3(Swedge90)	31	5.7.8
TB5.(Swedge70)	27	5.0.0
TB5(Swedge60)	30	33.8
TB6(Swedge5.5)	30	26.0

Figure 5.33 View of a wedge model

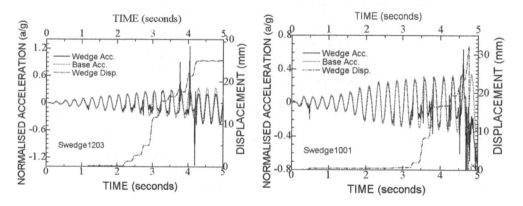

Figure 5.34 Dynamic response of wedge model: (a) TB1, (b) TB2

5.7.5.2 Shaking table tests

Three experiments were carried out on each wedge block configuration and dynamic displacement responses of the wedge blocks in addition to the acceleration responses were measured. Figures 5.35–5.36 shows typical acceleration time and displacement time responses for each wedge block configuration. As it is noted from the responses shown in Figures 5.35–5.36, the acceleration responses of the wedge block indicate some high-frequency waveforms on the overall trend of the acceleration imposed by the shaking table. When this type of waveform appears, the permanent displacement of the wedge block with respect to the base block takes place. Depending on the amplitude of the acceleration waves as well as its direction, the motion of the block may cease. In other words, a step-like behavior occurs.

The motion of the block starts when the amplitude of the input wave acts in the direction of the downside and exceeds the frictional resistance of the wedge block. When the

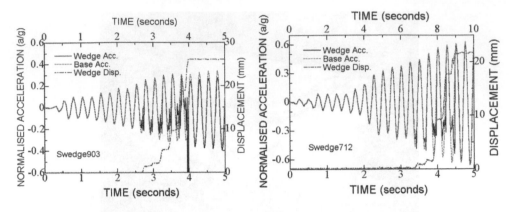

Figure 5.35 Dynamic response of wedge model: (a) TB3, (b) TB5

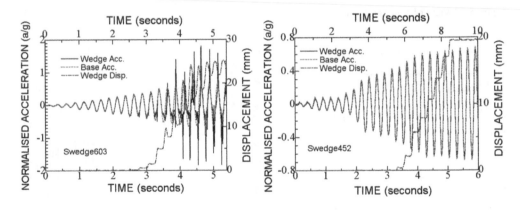

Figure 5.36 Dynamic response of wedge model: (a) TB5, (b) TB6

direction of the input acceleration is reversed, the motion of the block terminates after a certain amount of relative sliding. As a result, the overall displacement response is step-like.

Another important observation is that the frictional resistance between the wedge block and base block limits the inertial forces acting on the wedge block and the base block, even though the base block may undergo higher inertial forces. The sudden jumps in the acceleration response of the wedge block as seen in Figures 5.35–5.36 are due to the collision of the wedge block with the barrier. The initiation of the sliding of the wedge blocks was almost the same as those measured in the first series of the experiments.

5.7.6   Model experiments on shallow underground openings

The authors have been performing model experiments on underground openings for some time (Aydan *et al.*, 1994; Geniş and Aydan, 2002). The first series of experiments on shallow underground openings in discontinuous rock mass using nonbreakable blocks were reported by Aydan *et al.* (1994), in which a limit equilibrium method was developed for assessing

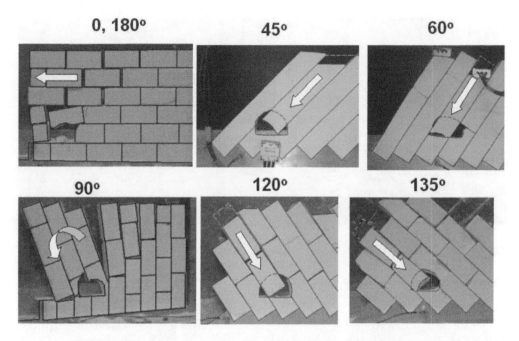

Figure 5.37 Failure modes of shallow tunnels adjacent to slopes with breakable material

their stability. These experiments have now been repeated using breakable material following the observations of damage to tunnels caused by the 2008 Wenchuan earthquake. The inclination of continuous discontinuity plane varied between 0 degrees and 180 degrees. Figure 5.37 shows views of some experiments. Unless the rock mass model itself failed, the failure modes were very similar to those of the model experiments using hard blocks. In some experiments with discontinuities dipping into the mountain side, flexural toppling of the rock mass model occurred. The comparison of the preliminary experimental results with the theoretical estimations based on Aydan's method (Aydan, 1989, Aydan *et al.*, 1994) are remarkably close to each other.

5.7.7 Monumental structures

5.7.7.1 Perry Banner Rock

A physical model of the Perry Banner Rock was prepared using model material, which has almost the same density of the original rock at a scale of 1/5.0 and total model height of 300 mm. The model was equipped with two laser displacement transducers (LDT-F, LDT-B), one acoustic emission sensor (AE-sensor) and two accelerometers (Acc. Top, Acc. B), as shown in Figure 5.38.

The model was fixed on the shaking table to receive same shaking during the dynamic test. The displacements, acoustic emissions (AE) and accelerations acting on the shaking table at the base and the model were recorded during the experiment and saved digitally in a data file.

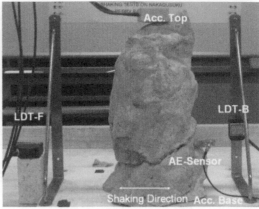

(a) General view   (b) Close-up view

Figure 5.38 General and close-up views of the model and its instrumentation

Figure 5.39 Views of orientations of shaking

As the shaking table can apply uniaxial accelerations, the model was shaken in three directions, namely 0, 45 and 90 degrees in order to investigate the effect of the inclination of the thoroughgoing discontinuity plane (Figure 5.39). The shaking direction angle is the acute angle between the strike of the thoroughgoing discontinuity plane and direction of shaking. Before the experiment leading to failure, the model was tested using the sweep testing procedure. The acceleration was fixed to 100 gals, and the frequency of wave was varied from 1 Hz to 50 Hz. The final experiment was concerned with the effect of grouting the gap in the model.

(A)   MODELS TESTS WITHOUT COUNTERMEASURES

Figures 5.40, 5.41 and 5.42 show the acceleration, AE and displacement responses measured for three directions, respectively. Depending upon the direction of the shaking, the blocks starts to exhibit nonlinear behavior at about 100 gals for the direction of 90 degrees while it is about 230 gals for the direction of 0 degree. The blocks become unstable when the base acceleration exceeds 350 gals.

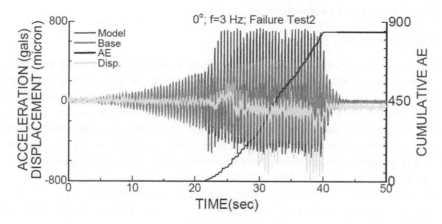

Figure 5.40 Responses for the shaking orientation of 0 degrees

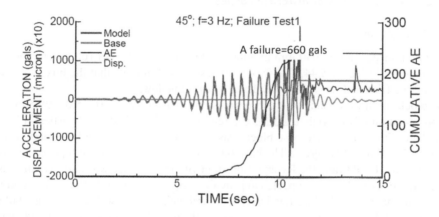

Figure 5.41 Responses for the shaking orientation of 45 degrees

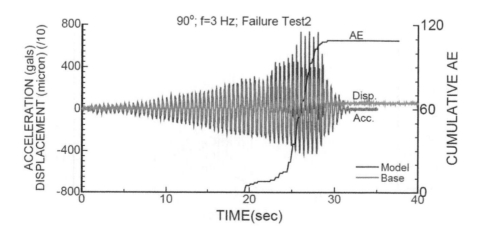

Figure 5.42 Responses for the shaking orientation of 90 degrees

(B)   MODELS TESTS WITH COUNTERMEASURES

Rock bolts and rock anchors are one of the effective ways of dealing with reinforcement issues in rock engineering (Aydan, 1989, 2018). However, the utilization of such counter-measures may not be attractive due to the disturbance of the appearance especially in archeological structures. As the resistance of the model was minimal for the orientation of 90 degrees, an experiment was carried out by introducing a bonding resistance to the through-going discontinuity plane. The bonding agent was double-sided bonding tape. Figure 5.43 shows the responses measured during the experiment. As noted from the figure and comparison with the response shown in Figure 5.42, the experimental results clearly indicated that the increase of the resistance was possible, and the overall seismic resistance of the block increases. This experimental finding was taken into account, and it was implemented during the remedial measures of the potentially unstable block.

5.7.7.2   Retaining walls of historical castles

(A)   MODEL SETUP

An acrylic transparent box 630 mm in length, 300 mm in height and 100 mm in width was used, as shown in Figure 5.44. The wall thickness was 10 mm so that the box was relative rigid, and the frictional resistance of sidewalls was quite low.

The blocks used were made of Ryukyu limestone with a size of 5.0 × 5.0 × 99.5 mm with the consideration of materials used for the retaining walls of historical castles in Ryukyu archipelago. Furthermore, the base block was such that the overall wall inclination can be chosen as 70, 83 and 90 degrees. The base block was fixed by two-sided tape to the base of the acrylic box. In addition, the Ryukyu limestone of the same size was laid over the base as seen in Figure 5.44. This was expected to provide a condition similar to the actual conditions observed in many historical castles in Ryukyu archipelago. The wall height was 240 mm, and the ratio of the height to width was 1/6. When the retaining wall inclination is 90 degrees without backfill material, the wall was expected to start rocking at an acceleration level of 167 gals.

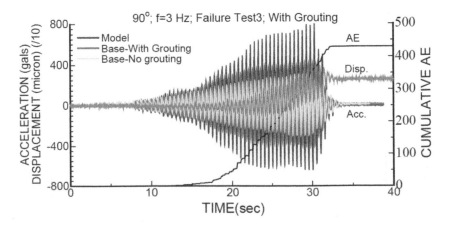

Figure 5.43 Responses of the model with countermeasures for the shaking orientation of 90 degrees

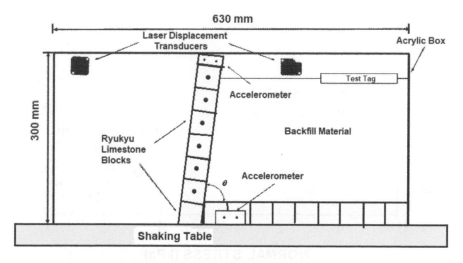

Figure 5.44 Illustration of model box

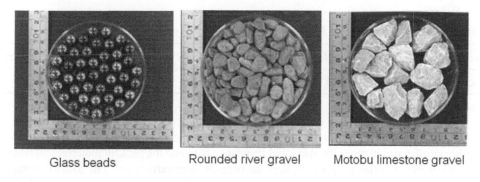

| Glass beads | Rounded river gravel | Motobu limestone gravel |

Figure 5.45 Views of backfill materials

(B)   BACKFILL MATERIALS AND THEIR PROPERTIES

Three different backfill materials were chosen (Figure 5.45). Glass beads were chosen to represent the lowest shear-resistant backfill material while the angular fragments of Motobu limestone was selected as the highest shear-resistant backfill material. The third backfill material was rounded river gravels having a shear resistance between those of the two other backfill materials.

A special shear testing setup was developed to obtain the shear strength characteristics of backfill materials under low normal stress levels, which are quite relevant to the model tests to be presented in this study. Figure 5.46 shows the shear strength envelopes for three backfill materials. As noted from the figure, the shear strength of rounded river gravel is in between the shear strength envelopes of glass-beads and Motobu limestone gravel. The strength of backfill materials is frictional, and the friction angle of the glass beads is about 21.68 degrees.

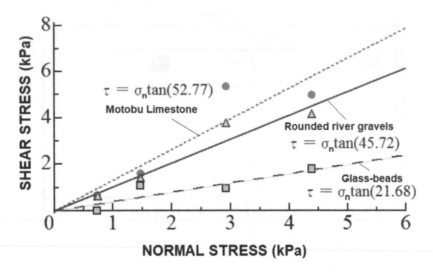

Figure 5.46 Shear strength envelopes for backfill materials

Table 5.4 Friction angle between Ryukyu limestone and backfill materials

Parameter	Glass-beads	Rounded River Gravel	Motobu Limestone Fragments
Friction angle	12.5–16.8	25.0–27.5	25.9–27.8

Another important factor for the stability of the retaining walls of historical castles as well as other similar structures is the frictional resistance between the backfill material and retaining wall blocks. For this purpose, tilting experiments were carried out. The backfill material contained in a box is put upon the Ryukyu limestone platens without any contact and tilted until it slides. This response of the backfill material contained in the box was measured using laser-displacement transducers. The inferred friction angles are given in Table 5.4. The lowest friction angle was obtained in the case of glass beads as expected.

(C)   SHAKING TABLE TESTS ON RETAINING WALLS WITH GLASS BEADS BACKFILL

A series of sweep tests were carried before the failure tests. Regarding the glass beads backfill material, the retaining walls were statically unstable for 90 degrees while they failed during the sweep test on the retaining walls with an inclination of 83 degrees. Therefore, we could show one example for retaining walls for the inclination of 70 degrees (Figure 5.47(a)). Its Fourier spectra analysis is shown in Figure 5.47(b). The results indicated no apparent natural frequency was dominant. The situation was quite similar in all experiments. Therefore, more emphasis will be given to the failure experiments.

Although the test on the retaining wall with an inclination of 83 degrees was intended for a sweep test, it resulted in failure. Figure 5.48 shows the displacement and base acceleration during the test. Failure tests on the retaining walls with an inclination of 70 degrees were

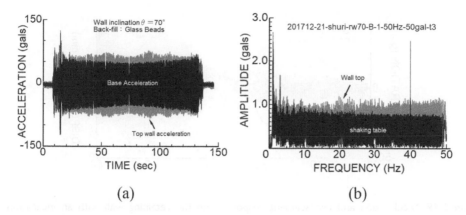

Figure 5.47 (a) Acceleration records of the shaking table and top of the retaining wall, (b) Fourier spectra of acceleration records

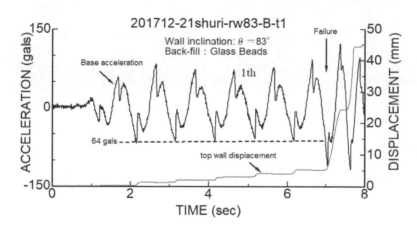

Figure 5.48 Acceleration and displacement responses on the retaining wall with an inclination of 83 degrees

carried out by applying sinusoidal waves with a frequency of 3 Hz. The amplitude waves were gradually increased until the failure occurred. Figure 4.49 shows an example of failure. The yielding initiated at about 110 gals, and the total failure occurred when the input acceleration reached 215 gals. Figure 5.50 shows the retaining wall before and after the failure test. The retaining wall failed due to toppling (rotation) failure, although some relative sliding occurred with the block at the toe of the model retaining wall.

(D) SHAKING TABLE TESTS ON RETAINING WALLS WITH RIVER GRAVEL BACKFILL

A series of sweep tests were carried before the failure tests as explained in the previous section. Regarding the rounded river gravel backfill material, the retaining walls were statically unstable for 90 degrees while the sweep test on the retaining walls with an inclination

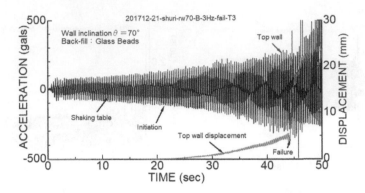

Figure 5.49 Acceleration and displacement responses on the retaining wall with an inclination of 70 degrees

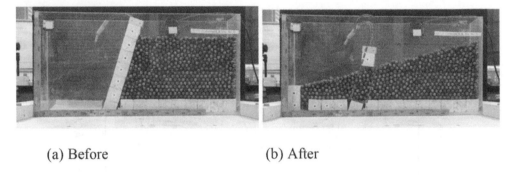

(a) Before   (b) After

Figure 5.50 Views of the model retaining wall with an inclination of 70 degrees before and after the test

of 83 and 70 degrees could be carried. We show one example for retaining walls for the inclination of 83 degrees in Figure 5.51(a) and its Fourier spectra analysis in Figure 5.51(b). The results indicated there was no dominant natural frequency for the given range of frequency. The situation was quite similar in all experiments for 83 and 70 degrees retaining wall models.

Failure tests on the retaining walls with inclinations of 83 and 70 degrees were carried out by applying sinusoidal waves with a frequency of 3 Hz. The amplitude waves were gradually increased until the failure occurred. Figures 5.52(a) and 5.52(b) show the acceleration and displacement responses of retaining walls with inclinations of 83 and 70 degrees as examples of failure tests. The yielding initiated at about 100 gals, and the total failure occurred when the input acceleration reached 210 gals for 83-degree retaining walls. On the other hand, the yielding initiated at 220 gals, and total failure occurred when the input acceleration was 430 gals for 70-degree retaining walls as seen in Figure 5.53. The retaining wall failed due to toppling (rotation) failure, although some relative sliding occurred with the block at the toe of the model retaining wall (Figure 5.53).

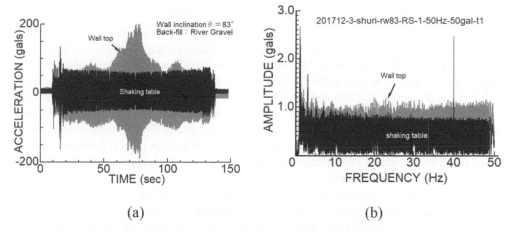

Figure 5.51 (a) Acceleration records of the shaking table and top of the 83 retaining wall with rounded river gravel backfill, (b) Fourier spectra of acceleration records

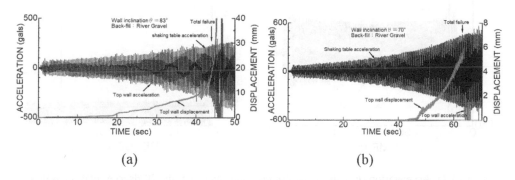

Figure 5.52 (a) Acceleration and displacement responses on the retaining wall with an inclination of 83 degrees, (b) acceleration and displacement responses on the retaining wall with an inclination of 70 degrees

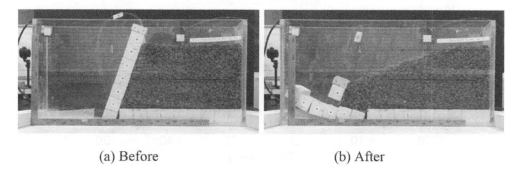

(a) Before   (b) After

Figure 5.53 Views of the model retaining wall with an inclination of 70 degrees before and after the test

(E)   SHAKING TABLE TESTS ON RETAINING WALLS WITH MOTOBU
       LIMESTONE GRAVEL BACKFILL

A series of sweep tests were carried out before the failure tests as explained in the previous section. Regarding the angular Motobu limestone gravel backfill material, the retaining walls were statically unstable for 90 degrees with a height of 25.0 mm. However, they were stable when the height was reduced to 160 mm. The sweep test on the retaining walls with inclinations of 90, 83 and 70 degrees were carried out. We show one example for retaining walls for the inclination of 70 degrees in Figure 5.54(a) and its Fourier spectra analysis in Figure 5.54(b). Again, the results indicated there was no dominant natural frequency for the given range of frequency. The situation was quite similar in all experiments for 90-, 83- and 70-degree retaining wall models.

Failure tests on the retaining walls with inclinations of 90, 83 and 70 degrees were carried out by applying sinusoidal waves with a frequency of 3 Hz. The procedure was the same as those in previous experiments. Figures 5.55, 5.56 and 5.57 show acceleration

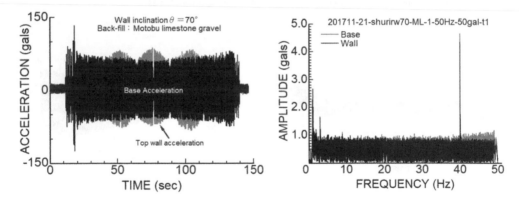

Figure 5.54 (a) Acceleration records of the shaking table and top of the 70-degree retaining wall with rounded river gravel backfill, (b) Fourier spectra of acceleration records

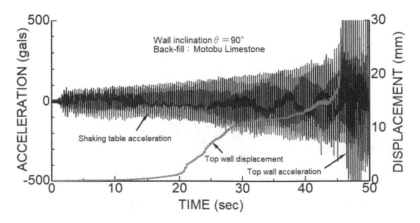

Figure 5.55 Acceleration and displacement responses on the retaining wall with an inclination of 90 degrees

and displacement responses of retaining walls with inclinations of 90, 83 and 70 degrees as examples of failure tests. The yielding initiated at about 110 gals, and the total failure occurred when the input acceleration reached 260 gals for 90-degree retaining walls. On the other hand, the yielding initiated at 130 gals, and the total failure occurred when the input acceleration was 300 gals for 83-degree retaining walls, as seen in Figure 5.58. The retaining walls failed due to toppling (rotation) failure.

The retaining walls with a 70-degree inclination and height of 25.0 mm did not fail during the entire test up to 400 gals as seen in Figures 5.59 and 5.60. Although some relative sliding occurred with the block at the toe of the model retaining wall when the base acceleration reached the level of 300 gals (Fig. 5.60). However, some settlement of the backfill occurred, and the retaining wall was pushed into passive sliding mode.

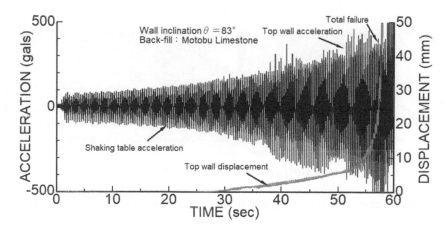

Figure 5.56 Acceleration and displacement responses on the retaining wall with an inclination of 83 degrees

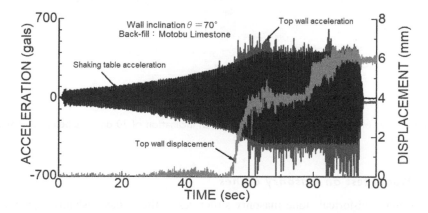

Figure 5.57 Acceleration and displacement responses on the retaining wall with an inclination of 70 degrees

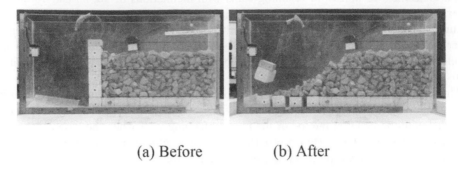

(a) Before   (b) After

Figure 5.58 Views of the model retaining wall with an inclination of 90 degrees before and after the test

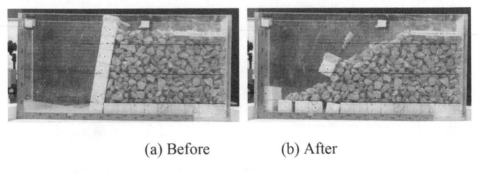

(a) Before   (b) After

Figure 5.59 Views of the model retaining wall with an inclination of 83 degrees before and after the test

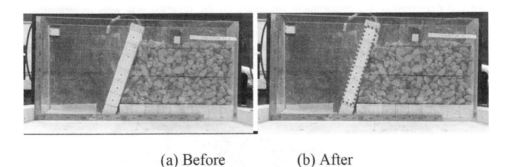

(a) Before   (b) After

Figure 5.60 Views of the model retaining wall with an inclination of 70 degrees before and after the test

5.7.8 Model test on masonry arches

There are many historical stone masonry structures in the Ryukyu islands. Ryukyu limestone blocks are generally used in the construction of the historical stone masonry structures. The major historical stone masonry structures are castles (i.e. Shuri Castle, Nakijin Castle,

Nakagusuku Castle, Gushikawa Castle etc.), burial tombs (Yodore) and imperial gardens. The authors were asked to assess the static and dynamic stability of walls and arch gates of Shuri Castle, a high retaining wall at Yodore imperial tomb and the arch bridge of Iedonchi imperial garden near the Shuri Castle. Some of experimental studies as well as numerical stability analyses are briefly reported in this subsection.

The Shuri Castle, whose ruins remain in Naha City of Okinawa Prefecture, is said to date back to the 12th century or earlier. The castle grounds and buildings were completely destroyed by the bombing of the U.S. Army during the Battle of Okinawa in 1945. Only some of castle and retaining walls remain with a certain degree of damage. The castle has been under restoration according to a map drawn during the Meiji period. In addition, the main buildings of the castle, castle walls, retaining walls and arches have been reconstructed. Since these structures are of masonry type without reinforcement, their seismic stability during earthquakes is of great concern.

Five arch configurations are denoted as Type-A, -B, -C, -D and -E and four of which (Type-A, -B, -D, -E) are commonly used in Shuri Castle in Okinawa island, Japan. All arch configurations were tested (Figure 5.61). The remaining arch form (Type-C) is quite common almost all over the world. The arches of Shuri Castle generally consist of two monolithic blocks in the form of a semicircle or an ovaloid shape while the Type-C arch consists of several blocks and has a semicircular shape. As the shaking table was uniaxial, the effect of the direction of input acceleration wave was investigated by changing the longitudinal axis of the arches (Figure 5.62).

The experimental results indicated that the common form of failure for all arch types for a shaking direction of 0 degrees is sliding at abutments and inward rotational fall of arch blocks subsequently. As for 90 degrees shaking, the arch failed in the form of toppling. The failure for 45 degrees shaking was a combination of sliding and toppling. The experiments clearly indicated that the amplitude of acceleration waves to cause failure was the lowest for 90 degrees shaking, while it was maximum for 0 degrees shaking.

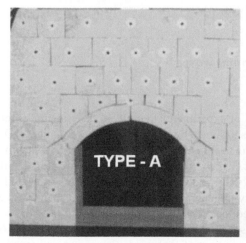 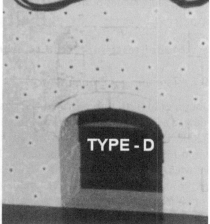

Figure 5.61 Common arch types used in Shuri Castle

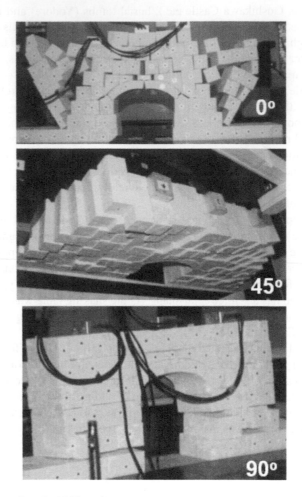

Figure 5.62 Failure modes of arch Type-A

5.7.9   Model tests subjected to tsunami waves

5.7.9.1   Tsunami generation device

Shimohira et al. (2019) performed a series of experiments using a model tsunami genera-tion device at the University of the Ryukyus designed by the author, and it is named as OA-TGD2000X to study the tsunami waves due to thrust and normal faulting event shown as shown in Figure 5.63. The dimensions and characteristics of the device were quite similar to that used in Tokai University except the wave-induction system. A tank was lowered or raised through pistons with a given velocity to generate rising or receding tsunami waves. The pressure and wave velocity at specified locations were measured using pressure sen-sors, and the amount of tank movement was measured using laser transducers. Figure 5.64. shows an example of record during the movement of tanks inducing rising and receding tsunami waves.

Figure 5.63 View of the tsunami generation device OA-TGD2000X

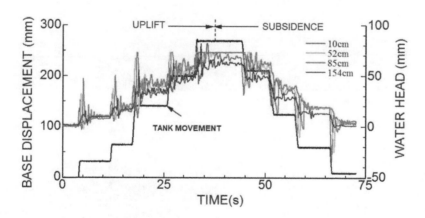

Figure 5.64 Water head response at specified locations in relation to the tank movement

5.7.9.2 Experiments

5.7.9.2.1 TRIANGULAR RYUKYU LIMESTONE BLOCKS

Triangular-shaped prismatic blocks shown in Figures 5.65 and 5.66 tested under the same condition. The longest side of the triangular prismatic block shown in Figure 5.65 was downward, while the longest side of the triangular prismatic block shown in Figure 5.66 was upward. While the downward triangular prismatic block was almost non-displaced, the upward triangular prismatic block was considerably displaced. One of the main reasons for such a big difference when they are subjected to the tsunami forces is that the tsunami wave applies a surging uplift force on the block. As for the downward triangular prism, the surging force increases the normal force on the block. We also put a rectangular prism of Ryukyu

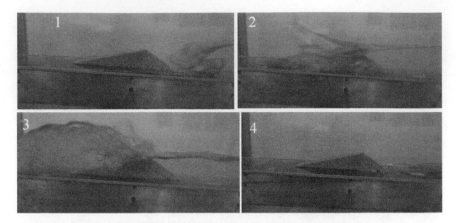

Figure 5.65 Views of the downward triangular prism block at different time steps

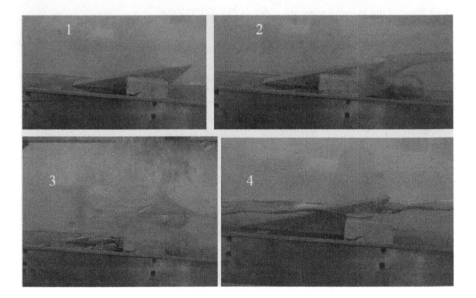

Figure 5.66 Views of the upward triangular prism block at different time steps

limestone next to the triangular prismatic block. The displacement of the rectangular block was quite small.

5.7.9.2.2 PLASTER BLOCKS

First a rectangular prismatic block made of plaster was subjected to rising tsunami waves as shown in Figure 5.67. The overall behavior is fundamentally similar to those tested in Tokai University. Nevertheless, the block was toppled toward the downstream side and displaced horizontally in the direction of receding tsunami waves as seen in Figure 5.67.

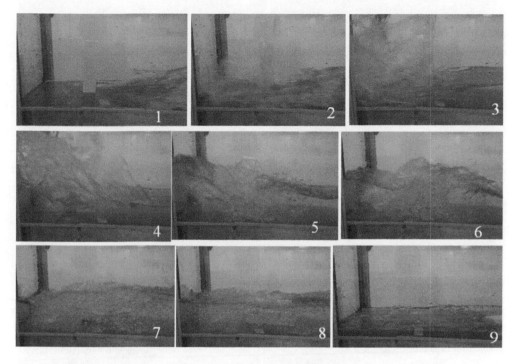

Figure 5.67 Views of the plaster block test at different time steps

Next two plaster prismatic blocks were laid over the Ryukyu limestone blocks as shown in Figure 5.68. The density of the plaster blocks is almost half of that of the Ryukyu limestone blocks. As seen from images 2 and 3 in Figure 5.68, the plaster blocks thrown upward and displaced in the direction of the tsunami waves. This experiment clearly demonstrates the importance of the density and overhanging degree of blocks when they are subjected to tsunami waves in nature.

5.7.9.2.3   BREAKABLE OVERHANGING CLIFFS

The next series of experiments involve the breakable blocks. Finding appropriate material for breakable blocks under the forces induced by tsunami forces by the experimental device was quite cumbersome. Although the materials had a very small density as compared to those in nature, it provided an insight view on the mechanism of formation of tsunami boulders, which was the main goal of this study. Figures 5.69 and 5.70 show the images of the models at different time steps. The surging tsunami wave enters under the overhanging blocks and applies upward forces. As a result, the overhanging block starts to bend upward, and breaks after a certain amount of displacement. In other words, the failure of the overhanging blocks is quite close to cantilever beams. However, the failure of the overhanging blocks is against gravity. Once the block is broken, it is dragged by the overflowing tsunami waves. This observation is in accordance with the mechanism proposed by Aydan and Toka-shiki (2019) for the formation of tsunami boulders. Our experiments clearly indicated that if

Figure 5.68 Top views of the plaster block overhanging the base Ryukyu limestone blocks at different time steps

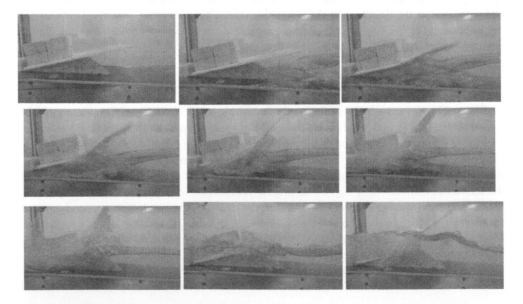

Figure 5.69 Views of the experiments using a breakable overhanging block at different time steps

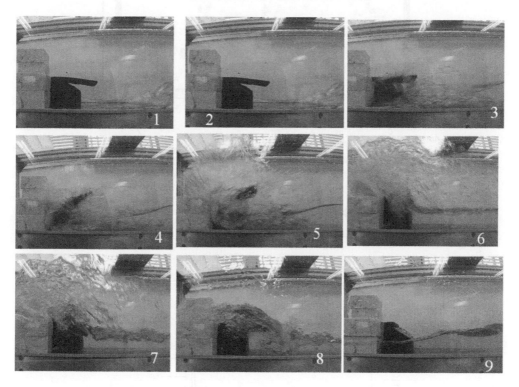

Figure 5.70 Views of the experiments using a breakable overhanging block at different time steps

the inclination of the lower side of the overhanging block ranges is 10–20 degrees, it is quite vulnerable to failure.

5.7.10 Experiments on tunnel models

Large rock samples of siliceous sandstone of Shizuoka third tunnel and Ryukyu limestone blocks having a circular hole with a diameter of 58 mm were tested. The height of the samples was 300 mm with a 200 mm width. In addition to acceleration measurements, multiparameter measurements in model tunnel experiments were performed on sandstone and limestone samples as illustrated in Figure 5.71.

Figure 5.72 shows the acceleration responses measured on a sandstone sample (300 × 200 × 138 mm) having a circular hole with a diameter of 58 mm. The overall acceleration responses are quite similar to those observed in uniaxial compression experiments. However, multiple acceleration responses with growing amplitudes occurred before that during the final rupture state. This phenomenon may be related to the ejection of small fragments from the perimeter of the model tunnel before the final failure of the sample. The observations of the experiments on the model tunnel in a brittle hard rock resemble the rockburst phenomenon experienced in rock engineering. Figure 5.73 shows a view of ejection of the rock fragments from the perimeter of the model tunnel.

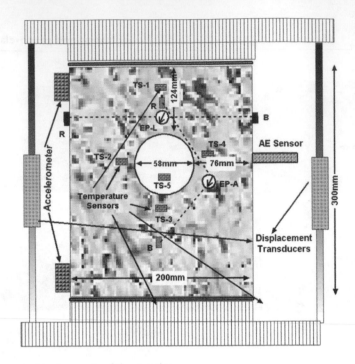

Figure 5.71 Layout of instruments of the samples

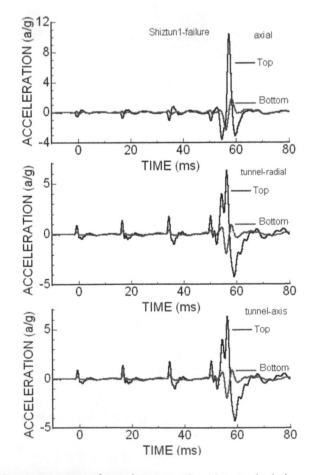

Figure 5.72 Acceleration response of a sandstone sample with a circular hole

Large rock samples of Ryukyu limestone blocks having a circular hole with a diameter of 58 mm were tested. The samples were 270 mm high, 160 mm wide and 140 mm thick. In addition to acceleration measurements, multiparameter measurements in model tunnel experiments were performed on limestone samples as illustrated in Figure 5.71.

Figure 5.74 shows the acceleration responses measured on a limestone sample denoted by BE-2–6-TUN. The overall acceleration responses are quite similar to those observed in

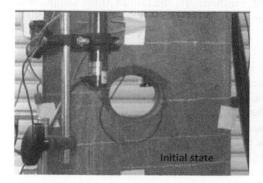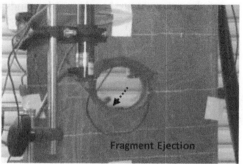

Figure 5.73 Views of the sample with a model tunnel during the experiment

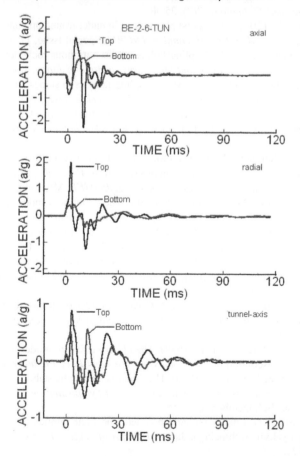

Figure 5.74 Acceleration response of a limestone sample with a circular hole

uniaxial compression experiments. The mobile part of the loading sample experiences larger accelerations. However, the amplitudes of acceleration responses of limestone sample are smaller than those measured in the experiment of the previous sandstone sample. The overall strength of limestone sample with a model tunnel is smaller than that of the sandstone sample, and the sandstone sample fails in a more brittle manner compared to the limestone sample. The compressive failure took place at sidewalls while tensile cracks appeared in the roof and floor of the model tunnel as expected. Furthermore, the failure of the sample was less violent as compared to that of the sandstone sample.

References

Aydan, Ö. (1989). The stabilisation of rock engineering structures by rock bolts. Geotechnical Engineering Department, Nagoya University, Nagoya, Doctorate Thesis.

Aydan, Ö. (2016). Issues on Rock Dynamics and Future Directions. Keynote, ARMS2016, Bali.

Aydan, Ö. (2018). *Rock Reinforcement and Rock Support*. CRC Press, Taylor and Francis Group, 486p.

Aydan, Ö. (2019). Some considerations on the static and dynamic shear testing on rock discontinuities. *Proceedings of 2019 Rock Dynamics Summit in Okinawa*, 7–11 May 2019, Okinawa, Japan, ISRM (Editors: Aydan, Ö., Ito, T., Seiki T., Kamemura, K., Iwata, N.), pp. 187–192.

Aydan, Ö. & Amini, M. G. (2009) An experimental study on rock slopes against flexural toppling failure under dynamic loading and some theoretical considerations for its stability assessment. *Journal of Marine Science and Technology*, 7(2), 25–40.

Aydan, Ö. and Geniş, M. (2010): Rockburst phenomena in underground openings and evaluation of its counter measures. *Journal of Rock Mechanics, TNGRM*, Special Issue, 17, 1–62.

Aydan, Ö. & Kawamoto, T. (1987) Toppling failure of discontinuous rock slopes and their stabilisation (in Japanese). *Journal of Mining and Metallurgy Institute of Japan*, Tokyo, 103(597), 763–770.

Aydan, Ö. & Kawamoto, T. (1992) The flexural toppling failures in slopes and underground openings and their stabilisation. *Rock Mechanics and Rock Engineering*, 25(3), 143–165.

Aydan, Ö. & Kumsar, H. (2010). An experimental and theoretical approach on the modeling of sliding response of rock wedges under dynamic loading. *Rock Mechanics and Rock Engineering*, 43(6), 821–830.

Aydan, Ö., Kyoya, T., Ichikawa, Y., Kawamoto, T. and Shimizu, Y. (1988). A model study on failure modes and mechanism of slopes in discontinuous rock mass. *Proceedings of the 23 National Conference on Soil Mechanics and Foundation Engineering, JSSMFE*, Miyazaki, 1, 1089–1092.

Aydan, Ö., Ohta, Y., Amini, M. & Shimizu, Y. (2019) The dynamic response and stability of discontinuous rock slopes. *Proceedings of 2019 Rock Dynamics Summit in Okinawa*, 7–11 May 2019, Okinawa, Japan, ISRM (Eds Aydan, Ö., Ito, T., Seiki T., Kamemura, K., Iwata, N.), pp. 519–524.

Aydan, Ö., Ohta, Y., Daido, M., Kumsar, H. Genis, M., Tokashiki, N., Ito, T. & Amini, M. (2011) Chapter 15: Earthquakes as a rock dynamic problem and their effects on rock engineering structures. *Advances in Rock Dynamics and Applications* (Eds Y. Zhou and J. Zhao), CRC Press, Taylor and Francis Group, pp. 341–422.

Aydan, Ö., Ohta, Y., Geniş, M., Tokashiki, N. & Ohkubo, K. (2010) Response and earthquake induced damage of underground structures in rock mass. *Journal of Rock Mechanics and Tunnelling Technology*, 16(1), 19–45.

Aydan, Ö., Shimizu, Y. & Ichikawa, Y. (1989). The effective failure modes and stability of slopes in rock mass with two discontinuity sets. *Rock Mechanics and Rock Engineering*, 22(3), 163–188.

Aydan, Ö., Shimizu, Y. & Karaca, M. (1994). The dynamic and static stability of shallow underground openings in jointed rock masses. *The 3rd International Symposium on Mine Planning and Equipment Selection*, Istanbul, October, 851–858.

Aydan, Ö. & Tokashiki, N. (2019) Tsunami boulders and their implications on a mega earthquake potential along Ryukyu Archipelago, Japan. *Bulletin of Engineering Geology and Environment*, 78(6), 3917–3925.

Bieniawski, Z.T. & van Tonder, C.P.G. (1969) A photoelastic-model study of stress distribution and rock around mining excavations. *Experimental Mechanics*, 9(2), 75–81.

Bray, J.W. & Goodman, R.E. (1981) The theory of base friction models. *International Journal of Rock Mechanics and Mining Science & Geomechanics* Abstracts, Oxford, 18, 553–568.

Brewster, D. (1815) Experiments on the depolarization of light as exhibited by various mineral, animal and vegetable bodies with a reference of the phenomena to the general principle of polarization, *Philosophical Transactions of the Royal Society*, 29–53.

Bucky, P.B. (1931) *The Use of Models for Study of Mining Problems*. Technical Publication 525, American Institute of Mineral and Metallurgical Engineering, NewYork.

Coker, E.G. & Filon, L.N.G. (1930/1957) *Treatise on Photoelasticity*. Cambridge Press, Cambridge.

Egger, P. (1979) A new development in the base friction technique. *Colloquium on Geomechanical Models,* ISMES, Bergamo. pp. 67–81.

Erguvanlı, K. & Goodman, R.E. (1972) Applications of models to engineering geology for rock excavations. *Bulletin of the Association of Engineering Geologist*, 9(1).

Everling, G. (1965) Model tests concerning the interaction of ground and roof support in gate roads. *International Journal of Rock Mechanics and Mining Science & Geomechanics* Abstracts, 1, 319–326.

Fumagalli, E. (1973) *Statical and Geomechanical Models*. Springer-Verlag, Vienna.

Geniş, M. & Aydan, Ö. (2002). Evaluation of dynamic response and stability of shallow underground openings in discontinuous rock masses using model tests. *Korea-Japan Joint Symposium on Rock Engineering*, Seoul, Korea, July, pp. 787–794.

Kumsar, H., Aydan, Ö. & Ulusay, R. (2000): Dynamic and static stability of rock slopes against wedge failures. *Rock Mechanics and Rock Engineering*, 33(1), 31–51.

Phillips, E. (1869) De l'equilibre des solides elastiques semblables. *Comptes rendus de l'Académie des Sciences*, Paris, 68, 75–79.

Shimohira, K., Aydan, Ö., Tokashiki, N., Watanabe, K. & Yokoyama, Y. (2019) *Proceedings of 2019 Rock Dynamics Summit in Okinawa*, 7–11 May 2019, Okinawa, Japan, ISRM (Eds: Aydan, Ö., Ito, T., Seiki T., Kamemura, K., Iwata, N.), pp. 193–198.

Stacey, T.R. (2006) Considerations of failure mechanisms associated with rock slope instability and consequences for stability analysis. *The Journal of the South African Institute of Mining and Metallurgy*, 106, 485–493.

Stephanson, O. (1971) Stability of single openings in horizontally bedded rock. *Engineering Geology*, 5(1), 5–71.

Sugawara, K., Akimoto, M., Kaneko, K. & Okamura, H. (1983) Experimental study on rock slope stability by the use of a centrifuge. *Proceedings of the Fifth International Society for Rock Mechanics Congress*, pp. C1–C4.

Chapter 6

Rock excavation techniques

6.1 Blasting

6.1.1 Background

Blasting is the most commonly used excavation technique in mining and civil engineering applications. Blasting induces strong ground motions and fracturing of rock mass in rock excavations.

The excavation of rocks in mining and civil engineering applications by blasting technique is the most common technique since chemical blasting agents were developed centuries ago (i.e. Hendron, 1977; Hoek and Bray, 1981). However, the development of modern blasting techniques is after the invention of dynamites. Blasting induces high ground motions and fracturing of rock mass adjacent to blast holes (i.e. Thoenen and Windes, 1942; Attewell *et al.*, 1966; Siskind *et al.*, 1980; Kutter and Fairhurst, 1971). Particularly, high ground motions may also induce some instability problems of rock mass and structures nearby (Northwood *et al.*, 1963; Tripathy and Gupta, 2002; Kesimal et al., 2008; Hao, 2002; Ak et al. 2009; Geniş et al. 2013). Furthermore, it may cause some environmental problems due to noise as well as vibrations of structures near populated areas (i.e. Aydan *et al.*, 2002).

Models for the attenuation of ground motions induced by blasting are generally based on velocity-type attenuation following the initial suggestions pioneered by United States Bureau of Mines (USBM) (Thoenen and Windes, 1942), and many models follow the footsteps of the USBM model (i.e. Attewell *et al.*, 1966; Tripathy and Gupta, 2002 etc.). These models are often used in an empirical manner to assess the environmental effects on structures and human beings (Northwood *et al.*, 1963; Hendron, 1977; Siskind *et al.*, 1980). Although it is mathematically possible to relate ground motion parameters with each other, it is not always straightforward to do so (Aydan *et al.*, 2002). Depending upon the sampling intervals of ground motion records, as well as superficial effects particularly during the integration process, the ground motion parameters and records may be different if they are measured by velocity meters or accelerometers. However, it should be noted that the acceleration records are essential when they are used in stability assessments.

6.1.2 Blasting agents

6.1.2.1 Dynamite

Dynamite is an explosive material of nitroglycerin, using diatomaceous earth or other absorbent substance such as powdered shells, clay, sawdust or wood pulp. The Swedish chemist and engineer Alfred Nobel invented dynamite in 1867. Dynamite is usually in the form of cylinders about 200 mm long and about 3.2 mm in diameter, with a weight of about 186 g. Dynamite is generally used in underground excavations as it produces less harmful gases during explosion.

6.1.2.2   Ammonium nitrate/fuel oil (ANFO)

ANFO (ammonium nitrate/fuel oil) is a widely used bulk industrial explosive mixture. It consists of 94% porous prilled ammonium nitrate (NH_4NO_3) (AN) that acts as the oxidizing agent and absorbent for the fuel and 6% fuel oil (FO). ANFO is widely used in open-cast coal mining, quarrying, metal mining and civil construction as it is a low-cost and ease-of-use matter among other conventional industrial explosives. The initiation of blasting is achieved using primer cartridges.

6.1.2.3   Blasting pressure for rock breakage

The detonation pressure is empirically related to the density (ρ_0) and detonation velocity (D) of explosives, and the following formula is generally used for estimating the intrinsic pressure of explosives:

$$p = \frac{1}{1+\gamma} \rho_o D^2 \tag{6.1}$$

where γ is parameter related to the intrinsic properties of explosives. Its value is generally 3 (Persson et al., 1994). Table 6.1 summarizes the detonation pressure of several explosives used in rock breakage. However, the actual blasting pressure acting on the wall of holes is much less than the detonation pressure due to the gap between explosives, thickness of the casing, deformability characteristics and fractures in rock mass.

6.1.3   Measurement of blasting vibrations in open-pit mines and quarries

The author has conducted measurements of blasting-induced motions in open-pit mines and quarries. The results of measurements conducted by the author are briefly explained in this subsection.

6.1.3.1   Orhaneli open-pit lignite mine

Measurements were carried out separately in Orhaneli open-pit and Gümüşpınar village (Aydan et al, 2002). Figure 6.1 shows a general view of open-pit mine and nearby

Table 6.1 Summary of basic parameters of commonly used explosives

Explosive	Detonation Velocity (m s⁻¹)	Density (g cm⁻³)	Detonation Pressure (GPa)
Dynamite	4500–6000 (7600)	1.3, 1.593 (1.51)	6–13.6 (22.0)
ANFO	2700–3600	0.882–1.10	0.7–9.0

Figure 6.1 General view of Orhaneli open-pit mine, Gümüşpınar village behind the pit and the major fault plan on the left side

Gümüşpınar village. The village is about 1 km from the open-pit mine. In the open-pit mine, a major normal fault and several minor strike-slip faults were observed. The minor strike-slip fault is observed near the ground surface, and they are limited to near-surface layers. The rock mass above the lignite seam consists of tuff, sandstone, breccia and marl.

Figure 6.2 shows the layout of the instrumentation employed at the open-pit mine blasting test. The blasting hole is 9.65 m deep with a diameter of 25 cm. It is a single hole with two deck charges. Both upper and lower deck charges consist of 50 kg of ANFO and 1 kg of cap-sensitive emulsion explosive (dynamite). Initiation was done by nonelectric shock tube (NONEL) detonator with 25 ms delay. The upper deck was fired first. Figure 6.3 shows a view of blasting operation experiment. The distance to the blasting hole was 66 m from Acc-3.

Figure 6.3 shows the records of accelerometer denoted as Acc-1 during blasting. As seen from the figure, the magnitude of the vertical component of the acceleration waves is the largest, while that of the traverse components is the least among other components. Furthermore, the peak of the vertical component appears a few milliseconds before the others. The reason for the difference between vertical and horizontal components may be related to the damage and weakening caused by the previous blasts to the top 1–1.5 m of the bench on which the instruments were located. Therefore the weakened top part of the bench can be regarded as a low-velocity layer as compared with the rest of the bench below. This low-velocity (damaged) layer may cause the attenuation of horizontal components, while its effect on the vertical component is less since the vertical component wave travels mostly through undisturbed marl beds. Furthermore, the interface between coal and marl, which is just 8–10 m below the ground surface, may act as a good reflecting surface so that the vertical component is enhanced in amplitude.

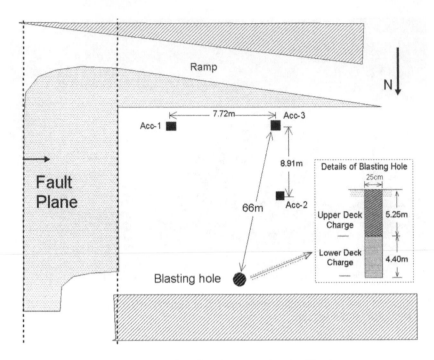

Figure 6.2 Layout of instrumentation employed in the open-pit mine blasting test

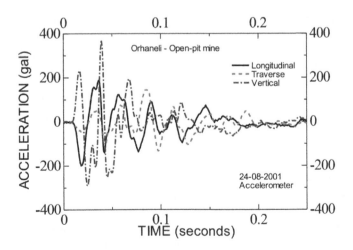

Figure 6.3 Acceleration records from Acc-1 accelerometer

Figure 6.4 shows the vertical acceleration records of accelerometers denoted as Acc-1, Acc-2 and Acc-3 during blasting. Although the magnitudes of initial peaks follow the order of distance to the blasting hole, the magnitude of the farthest accelerometer denoted as Acc-3 is larger than the others. This may be caused by some slight variation of fixation and ground

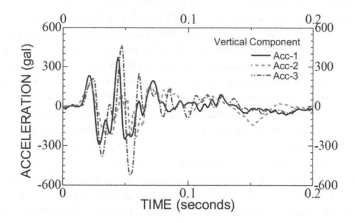

Figure 6.4 Vertical components from accelerometers denoted as Acc-1, Acc-2 and Acc-3

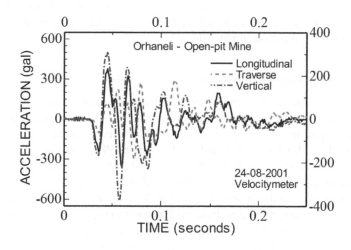

Figure 6.5 Acceleration components obtained from the numerical derivation of velocity records

conditions beneath the accelerometers. The peak value exceeds 500 gal, and the waves attenuate as time goes by within 0.2 s following the blasting. The ground near accelerometers was blasted previously. The accelerometer locations could be disturbed at varying degrees depending upon the distance to the previous blast holes. Therefore, the attenuation of records of accelerometer Acc-2 is greater than the others, and long-period waves become dominant.

Figure 6.5 shows the acceleration obtained from the numerical derivation of velocity records using sampling interval of the velocity-meter (denoted Vel-3) next to the accelerometer Acc-3 (see Figure 6.2 for location). Although the records are very similar to each other, the accelerations computed from velocity meters are larger in amplitude as compared with those from accelerometers. For example, for the vertical component, the maximum amplitude of the acceleration wave was 525 gals from the accelerometer as compared with 609 gals computed from the velocity record of the velocity meter. The amplitude obtained from the velocity meter records is about 1.16 times that of the true acceleration records. The difference may arise from one or all of the following reasons: slight variation of fixation of instruments, the errors inherent in numerical derivation arising from digital wave forms or the frequency dependence of directly measured acceleration value. The validity of the first and second reasons must be checked by further measurements and analysis. If, especially, the second reason holds true, the empirical damage criteria based on the peak particle velocity of the ground could not be employed straightforwardly in the case integration of directly measured acceleration values, or vice versa. Hence a different damage criterion should be developed for the case of direct monitoring of acceleration.

The amplitude of vertical component of acceleration waves caused by blasting is larger than that of other components. The amplitude of the acceleration waves is in the order of vertical, longitudinal (radial) and traverse (tangential). However, the response spectra imply that amplifications are in the reverse order. Fourier spectra of longitudinal, traverse and vertical components of the acceleration records of the accelerometer and velocity meter indicated that dominant frequencies of the waves observed at 8–10 Hz and 30–40 Hz account for the fundamental vibration mode. The results indicate that structures having a natural period less than 0.06 s could be very much influenced. The effect of blasting should be smaller for structures having natural periods greater than 0.1 s.

6.1.3.2   Demirbilek open-pit lignite mine

Aydan *et al.* (2014a) performed ground motion measurements in the open-pit mine near Demirbilek village during blasting, in which the attenuation of ground motions and the effect of existing faults were investigated, using both velocity meters and accelerometers simultaneously. The YOKOGAWA WE7000 modular high-speed PC-based data-acquisition system was used. It can handle 16 channels simultaneously. The sampling interval was set to 10 ms during measurements. AR-10TF accelerometers with three components were used, and this device can measure accelerations up to 10 g. X and Y components of the accelerometer were aligned radially and in tangential directions with respect to the blasting location. Z-direction measured the up and down (UD) component of ground motions. Figure 6.6 shows one of the layouts for ground motion observations and a view of blasting.

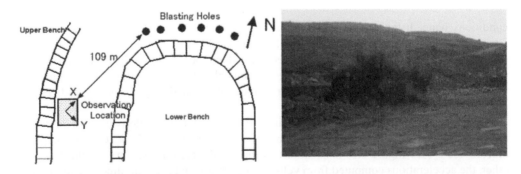

Figure 6.6 Layout of ground motion observations and a view of blasting

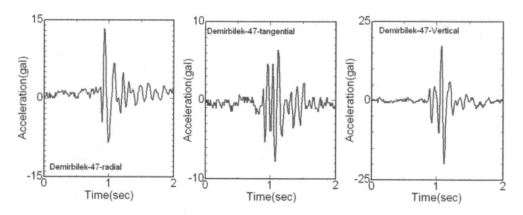

Figure 6.7 Acceleration responses measured during the *in-situ* blasting experiment number 47

A typical blast hole in the lignite mines of Turkey is generally 8–10 m deep with a diameter of 25 cm. It consists of 50–75 kg of ANFO and 0.5–1 kg of cap-sensitive emulsion explosive (dynamite). One third of each blast hole stemmed with soil, and initiation was done by nonelectric shock tube detonator with a 25–50 ms delay.

The characteristics of ground motions induced by blasting depends upon the amount of explosive, the layout of blast holes and benches, delays and geomechanical properties of rock mass. Figure 6.7 shows an example of acceleration responses during the blasting experiment numbered 47 with a 15.5 kgf ANFO explosive. The distance of the blasted hole was approximately 40 m away from the monitoring location on the same bench level. Typical rocks observed in the lignite mine are lignite itself, marl, sandstone, mudstone and siltstone.

As mentioned in the introduction, most of the attenuation relations used in the evaluation of the effects of blasting are of the velocity type. There are very few attenuation relations for blasting-induced accelerations. Dowding (1985) proposed an empirical attenuation relation for maximum acceleration. Wu *et al.* (2003) also developed an empirical relation using the results of small-scale field blast tests involving soil and granite. However, the empirical relations particularly overestimate maximum accelerations within a distance of 100 m of the blast location. Nevertheless, the empirical relation proposed by Dowding performs better than that by Wu *et al.* (2003). Therefore, the authors develop their attenuation relations for ground conditions typical in the lignite mines of Turkey.

The attenuation of ground acceleration may be given in the following form as a convolution of three functions F, G and H in analogy to the attenuation relation of ground motions induced by earthquakes (Aydan 2012; Aydan et al. 2011):

$$a_{max} = F(V_p)G(R_e)H(W) \tag{6.2}$$

where V_p, R_e and W are elastic wave velocity, distance from the explosion location and weight of explosives. The units of V_p, R_e and W are m s^{-1}, meter (m) and kilogram force (kgf) while the unit of acceleration is gal. In analogy to the spherical attenuation relation proposed

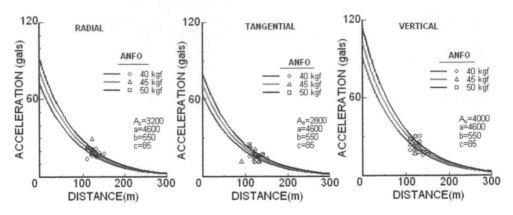

Figure 6.8 Attenuation of maximum ground acceleration with distance for a single-hole blasting experiment

by Aydan (1997, 2001, 2007, 2012; Aydan and Ohta, 2011), the functions F, G and H may be assumed to be of the following forms:

$$F(V_p) = A_o(e^{V_p/a} - 1) \tag{6.3a}$$

$$H(W) = (e^{W/b} - 1) \tag{6.3b}$$

$$G(R_e) = Ae^{-R_e/c} \tag{6.3c}$$

The coefficients of a, b and c are found to be approximately 4600, 550 and 85, respectively, for the observation data obtained from single-hole blasting experiments at Demirbilek open-pit mine as shown in Figure 6.8, together with the values of the coefficients of Equation (6.3) for each component. The values of coefficient A_o are found to be 3200, 2800 and 4000 for radial, tangential and vertical components, respectively. However, the value of coefficient c may be different for each component of ground acceleration in relation to explosive type. The value of coefficient c is applicable to ANFO explosives, which are commonly used in Turkish lignite mines. The empirical relation by Dowding (1985) follows a similar approach in order to take into account the effect of ground conditions in attenuation relations.

6.1.3.3 ELI Işıkdere open-pit mine

(A) 2010 MEASUREMENTS

2010 measurements were carried out using a single accelerometer of G-MEN type. The blasting holes were 7 m deep, and two rows of the holes with a separation distance of 7 m were drilled parallel to the bench face. For each hole, the amount of ANFO was 50–75 kg, and stemming was 1/3 of the total depth of the hole. Each sag has 25 kg ANFO. The delay between the front and back rows of the holes was 25 ms. Blasted rock mass consists mainly of marl. Figure 6.9 shows views of blasts while Figure 6.10 shows the acceleration records for different blasts. As noted from the figure, the acceleration records are not symmetric with respect to time axis.

Figure 6.9 View of blasting on 23 August 2010 at ELI Işıkdere open-pit mine

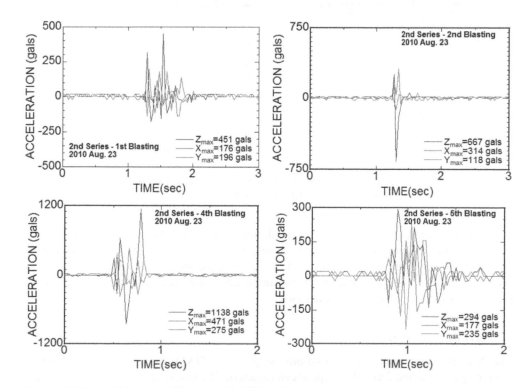

Figure 6.10 View of blasting on 23 August 23 2010 at ELI Işıkdere open-pit mine

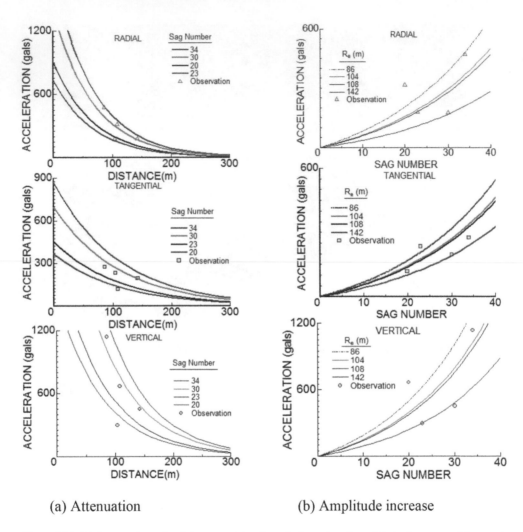

(a) Attenuation　　　　　　　　(b) Amplitude increase

Figure 6.11　(a) Attenuation of radial, tangential and vertical accelerations, (b) the increase of the maximum acceleration with respect to sag number

Figure 6.11 shows the attenuation of radial, tangential and vertical accelerations and the increase of the maximum acceleration with respect to the sag number in accordance with functional forms of Equations (6.3b) and (6.3c). As expected, the maximum acceleration decreases as a function of distance while the amplitude of the maximum ground acceleration increases as the amount of ANFO increases.

(B)　2011 MEASUREMENTS

The 2011 measurements were carried out using 7 QV3-OAM stand-alone accelerometers with trigger mode and 2 G-MEN-type accelerometers. The first series of investigations were aimed to see the attenuation of accelerometers. The nearest station to the blasting point was 10 m (Figure 6.12). Figure 6.13 shows some of the acceleration records.

(a) First series   (b) Second series

Figure 6.12 Views of blasting experiments before and during blasts

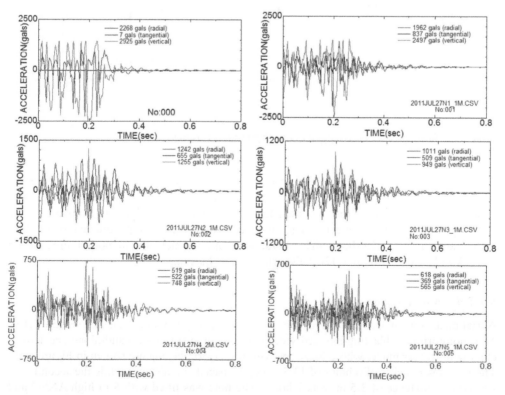

Figure 6.13 Acceleration records for the first series of blasting experiments

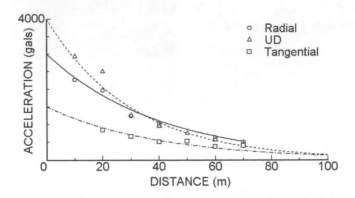

Figure 6.14 Attenuation of maximum ground acceleration with respect to distance

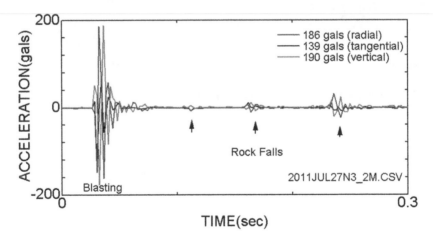

Figure 6.15 Rock fall-induced ground motions triggered by blasting

Figure 6.14 shows the attenuation of maximum acceleration as a function of distance. As indicated previously, the vertical component is the largest, while the tangential component is the smallest. However, the attenuations with distance are different from each other and tangential component attenuates gradually compared with the rapid attenuation of other components. Figure 6.15 shows an acceleration record in which ground motions are induced by rock falls seen in Figure 6.12(a) also triggered by blasting.

6.1.3.4 Motobu quarry

A trial measurement was done at Motobu limestone quarry in Okinawa island (Figure 6.16). The monitoring of blasting-induced vibrations is done using four stand-alone accelerometers with trigger mode, whose locations with respect to blasting are shown in Figure 6.16. The first row consists of 6 holes of 13.5 m depth spaced at a distance while the second row, spaced at a distance of 3.5 m, was 7 holes. The hole was filled with 5 m high ANFO and

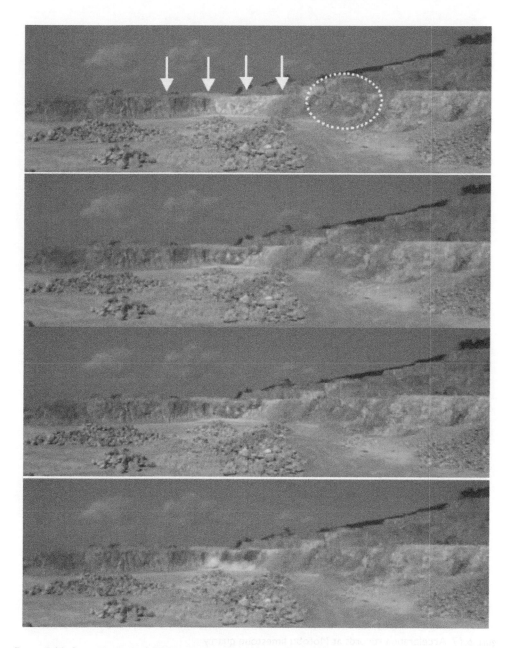

Figure 6.16 Several views of the blasting and location of instruments

8.6. m stemming material. However, one of the accelerometers did not function during the blasting.

Figure 6.17 shows the acceleration records. The axial component was larger than the UD component, and they attenuate with distance from the blasting location. In the same records, some ground motions induced by the falling rock blocks are also noticed.

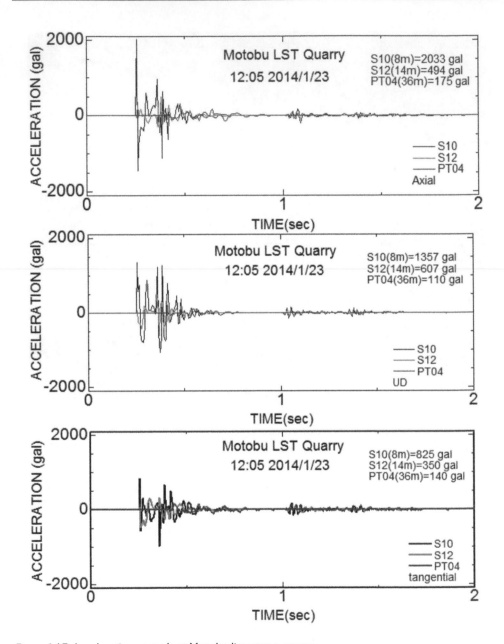

Figure 6.17 Acceleration records at Motobu limestone quarry

6.1.4   Measurements at underground openings

The author has conducted measurements of blasting-induced motions in several tunnels in Japan and Turkey. The results of measurements conducted by the author are briefly explained in this subsection.

6.1.4.1 Kuriko Tunnel

The Kuriko Tunnel in Fukushima prefecture is excavated through Mt. Kuriko (Watanabe et al. 2013). A geology of the tunnel consists of granite, rhyolite, tuff, andesitic dykes and intercalated sedimentary rocks such as sandstone, mudstone and conglomerate. While granite is exposed in the Fukushima side (east), folded sedimentary rocks with folding axis aligned in north–south outcrops on Yonezawa side (west). Sedimentary rocks are covered with tuff and rhyolite and intruded with andesitic dykes (Figure 6.18).

The length of blast hole rounds was 1.5 m, and a total of 156 kg dynamite was used at the location where measurement was done. Figure 6.19 shows the layout of blast holes,

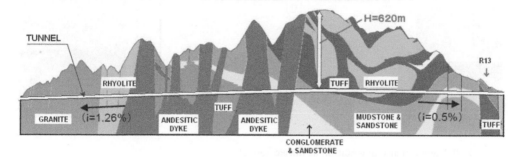

Figure 6.18 Geological cross section of Kuriko evacuation tunnel beneath Mt. Kuriko

Source: Modified from Haruyama and Narita (2009)

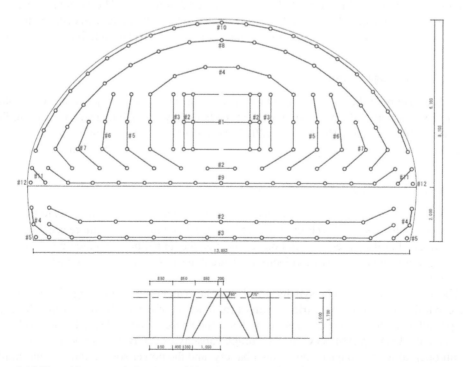

Figure 6.19 Typical layout and plan view of blast holes used at Kuriko Tunnel

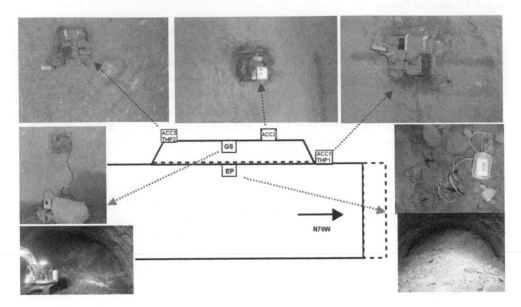

Figure 6.20 Layout of instrumentation at Kuriko Tunnel

while Figure 6.20 shows the layout of instrumentation and their views with respect to tunnel advance direction. Figure 6.21 shows the acceleration records. The axial and tangential components were larger than the radial component, and they attenuate with distance from the blasting location. Furthermore, several shocks are recorded in relation to blasting sequence.

6.1.4.2   Taru-Toge Tunnel

Taru-Toge Tunnel is being constructed as a part of an expressway project connecting Shin-Tomei Expressway and Chuo Expressway at the boundary of Shizuoka and Yamanashi Prefectures in Central Japan (Imazu et al. 2014; Aydan et al. 2016). The tunnel passes through a series of mudstone, sandstone, conglomerate layers with folding axes aligned north–south.

(A)   INSTRUMENTATION AND INSTALLATION

The total number of accelerometers was 13, and the accelerometers were fixed to the plates of the rock bolts or steel ribs (Figures 6.22 and 6.23). In addition, three more accelerometers were attached to the plates of rock bolts at the passage tunnel between the main tunnel and the evacuation tunnel for wave velocity measurements during the blasting operation at the main tunnel.

The accelerometers can be synchronized, and they can be set to the triggering mode with the capability of recording pretrigger waves for a period of 0.5 s. The trigger threshold and the period of each record can be set to any level and chosen time as desired. The accelerometer is named QV3-OAM-SYC, has a storage capacity of 2 GB and is a stand-alone type. It can operate for two days using its own battery, and the power source can be solar light if appropriate equipment is used. In other words, it is an eco-friendly acceleration monitoring

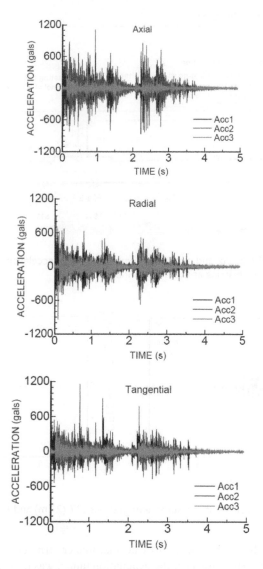

Figure 6.21 Acceleration records for each component

Figure 6.22 View of the accelerometer and its fixation in the main tunnel

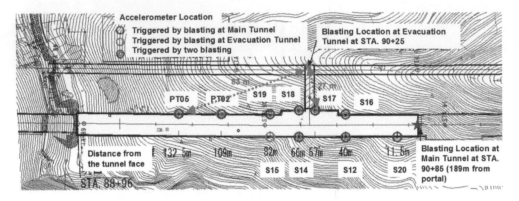

Figure 6.23 Locations of blasting and accelerometers (Note that accelerometer S15 had a battery problem during blasting at the main tunnel.)

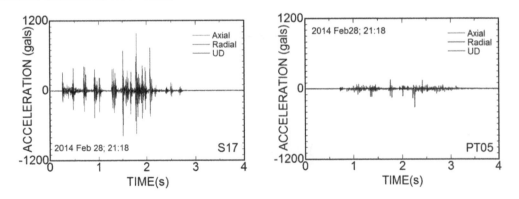

Figure 6.24 Acceleration records at measurements stations S17 (27 m) and PT06 (83 m)

system. Figure 6.23 shows the location of accelerometers triggered during each blasting operation. The blasting (Blasting-1) at the evacuation tunnel was done at 21:18 on 28 February 2014. The second blasting (Blasting-2) was carried out at 10:54 on 5 March 2014.

(B)   CHARACTERISTICS OF GROUND VIBRATION DURING BLASTING

Blasting-1   The blasting (Blasting-1) at the evacuation tunnel was done at 21:18 on 28 February 2014, and the amount of explosive was 75 kgf with 10 rounds with a delay of 0.4–0.5 s. The threshold value for triggering was set to 10 gals and the total number of the triggered accelerometers was 9. The most distant accelerometer was PT06, and its distance from the blasting location was 83 m. The highest acceleration was recorded at the accelerometer denoted as S17, and its value was about 1000 gals. Figure 6.24 shows the acceleration records at the stations denoted S17 and PT05. As noted from the figure, the amplitude of accelerations decreases with distance as expected. Another interesting observations is that the acceleration records are not symmetric with respect to the time-axis. Furthermore, acceleration wave amplitudes differ depending upon the direction.

Figure 6.25 also shows the Fourier spectra of each component of acceleration waves. As noted, the figures for the Fourier spectra of the radial and axial components of the accelerometer S17 consist of higher-amplitude and higher-frequency content. As for the distant accelerometer PT05, the vice-versa condition is observed, as expected.

Figure 6.26 shows the acceleration response spectra of acceleration records taken at S17 and PT05 for the respective directions. As noted from the figure, the acceleration response spectra have very short natural periods, as expected.

Figure 6.27 shows the attenuation of maximum acceleration at all stations. As the wave forms are unsymmetric with respect to time-axis, peak values are plotted as positive peak (PP) and negative peak (NP) with the consideration of their position in relation to the blasting location. As noted from the figure, the attenuation is quicker at the unblasted side compared with those on the blasted side. The data is somewhat scattered, and this may be related to the existence of structural weakness zones in the rock mass.

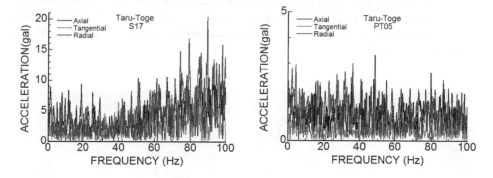

Figure 6.25 Fourier spectra of acceleration records shown in Figure 6.24

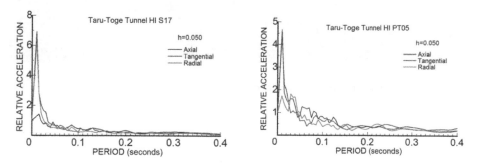

Figure 6.26 Acceleration responses spectra of acceleration records shown in Figure 16.3.

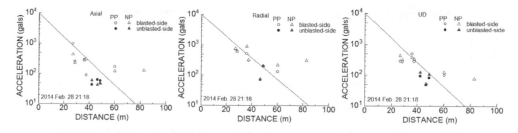

Figure 6.27 Attenuation of maximum ground acceleration with distance for Blasting-1

Blasting-2   The second blasting (Blasting-2) at the main tunnel was carried out at 10:54 on 5 March 2014, and the total amount of blasting was 30.4 kgf with 0.4–0.5 ms delays per round. Figure 6.28 shows the tunnel face before and after blasting of the lower bench. The threshold value for triggering was set to 10 gals, and the total number of the triggered accelerometers was 14. The most distant accelerometer was PT05 and its distance from the blasting location was 132.5 m. The highest acceleration was recorded at the accelerometer denoted as S20, and its value was about 1030 gals.

Figure 6.29 shows the acceleration records at the stations denoted S20 and PT05. As also noted from the figure, the amplitude of accelerations decreases with distance as expected. Another interesting observations is that the acceleration records are not symmetric with respect to the time-axis. Furthermore, acceleration wave amplitude differs depending upon the direction of measurements. The comments for Fourier spectra basically would be the same except those for the tangential component. Lower-frequency content waves become dominant for the tangential component as seen in Figure 6.30. Figure 6.31 shows the acceleration response spectra of acceleration records taken at S20 and PT05 for the respective directions. As noted from the figure, the acceleration response spectra have very short natural periods, as expected.

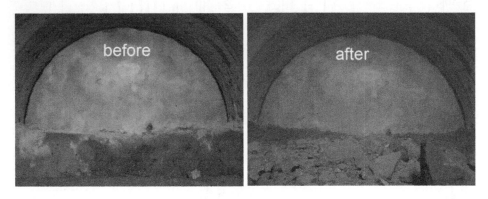

Figure 6.28 Views of tunnel face before and after blasting

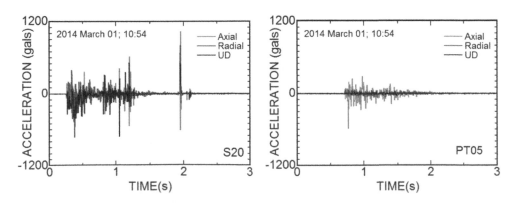

Figure 6.29 Acceleration records at measurements stations S20 (11.6 m) and PT05 (132.5 m)

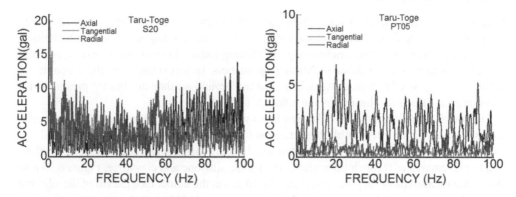

Figure 6.30 Fourier spectra of acceleration records shown in Figure 6.29

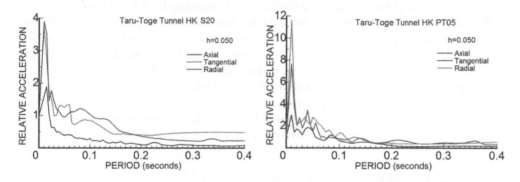

Figure 6.31 Acceleration response spectra of acceleration records shown in Figure 6.29

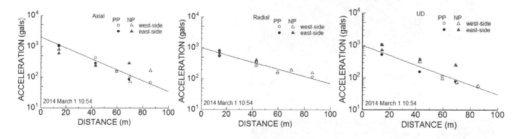

Figure 6.32 Attenuation of maximum ground acceleration with distance for Blasting-2

Figure 6.32 shows the attenuation of maximum acceleration at all stations triggered. As the wave forms are unsymmetric with respect to time axis, peak values are plotted as positive peak (PP) and negative peak (NP) with the consideration of their position in relation to the blasting location. As the passage tunnel exists on the west side between the main tunnel and the evacuation tunnel, the maximum accelerations are somewhat smaller at the station S18. Despite some scattering of measured results, the attenuation of maximum acceleration decreases with the increase of distance exponentially. The data scattering may also be related to the existence of structural weakness zones in the rock mass.

6.1.4.3   Zonguldak tunnels

Several tunnels in association with rehabilitating the intercity roadways in Turkey have been excavated in Zonguldak and its close vicinity, using the drilling and blasting technique. Genis *et al.* (2013) have been monitoring the blasting-induced vibrations in several adjacent tunnels near at Sapça, Üzülmez ve Mithatpaşa tunnels. In this subsection, the outcomes of the monitoring of vibrations in tunnels and at the ground surface are briefly presented for assessing the effects of blasting on the adjacent structures.

The Sapça tunnels are double-tube two-lanes tunnels. The main purposes of the measurements were to see the effect of blasting at a new tunnel on the adjacent tunnel and ground surface. Figure 6.33 shows the position of tunnels, blasting location and measurement locations of a tunnel excavated through intercalated sandstone, siltstone and claystone. Figure 6.33 also shows the measurements at ground surface (70 m) and at the tunnel face (32 m) of the adjacent tunnel. The amount of blasting was 6–12 kg for each round. Although the ground motions are less on the ground surface than those at the adjacent tunnel due to the distance from the location of blasting, the ground motions are relatively high regarding the UD component.

The Üzülmez tunnels are also two-lanes double-tube tunnels. The pillar distance between the tunnels is about 11 m. Three accelerometers were installed at the pillar side of the adjacent tunnel and two accelerometers at the mountainside. One more accelerometer was installed at the tunnel where blasting was carried out. Figure 6.34 shows the measurement results. Despite the distance, the measurements were highest in the tunnel of blasting, and the accelerations were high at the pillar side.

The Mithatpaşa tunnels fundamentally have similar geometrical features, while rock mass is limestone, and some karstic caves were encountered during excavation.

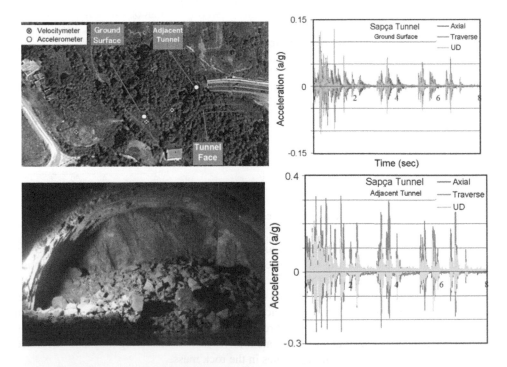

Figure 6.33 Position of tunnels, blasting location and measurement locations and measured responses

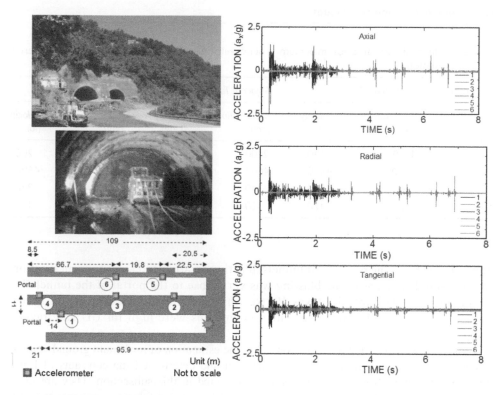

Figure 6.34 Position of tunnels, blasting location and measurement locations and measured responses

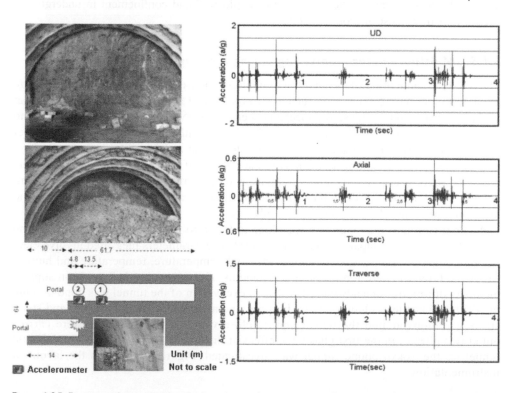

Figure 6.35 Position of tunnels, blasting location and measurement locations and measured responses

Table 6.2 Values of constants for maximum ground velocity and ground acceleration for different situations

Coefficient	Inside Tunnels		Adjacent Tunnels		Ground Surface	
	Acceleration (gal)	Velocity (kine)	Acceleration (gal)	Velocity (kine)	Acceleration (gal)	Velocity (kine)
A_0	4000	140	2000	80	6000	200
a	4600	4600	4600	4600	4600	4600
b	240	240	240	240	240	240
c	100	120	120	100	100	120

Figure 6.35 shows the position of tunnels, blasting location and measurement locations and measured responses. The blasting was very close to the portal of the tunnel. The total amount of the explosives was 192 kg while the amount of explosive for rounds changed between 10 kg to 25 kg. The accelerations were high for UD and traverse components.

Genis *et al.* (2013) extended Equation 6.2 to the estimation of maximum ground velocity in addition to the maximum ground acceleration and determined the constants of Equation 6.3 for tunnels from the measurements presented in this subsection. They are given in Table 6.2. The coefficients are slightly different from those for measurements at lignite mines. The reason may be the difference of explosives and confinement in underground excavations from those at ground surface.

6.1.5   *Multiparameter monitoring during blasting*

The real-time monitoring of the stability of tunnels is of great importance when tunnels are prone to failure during excavation such as rockbursting or squeezing. It is also known that when rock starts to fail, the stored mechanical energy in rock tends to transform itself into different forms of energy. Experimental studies by the authors showed that rock indicates distinct variations of multiparameters during deformation and fracturing processes. These may be used for the real-time assessment of the stability of rock structures. The parameters measured involve electric potential variations, acoustic emissions, rock temperature, temperature and humidity of the tunnel in addition to the measurements of convergence and loads on support members during the face advance.

An application of this approach was done to the third Shizuoka tunnel of the second Tomei Expressway in Japan (Aydan *et al.*, 2005). The parameters measured involve electric potential variations, acoustic emissions, rock temperature, temperature and humidity of the tunnel in addition to the measurements of convergence and loads on support members during the face advance. The tunnel excavation of the tunnel was done through the drilling-blasting technique. Each blasting operation causes both dynamic and static stress variations around the tunnel and its close vicinity. Measurements were carried out at two phases. In the first phase, the effect of the face advance of the Nagoya-bound tunnel on the Tokyo-bound tunnel was investigated. Figure 6.36 shows the layout of instrumentation.

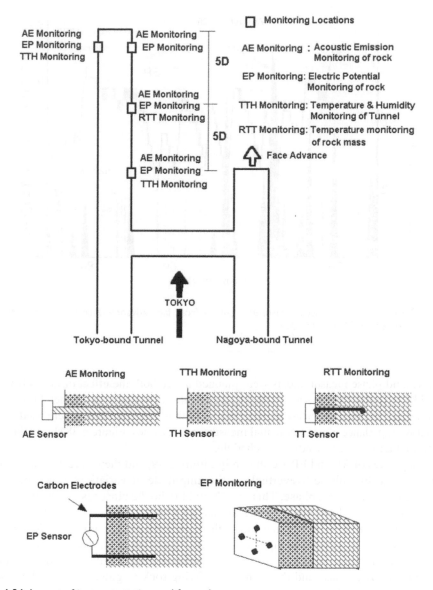

Figure 6.36 Layout of instrumentation and face advance

Figure 6.37 shows the measured AE and electric potential responses as a function of time. Vertical bars in the same figure indicate the blasting operations. After each blasting operation, distinct AE and EP variations were observed. These variations cease after a certain period of time. The electric potential increase simultaneously and tends to decrease as time goes by. AE response also showed the same type of response. When the tunnel is stable, it is expected that these variations should disappear on the basis of experimental observations and theoretical considerations by the author.

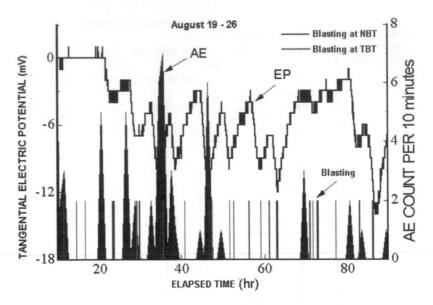

Figure 6.37 AE and EP responses measured resulting from face advance during the period from 19 August to 26 August 2004

The second phase measurements were planned to see both the effect of face advance on AE and EP responses in the same tunnel as well as that of the adjacent tunnel. Furthermore, two new electric potential measurement devices with higher impedance were used in addition to low-impedance electric potential measurement devices. Figure 6.38 shows the layout of instrumentation and face advance schedule.

The responses of AE and EP are shown in Figure 6.39, and they were basically similar to those of the first phase. Nevertheless, the amplitude of variations was much larger than those of the previous phase. This was thought to be the closeness of the instruments to the tunnel face where blasting operations were carried out. The electric potential measurements with low-impedance and high-impedance electric potential devices were almost the same. However, the high-impedance electric potential measurement devices are desirable. In addition to that, some problems were noted with the fixation of electrodes into the rock mass and the damage by flying rock fragments during the blasting operation. However, the electric potential variations may sometimes include the far-field effects from the deformation of the Earth's crust associated with earthquakes. During the measurement of electric potentials, an additional device would be necessary outside the tunnel. In this particular case, such a device was installed about 2 km west of the tunnel.

The same instrumentation was repeated at Tarutoge Tunnel. Figure 6.40 shows the variations of acoustic emissions (AE) and electric potential variations in relation to blasting at the main tunnel and evacuation tunnel. The overall responses are similar to those measured at Shizuoka Third Tunnel.

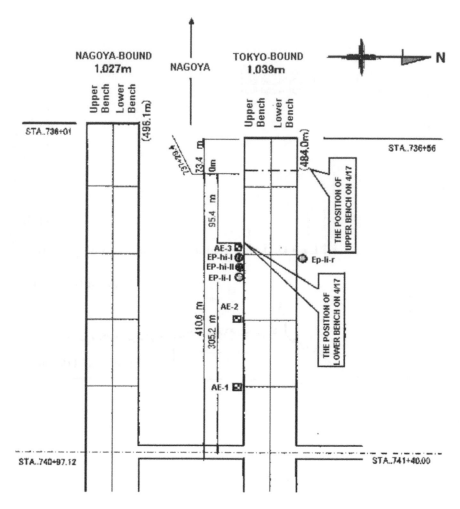

Figure 6.38 Layout of instrumentation and face advance schedule

A trial measurement was carried out on the groundwater pressure variation during blasting operation at Demirbilek open-pit lignite mine. Figure 6.41 shows the monitoring results during the blasting operation numbered Demirbilek-07-blasting. The 1500 mm deep borehole, with a diameter of 160 mm, was drilled and filled with water. A water pressure sensor was installed in the borehole. The water pressure was measured using the same monitoring system. Although a very slight time lag exists between the peaks of ground acceleration and groundwater, the water pressure fluctuation occurred in response to the ground acceleration fluctuation. Furthermore, the groundwater level deceases thereafter due to seepage into surrounding rock as well as the increased permeability due to the damage rock by blasting.

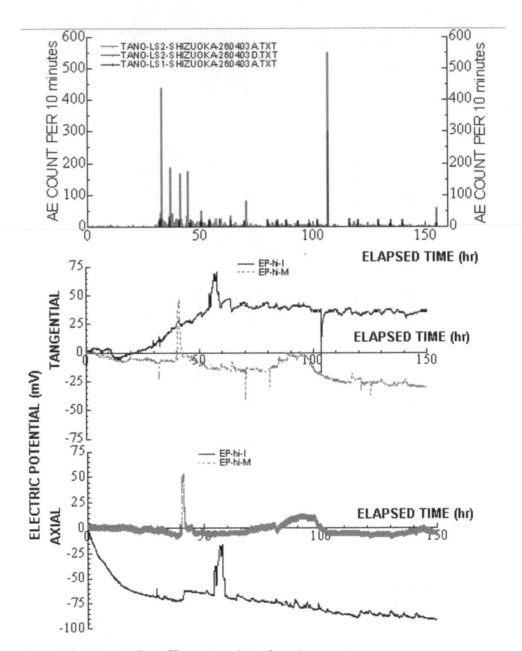

Figure 6.39 Measured AE and EP responses during face advance

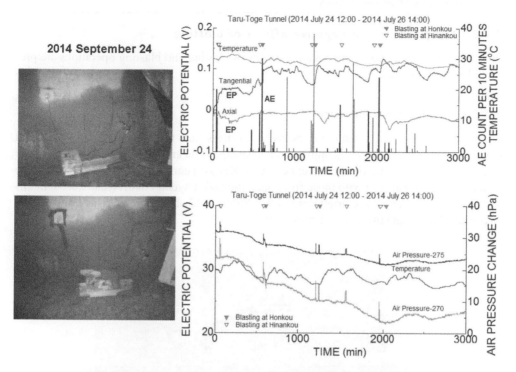

Figure 6.40 AE and EP responses measured resulting from face advance during the period from 24 July to 26 July 2014

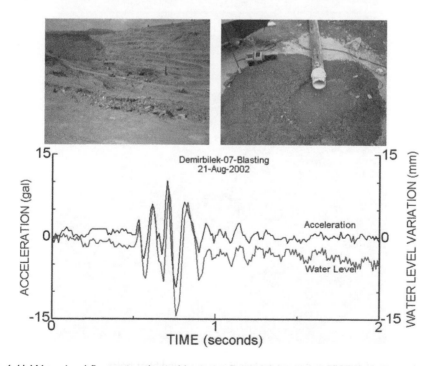

Figure 6.41 Water level fluctuation during blasting at Demirbilek open-pit mine

6.1.6 *The positive and negative effects of blasting*

In this section, the positive and negative effects of blasting and blasting operations are presented and discussed.

6.1.6.1 In-situ *stress inference*

Aydan (2013) proposed a method to estimate the stress state from the damage zone around blasted holes. This method was applied to the damage zone around blasted holes, and some stress inferences were made for the tunnel face at Kuriko Tunnel and Taru-Toge Tunnel. The estimations are compared with those from other methods (Aydan 2000: Aydan 2003). In this section, these examples are briefly presented. Figure 6.42 shows the damage zone around two blasted holes and the inferred stress state.

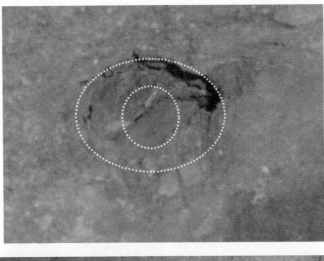

Figure 6.42 Views of damage and estimated yield zones around blast holes

(A)   KURIKO TUNNEL

The damage around blast holes are shown in Figure 6.42. The fracture zone formation around the blasted holes are almost elliptical with the longitudinal axis almost horizontal. These results imply that lateral stress is higher than the vertical stress on the tunnel face plane.

The lateral stress coefficient was taken as 2.2, and the inclination of the maximum principal stress was assumed to be 0 from horizontal by taking into account the actual shape of the damage zone around blast holes. The results are given in Table 6.3 for a blast hole pressure of 150 MPa; the inferred plastic zones for different yield criteria and actual plastic zone are shown in Figure 6.43. The estimated damage zones are quite close to those shown in Figure 6.4. The *in-situ* stress estimations are also similar to the *in-situ* stress measurements by the AE method, in which the lateral stress coefficient was found to be 1.7.

Table 6.3 Inferred *in-situ* stress parameters and yield function parameters used in computations

σ_{10} (MPa)	k	σ_c (MPa)	σ_t (MPa)	$\phi(°)$	m	S_∞ (MPa)	b_1 (1/MPa)
9.0	2.2	90	6	60	14.93	360	0.045

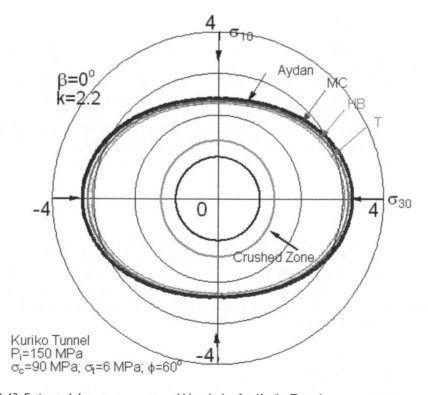

Figure 6.43 Estimated damage zones around blast holes for Kuriko Tunnel

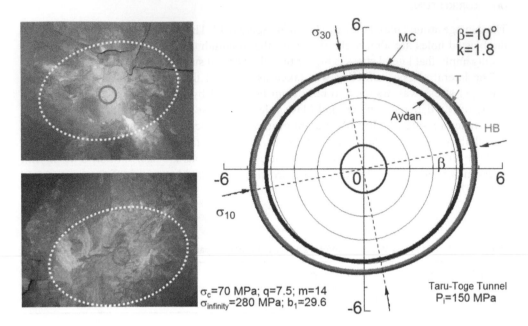

Figure 6.44 Estimated stress state from the damage zone around blasted holes

(B)   TARU-TOGE TUNNEL

Imazu *et al.* (2014) utilized the fault striation method (Aydan, 2000) and blasted hole damage method (BHMD) (Aydan, 2013) at Taru-Toge Tunnel. The estimated lateral stress coefficient from the fault striation method at the tunnel cross section ranges between 1.6 and 1.7. The BHDM was applied to the damage zone around blasted holes, and some stress inferences were made for the tunnel face. Figure 6.44 shows the damage zone around two blasted holes and the inferred stress state. The estimations indicate that the lateral stress coefficient is about 1.8 and that it is inclined with an angle of 10 degrees to the west. It is also interesting to note that this ratio is also very close to that estimated from the fault striation method.

6.1.6.2   Rock mass property estimation from wave velocity using blasting-induced waves

The mechanical properties of rock mass may be estimated using the elastic wave velocity. Direct relations or normalized relations exist in the literature (i.e. Ikeda, 1970; Aydan *et al.*, 1993, 1997; Sezaki *et al.*, 1990). The uniaxial compressive strength (UCS) of intact rock ranges between 40–70 MPa, and its P-wave velocity is in the range of 3–4 km s^{-1}. The direct relations between the UCS of rock mass and elastic wave velocity proposed by Aydan *et al.* (1993, 2014), and Sezaki *et al.* (1990) are, respectively:

$$\sigma_{cm} = 5\left(V_{pm} - 1.4\right)^{1.43} \tag{6.4}$$

$$\sigma_{cm} = 1.67\left(V_{pm} - 0.33\right)^{2} \tag{6.5}$$

$$\sigma_{cm} = 0.98V_{pm}^{2.7} \tag{6.6}$$

where σ_{cm} and V_{Pm} are uniaxial compressive strength (UCS) and P-wave velocity of rock mass.

Another relation, which is called "rock mass strength ratio," is as follows:

$$\sigma_{cm} = \left(\frac{V_{pm}}{V_{pi}}\right)^2 \sigma_{ci} \qquad (6.7)$$

where σ_{ci} and V_{pi} are uniaxial compressive strength (UCS) and P-wave velocity of intact rock.

Aydan *et al.* (2016) developed a portable system to measure the P-wave and S-wave velocities of the surrounding rock mass. The measurement system consists of five accelerometers connected to one another with wire and operated through a "start-stop" switch. This system was first used on 5 March 2014 and 1–3 September 2016 at Tarutoge Tunnel. The distance between accelerometers in 1–3present system can be up to 6 m (measurement length is about 30 m), and the sampling interval is 50 μs. Figure 6.45 shows the installation of the device in Tarutoge Tunnel, and measured velocity responses are shown in Figure 6.46.

The estimated wave velocity of rock mass in the instrumented zone is estimated to be 1.9–2.6, 2.0–2.7 and 1.8–2.4 km s⁻¹ during 6 March 2015. The normalized UCS of rock mass is estimated to be about 0.2–0.46 times that of intact rock. If direct relations are used and normalized by that of intact rock (Aydan *et al.*, 2014b), the normalized UCS of rock mass would be 0.1–0.2 times that of intact rock.

The measurement during 1–3 September 2016 yielded that the P-wave velocity was about 2.7 km s⁻¹. In view of the P-wave velocity of intact rock, the results are quite close to those

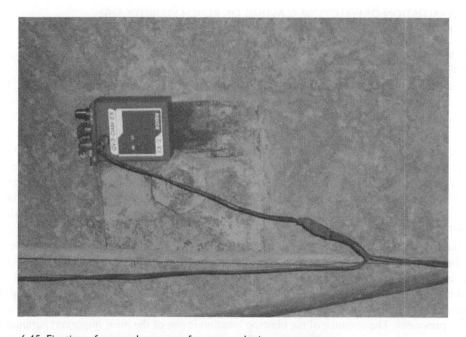

Figure 6.45 Fixation of an accelerometer for wave velocity measurement

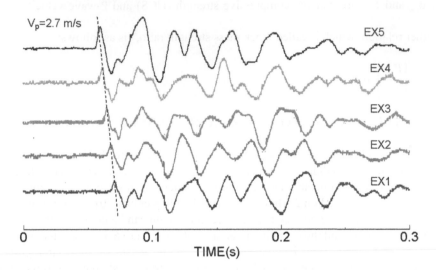

Figure 6.46 Acceleration records during the arrival of P-waves induced by blasting

of 5 March 2016 measurements. The normalized UCS of rock mass is estimated to be about 0.46 times that of intact rock.

6.1.6.3   Instability problems

(A)   EVALUATION OF EFFECTS OF BLASTING ON BENCH STABILITY AND
RESPONSES

The width and height of benches range between 16 to 46 m and 8 to 10m, respectively, at Demirbilek open-pit lignite mines (Aydan *et al.*, 2014a). The slope angle of the benches also varies between 50 and 60 degrees depending upon the mining operations and layouts. The failure of benches was observed in the upper marl unit, and they were of planar type on the east slope of the open-pit mine. Figure 6.47 shows the failure of benches at different locations at the east side of the open-pit mine. As seen in the pictures, the planar sliding failure is the dominant failure mode. The engineers of the open-pit mine also reported that the stability issues become very important following heavy rains. On the other hand, benches on the west slope of the open-pit mine were more stable, although some cracking and opening were observed. The instability problems on the benches of the west slope were associated with normal fault. Unless the fault plane cut through the benches, there were no major slope stability problems on the west slope benches.

Aydan and his coworkers (Aydan *et al.*, 1996; Aydan and Ulusay, 2002) proposed a method to estimate the movements of slopes involving sliding failure. This method was further elaborated in subsequent studies (Aydan *et al.*, 2006, 2008, 2009a, 2011; Tokashiki and Aydan, 2010; Aydan and Kumsar, 2010). The author utilized this technique for assessing the response and stability of the slope subjected to dynamic forces as well as gravity and pore water pressures. The distance of the blasting location is one of the most important parameters for analyzing the response and stability of bench slopes.

Figure 6.47 Views of stability problems on the benches of the east slope

First the effect of vibrations measured by a single blast hole experiment with a distance of 100 m and an ANFO explosive of 50 kgf on the benches consisting of the upper marl unit was investigated. The computations indicated that no movement would occur under dry and fully saturated conditions. Then, the number of blast holes was assumed to be 18, which is commonly used during blasting operations in the open-pit mining of lignite mines in Turkey. The amplitude of the accelerations was increased by 4.7 times that of the single blast hole with the consideration of 18 blast holes with an ANFO explosive of 50 kgf. In the computations, both horizontal and vertical accelerations were considered. If the rock mass is assumed to be dry, the computations indicated that no relative movement along bedding planes emanating from the toe of the benches would occur. If the groundwater coefficient is more than 0.76, relative movement along the bedding planes occurs. Figure 6.48 shows the computed relative displacement responses in relation to the horizontal base acceleration. We introduce a water force coefficient denoted as r_u to count the effect of groundwater in the body subjected to slide. It is defined as the volume of water to the total volume of the body prone to sliding. As noted from the figure, if rock mass is fully saturated ($r_u = 1.0$) or nearly fully saturated ($r_u = 0.8$), some permanent displacement would occur after each blasting operation. The sliding body in benches would gradually be displaced, and the separation of the sliding body from the rest of benches would occur at the upper levels and relative offsets at the toe of the slope. When the relative displacement becomes more than the half of the typical block size, blocks from the benches would fall to the lower level benches (Aydan *et al.*, 2009b). This process would repeat itself successively at a given time interval.

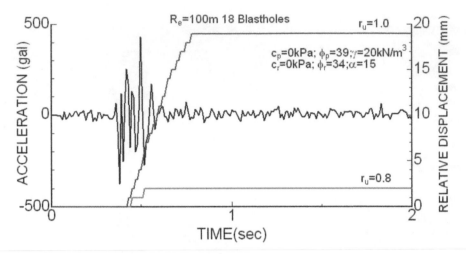

Figure 6.48 Relative displacement responses of the benches of the east slope for each blasting operation

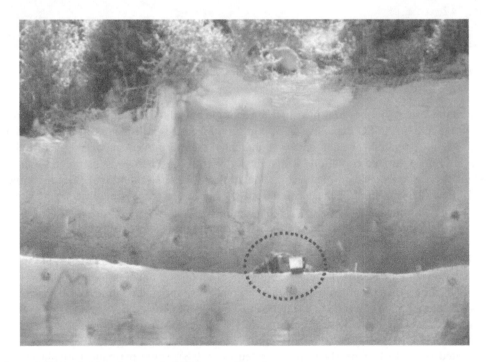

Figure 6.49 Rock fall at the portal of Mithatpaşa Tunnel

It is well-known that the blasting may cause the individual blocks to topple or sliding. Figure 6.49 shows an example of rockfall at the portal of Mithatpaşa Tunnel described in Subsection 6.3.4. Such events may cause casualties as well as property damage. The stability of such blocks can be evaluated using some dynamic limiting equilibrium approaches already

Figure 6.50 Blasting-induced failure at Gökgöl cave

mentioned. Similar events can also be found at karstic caves as seen at Gökgöl karstic cave in Zonguldak (Figure 6.50). The major collapse of the hall at the cave was caused by uncontrolled blasting operations in 1960 for enlarging the national highway

6.1.6.4   Vibration effects on buildings

Using the acceleration records of accelerometer Acc-1 and accelerations computed from the records of velocity meter Vel-3, a series of response analyses are carried out. Figure 6.51 shows the normalized acceleration response spectra of each component of acceleration records with damping coefficient (*h)* values of 0.000, 0.025, 0.050, respectively.

The results indicate that structures having a natural period less than 0.06 s could be very much influenced. The effect of blasting should be smaller for structures having natural periods greater than 0.1 s. Figure 6.52 and 6.53 show the responses of a structure with a natural period of less than 0.06 s and damping coefficient of 0.05 (h= 0.05) for each acceleration component from the accelerometer measurements and velocity meter measurements. As understood from Figures 6.52 and 6.53, the absolute acceleration acting on structures should be less than the accelerations of input motion. It would be safe to assume that the induced accelerations caused by blasting should act on the ground and structure system without any reduction.

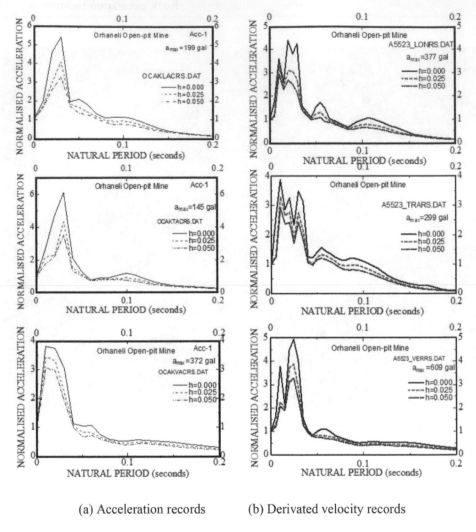

(a) Acceleration records   (b) Derivated velocity records

Figure 6.51 Normalized acceleration response spectra of the records of accelerometer Acc-1 and records of velocity meter Vel-3

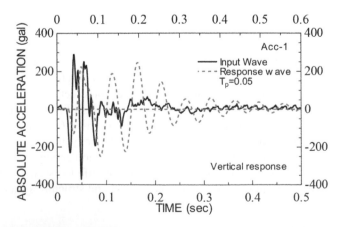

Figure 6.52 Absolute vertical acceleration response of a structure system (Acc-1)

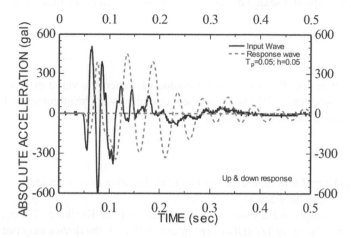

Figure 6.53 Absolute vertical acceleration response of a structure system using the acceleration computed from the velocity records of velocity meter Vel-3

6.1.6.5   Air pressure due to blasting

A typical pressure–time profile for a blast wave in free air is shown in Figure 6.54. It is characterized by an abrupt pressure increase at the shock front, followed by a quasi exponential decay back to ambient pressure p_o and a negative phase in which the pressure is less than ambient. The pressure-time history of a blast wave is often described by exponential functions such as Frielander's equation (Smith and Etherington, 1994)

$$p = p_o + p_s \left(1 - \frac{t}{T_s}\right) e^{-bt/T_s} \tag{6.8}$$

where p_s is peak overpressure, T_s is duration of the positive phase, and i_s is specific impulse of the wave that is the area beneath the pressure–time curve from the arrival at time to t_o the end of the positive phase.

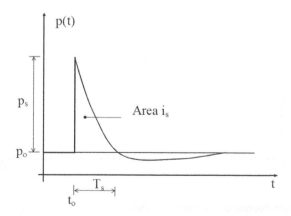

Figure 6.54 Air pressure variation during blasting

It is common to use the scaled distance to evaluate the effects of blasting given as:

$$Z = \frac{R}{\sqrt[n]{W}} \qquad (6.9)$$

where R is distance and W is mass of the explosive. When the value of n is 3, it is called Hopkinson blast scaling. USBM suggests the value of n as 2. The attenuation of air pressure is generally given in the following form:

$$p = AZ^{-\alpha} \qquad (6.10)$$

If W is given in kg and R in m, the unit of p is kPa, the value of A is generally about 186, and the power of α ranges between 1.2 and 1.5.

Air pressure changes were measured at Takamaruyama, Kuriko and Tarutoge Tunnels using TR-73U produced by TANDD Corporation. Although the device may not be appropriate for very sensitive air pressure changes induced by blasting, the measurements are quite meaningful, as seen in Figures 6.55–6.57. The cross sections of the tunnels are almost the

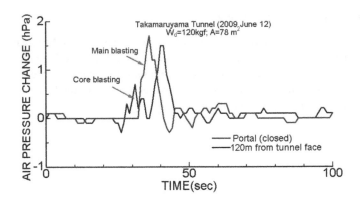

Figure 6.55 Air pressure fluctuations at Takamaruyama Tunnel

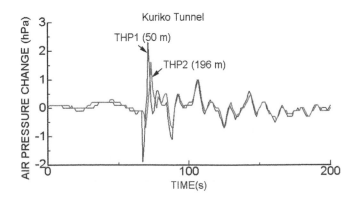

Figure 6.56 Air pressure fluctuations at Kuriko Tunnel

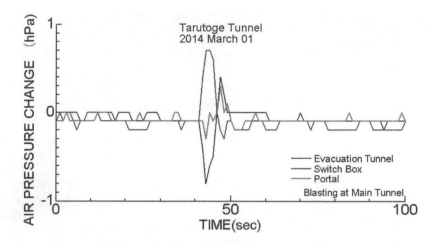

Figure 6.57 Air pressure fluctuations at Tarutoge Tunnel

same. Takamaruyama and Kuriko Tunnels are single tubes, while the Taru-Toge Tunnel has a branch connecting to the evacuation tunnel. The air pressure decreases at the evacuation tunnel when the shock wave passes by the branch and then increases as the air pressure is confronted by the door at the tunnel portal. Furthermore, the amplitude of the waves decreases as the distance increases.

6.1.6.6 Flyrock distance

Flyrocks are another major issue for the safety of people and damage to machinery and equipment when blasting is employed (Figure 6.58). The flyrock issue was caused by either inappropriate stemming, blasting sequence or charges or the existence of some weak zones in rock mass. It is reported that the flyrock may travel up to a distance of 900 m in worst cases. The ejection velocity may reach up to 300 m s⁻¹. Flyrock distance in common practice is less than 50 m, and it may sometimes reach a distance of 95 m. There are some empirical relations to estimate the maximum distance of flyrocks from the blasting relations.

One of the empirical relations is given by Lundborg (1981):

$$L_{\max} = 260D^{2/3} \tag{6.11}$$

The unit of fly distance is m, and D is given in inches.

The fly distance may be obtained from the simple physical laws of an object thrown with an initial angle (β_0) and velocity (v_0) at a given height (h_0). The air resistance may be taken into account as a viscous drag F_d. The fundamental equations of the flying object may be written as:

$$\frac{d^2x}{dt^2} = -\frac{F_d}{m}; \frac{d^2y}{dt^2} = -g - \frac{F_d}{m} \tag{6.12}$$

Figure 6.58 Flyrock during blasting operation at Işıkdere open-pit lignite mine

where g is gravitational acceleration. If the viscous resistance is given in the following form:

$$F_d = \eta m v^n \tag{6.13}$$

the governing equation becomes:

$$\frac{d^2 x}{dt^2} = -\eta v^n, \frac{d^2 y}{dt^2} = -g - \eta v^n \tag{6.14}$$

where η is viscous drag coefficient. When $n = 2$, it is called Stoke's law. If $n = 1$, it is possible to solve the preceding differential equation. Otherwise the differential equations become nonlinear. The solution of Equation 6.14 without drag force would yield the trajectory of a flyrock with conditions illustrated in Figure 6.59:

$$x = v_o \cos \beta t \text{ and } y = y_o + v_o \sin \beta t - \frac{g}{2} t^2 \tag{6.15}$$

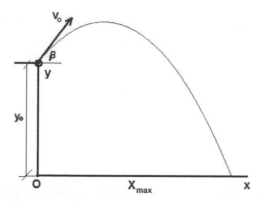

Figure 6.59 Trajectory of a flyrock and its conditions

The maximum fly distance of flyrock in air can be estimated from Equation (6.15) with condition of $y = 0$:

$$X_{max} = \frac{v_o \cos \beta}{g} \left(v_o \sin \beta + \sqrt{(v_o \sin \beta)^2 + 2 y_o g} \right) \tag{6.16}$$

When $y_o = 0$, then the travel distance of a flyrock takes the following form:

$$X_{max} = \frac{(v_o)^2 \sin 2\beta}{2g} \tag{6.17}$$

The maximum fly distance would be obtained when $\beta = 45^o$. If $\beta = 0^o$, the travel distance of a flyrock ejected from a given height (y_o) can be obtained from Equation 6.16:

$$X_{max} = v_o \sqrt{\frac{2 y_o}{g}} \tag{6.18}$$

As noted from these relations, the initial conditions are very important for the fly distance of flyrocks.

If drag forces are taken into account, the solution of Equation 6.12 yields the trajectory of the flyrock:

$$x = \frac{v_o \cos \beta}{\eta} \left(1 - e^{-\eta t} \right) \text{ and } y = y_o + \frac{1}{\eta} \left(v_o \sin \beta + \frac{g}{\eta} \right) \left(1 - e^{-\eta t} \right) - \frac{g}{\eta} t \tag{6.19}$$

The solution of Equation 6.12 for $n > 1$ requires the utilization of numerical techniques because it is difficult to obtain the closed-form solutions. Figure 6.60 compares the effect of viscous drag resistance of air on the fly trajectory of the flyrock.

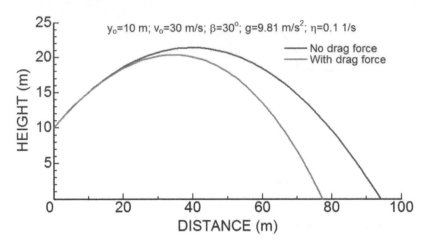

Figure 6.60 Comparison of the effect of viscous drag resistance of air on the fly trajectories of the flyrock

6.2   Machine excavations

6.2.1   Road headers

A roadheader is a piece of excavation equipment consisting of a boom-mounted cutting head, a loading device usually involving a conveyor and a crawler traveling track to move the entire machine forward into the rock face (Figure 6.61). The cutting head can be a general purpose rotating drum mounted in line or perpendicular to the boom together with picks. Roadheaders were first developed for the coal mining industry in the early 1960s (Copur *et al.*, 1998). In general, roadheaders can be divided into two types: milling (axial) type, with the cutter head rotating around the boom axis, and ripping (transverse) type, with the head rotating perpendicularly to the boom axis (Copur *et al.*, 1998).

6.2.2   Tunnel boring machines (TBMs)

A tunnel boring machine (TBM) is used to excavate tunnels, shafts and raise-bores with a circular cross section through various ground conditions. Tunnel diameters can range from 1 m to 17.6 meters to date (Figure 6.62). Tunnel boring machines are used as an alternative to drilling and blasting (D&B) methods in rock. The first successful application of the TBM was during the construction of the tunnels beneath Thames in 1825. TBMs have the advantages of limiting the disturbance to the surrounding ground and producing a smooth tunnel wall. This significantly reduces the cost of lining the tunnel. When surrounding rock is heavily fractured and sheared, the TBM may get stuck. Modern TBMs typically consist of the rotating cutting head, followed by a main bearing, a thrust system and trailing support mechanisms. The type of machine used depends on the particular geology of the project, the amount of groundwater present and other factors. TBMs can be shielded or open depending upon the ground condition. When the face is unstable or the groundwater condition is bad, earth-pressure balanced (EBP)–type TBMs are used.

Figure 6.61 Excavation of storage rooms by roadheaders at Cappadocia

(a) TBM at Hida Tunnel   (b) Stuck TBM at Pinglin Tunnel

Figure 6.62 Views of Hida and Pinglin Tunnels

6.3   Impact excavation

Breaker is a powerful percussion hammer fitted to an excavator for breaking rocks. It is powered by an auxiliary hydraulic or pneumatic system from the excavator, which is fitted with a foot-operated valve for this purpose. They are generally used when blasting cannot be used due to safety or environmental issues (Figure 6.63).

Figure 6.63 Views of breakers

Figure 6.64 Trimming basaltic rock blocks at Iguassu Falls in Brazil.

6.4   Chemical demolition

Chemical demolition is a technique in use since the 1970s. This technique is used where blasting could not be implemented due to safety and environmental concerns. The method itself is generally expensive compared to other methods. The basic principle is based on injecting expansive grout into the holes and splitting rock through the arrangement of holes in a given pattern and spacing. Figure 6.64 shows an example of chemical demolition to trim basaltic rock blocks at Iguassu Fall in Brazil.

References

Ak, H., Iphar, M., Yavuz, M. & Konuk, A. (2009) Evaluation of ground vibration effect of blasting operations in a magnesite mine. *Soil Dynamics and Earthquake Engineering*, 29(4), 669–676.

Attewell, P.B., Farmer, I.W. & Haslam, D. (1966) Prediction of ground vibration from major quarry blasts. *The International Journal of Mining and Mineral Engineering*, 621–626.

Aydan, Ö. (1997) Seismic characteristics of Turkish earthquakes. *Turkish Earthquake Foundation*, TDV/TR 97–007, 41 pages.

Aydan, Ö. (2000) A stress inference method based on structural geological features for the full-stress components in the earth' crust, *Yerbilimleri*, Ankara, 22, 223–236.

Aydan, Ö. (2001) Comparison of suitability of submerged tunnel and shield tunnel for subsea passage of Bosphorus. *Geological Engineering Journal*, 26(1), 1–17.

Aydan, Ö. (2003) The Inference of crustal stresses in Japan with a particular emphasis on Tokai region. *International Symposium on Rock Stress*. Kumamoto, 343–348.

Aydan, Ö. (2007) Inference of seismic characteristics of possible earthquakes and liquefaction and landslide risks from active faults (in Turkish). *The 6th National Conference on Earthquake Engineering of Turkey, Istanbul*, 1, 563–574.

Aydan, Ö. (2012) Ground motions and deformations associated with earthquake faulting and their effects on the safety of engineering structures. *Encyclopaedia of Sustainability Science and Technology*, Springer, New York, R. Meyers (Ed.), 3233–326.3.

Aydan, Ö (2013) In-situ stress inference from damage around blasted holes. *Journal of Geo-System Engineering*, Taylor and Francis, 16(1), 83–91.

Aydan, Ö., Akagi, T. & Kawamoto, T. (1993) Squeezing potential of rocks around tunnels: Theory and prediction. *Rock Mechanics and Rock Eng*ineering, Vienna, 26(2), 137–163.

Aydan, Ö., Bilgin, H.A. & Aldas, U.G. (2002) The dynamic response of structures induced by blasting. *Int. Workshop on Wave Propagation, Moving load and Vibration Reduction Okayama, Japan, Balkema*. pp. 3–10.

Aydan, Ö., Daido, M., Tano, H., Tokashiki, N. & Ohkubo, K. (2005) A real-time multi-parameter monitoring system for assessing the stability of tunnels during excavation. *ITA Conference, Istanbul*. pp. 663–669.

Aydan, Ö., Geniş, M. & Bilgin, H.A. (2014a) The effect of blasting on the stability of benches and their responses at Demirbilek open-pit mine. *Environmental Geotechnics*. ICE, 1(4), 240–248.

Aydan, Ö. & Ohta, Y. (2011) A new proposal for strong ground motion estimations with the consideration of charactcristics of earthquake fault. *Seventh National Conference on Earthquake Engineering, Istanbul*, Paper No. 66, 1–10 pages.

Aydan, Ö., Ohta, Y., Daido, M., Kumsar, H., Genis, M., Tokashiki, N., Ito, T. & Amini, M. (2011) Chapter 16.: Earthquakes as a rock dynamic problem and their effects on rock engineering structures. In: Zhou, Y. & Zhao, J. (ed.) *Advances in Rock Dynamics and Applications*. CRC Press, Taylor and Francis Group, London. pp. 341–422.

Aydan, Ö., Tano, H., Ideura, H., Asano, A., Takaoka, H., Soya, M. & Imazu M. (2016) Monitoring of the dynamic response of the surrounding rock mass at the excavation face of Tarutoge Tunnel, Japan. EUROCK2016, Ürgüp, 1261–1266.

Aydan, Ö., Üçpırtı, H. & Kumsar, H. (1996) The stability of a rock slope having a visco-plastic sliding surface. *Rock Mechanics Bulletin (Kaya Mekaniği Bülteni)*, 6, 39–49.

Aydan, Ö., Ulusay, R. & Kawamoto, T. (1997) Assessment of rock mass strength for underground excavations, *Proc. of the 36th US Rock Mechanics Symposium, New York*. pp. 777–786.

Aydan, Ö., Ulusay, R. & Tokashiki, N. (2014b) A new Rock Mass Quality Rating System: Rock Mass Quality Rating (RMQR) and its application to the estimation of geomechanical characteristics of rock masses. *Rock Mechanics and Rock Eng*ineering, Springer, Vienna, 47(4), 666–676.

Copur, H., Ozdemir, L. & Rostami, J. (1998) *Road-Header Applications in Mining and Tunnelling Industries*. Society for Mining, Metallurgy and Exploration, Orlando, FL.

Dowding, C.H. (1986.) *Blast Vibration Monitoring and Control*. Englewood Cliffs, NJ, Prentice-Hall.

Genis, M., Aydan, Ö. & Derin, Z. (2013) Monitoring blasting-induced vibrations during tunnelling and its effects on adjacent tunnels. *Proc. of the 3rd Int. Symp. on Underground Excavations for Transportation, Istanbul*. pp. 210–217.

Hao, H. (2002) Characteristics of non-linear response of structures and damage of RC structures to high frequency blast ground motion. *Wave2002*, Okayama.

Hendron, A.J. (1977) Engineering of rock blasting on civil projects. In: *Structural and Geotechnical Mechanics*. Prentice-Hall, Englewood Cliffs, NJ.

Ikeda, K. (1970) A classification of rock conditions for tunnelling. *Proceedings of the 1st Int. Congr. on Engineering Geology, IAEG, Paris*. pp. 6.6.8–6.66.

Imazu, M., Ideura, H. & Aydan, Ö. (2014) A Monitoring System for Blasting-induced Vibrations in Tunneling and Its Possible Uses for The Assessment of Rock Mass Properties and In-situ Stress Inferences. *Proceedings of the 8th Asian Rock Mechanics Symposium*, Sapporo, 881–890.

Kesimal, A., Ercikdi, B. & Cihangir, F. (2008) Environmental impacts of blast-induced acceleration on slope instability at a limestone quarry. *Environmental Geology*, 6.4(2), 381–389.

Kutter, H.K., Fairhurst, C. (1971) On the fracture process in blasting. *International Journal of Rock Mechanics Min. Sci., & Geomech.*, Abstr., Pergamon Press, 8., 181-188.

Lundborg, N. (1981) The probability of flyrock damages. Swedish Detoni Research Foundation, Stockholm, D.S. 6., 39 pp.

Northwood, T.D., Crawford, R. & Edwards, A.T. (1963) Blasting vibrations and building damage. *The Engineer*, 216.

Persson, P.A., Holmberg, R. & Lee, J. (1994). *Rock Blasting and Explosives Engineering*. Boca Raton, FL: CRC Press, 540 pp.

Sezaki, M., Aydan, Ö., Ichikawa, Y. & Kawamoto, T. (1990) Mechanical properties of rock mass for the pre-design of tunnels by NATM using a rock mass data-base (in Japanese). *Journal of Civil Engineers of Japan, Construction Division*, 421-VI-13, 6.6.–133.

Siskind, D.E., Stagg, M.S., Koop, J.W. & Dowding, C.H. (1980) Structure response and damage produced by ground vibration from surface mine blasting. United States Bureau of Mines, Report of Investigations, No. 86.07.

Smith, P.D. & Hetherington, J.G. (1994) *Blast and Ballistic Loading of Structures*. Butterworth-Heinemann Ltd, Great Britain.

Thoenen, J.R. & Windes, S.L. (1942) Seismic effects of quarry blasting. *U.S. Bureau of Mines Bulletin*, 442.

Tripathy, G.R. & Gupta, I.D. (2002) Prediction of ground vibrations due to construction blasts in different types of rock. *Rock Mechanics and Rock Engineering*, 36(3), 196–204.

Watanabe, H., Aydan, Ö. & Imazu, M. (2013) An integrated study on the stress state of the vicinity of Mt. Kuriko. *The 6th International Symposium on In-Situ Rock Stress (SENDAI)*. pp. 831–838.

Wu, C., Hao, H., Lu, Y. & Zhou, Y. (2003) Characteristics of stress waves recorded in small-scale field blast tests on a layered rock-soil site. *Geotechnique*, 63(6), 687–699.

Chapter 7

Vibrations and vibration measurement techniques

7.1 Vibration sources

Vibrations are caused by different processes such as blasting, machinery, impact hammers, earthquakes, rockburst, bombs (including missiles), traffic, winds, lightning, weight drop and meteorites. Vibrations induced by impact hammers, blasting with small explosives, TBM may be used to evaluate wave velocity characteristics of rocks and rock masses for assessing the rock mass properties for design purposes. In recent years, it is also used to assess the rock mass conditions such as the existence of weak/fracture zones or cavities. They are used to infer the rock mass properties as well as some yielding or loosening around rock structures.

In addition, vibrations may be used to investigate the soundness of support or reinforcement members. Off-course, they are used to investigate the effect of shaking caused by earthquakes or other large-scale vibration sources.

7.2 Vibration measurement devices

Vibration measurements devices may be of acceleration, velocity or displacement types. The devices may utilize piezoelectric sensors, servo-acceleration sensor, electro-kinetic velocity sensors and noncontact displacement sensor. The devices used for measuring vibrations are called accelerometers, velocity meters, displacement meters.

A piezoelectric sensor utilizes piezoceramics or single crystals such as quartz or tourmaline. The basic concept is to convert the mechanical motion into electric signals.

Servo-acceleration sensors utilize a displacement detector; a current is fed to the coil to get the pendulum mass back to the original position when it is subjected to a motion. The current will be proportional to the acceleration that is converted to an output voltage. The servo-type accelerometer has higher sensitivity, stability and more accurate phase responses in the lower frequency range than those of other vibration transducers. These sensors are commonly used in micro tremor measurements.

Electro-kinetic sensors convert vibrations into electric signals through the measurement of a streaming potential induced by the passage of polar fluid through a permeable refractory-ceramic or fritted-glass member between two chambers.

7.3   Theory of wave velocity measurement in layered medium

7.3.1   *Principles*

Let us consider a two-layered medium as shown in Figure 7.1. S denotes the source geophone, and R denotes the receiver geophone. It is assumed that the wave velocity V_1 of layer 1 is smaller than that of layer 2 ($V_2 > V_1$). For this particular situation, there will be numerous wave paths. Among them, three wave paths would be of particular importance. These wave paths are called direct wave (S-R), reflected wave path (S-C-R) and refracted wave path (S-A-B-R).

If the medium is assumed to be elastic and its density remains the same within the layer, Snell's law holds for incidence angle and refraction angle:

$$\frac{\sin i}{\sin r} = \frac{V_1}{V_2} \tag{7.1}$$

The refraction angle r of the refracted wave path shown in Figure 7.1 is 90 degrees. Therefore, the critical incidence wave angle i_c can be easily obtained from Equation (7.1) as follows:

$$\sin i_c = \frac{V_1}{V_2} \tag{7.2}$$

One can easily write the following relation between distance and arrival time for the direct wave path:

$$t = \frac{X}{V_1} \tag{7.3}$$

As for the reflected wave path, the relations between distance and arrival time is given by:

$$t = \frac{SC}{V_1} + \frac{CR}{V_1} \tag{7.4}$$

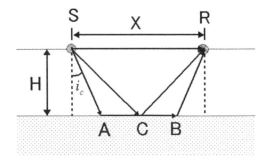

Figure 7.1 Wave paths in a two-layered medium

As $SC = CR$ and given by:

$$SC = CR = \sqrt{H^2 + \frac{X^2}{4}}$$

Equation (7.4) becomes:

$$t = \frac{2}{V_1}\sqrt{H^2 + \frac{X^2}{4}} \tag{7.5}$$

The relation between distance and arrival time for the refracted wave path shown in Figure 7.1 can be written as follows:

$$t = \frac{SA}{V_1} + \frac{AB}{V_2} + \frac{BR}{V_1} \tag{7.6}$$

From the geometry of the path, one can write the following relations:

$$SA = BR = \frac{1}{V_1} \cdot \frac{H}{\cos i_c}, \; AB = X - 2H \tan i_c \tag{7.7}$$

as

$$\cos i_c = \sqrt{1 - \left(\frac{V_1}{V_2}\right)^2}$$

Equation (7.6) takes the following form:

$$t = \frac{2H}{V_1}\sqrt{1 - \left(\frac{V_1}{V_2}\right)^2} + \frac{X}{V_2} \tag{7.8}$$

The thickness of layer 1 can be obtained by equating the arrival times of the direct wave and refracted wave if the critical distance X_c and wave velocities V_1 and V_2 are obtained from the records as follows:

$$H = \frac{X_c}{V_2} \frac{1 - \frac{V_1}{V_2}}{\sqrt{1 - \left(\frac{V_1}{V_2}\right)^2}} \tag{7.9}$$

As an application of this theory, a computation example was carried out, and the results are shown in Figure 7.2, together with the assumed wave velocities and the thickness of layer 1.

The technique described is generally used to identify the loosed zones, excavation-induced damage zone (EDMZ) around the excavations or velocity structure of the rock mass. Several practical applications of this method are given in the following subsections.

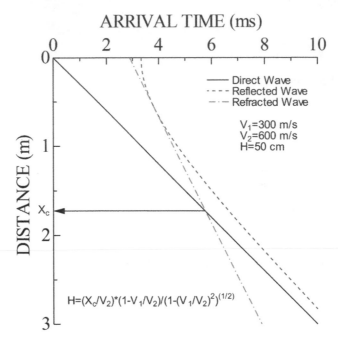

ARRIVAL TIME (ms)

— Direct Wave
--- Reflected Wave
—·— Refracted Wave

V_1=300 m/s
V_2=600 m/s
H=50 cm

X_c

$$H=(X_c/V_2)*(1-V_1/V_2)/(1-(V_1/V_2)^2)^{(1/2)}$$

Figure 7.2 Relations between distance and arrival time for different wave paths

Figure 7.3 Some views of measurements and instruments

7.3.2 Elastic wave velocity measurements

7.3.2.1 Measurements at Amenophis III pharaoh underground tomb

The instrument called Mc SEIS III developed by Oyo Corporation of Japan was used for this purpose (Figure 7.3). The device has three geophones and one hammer equipped with

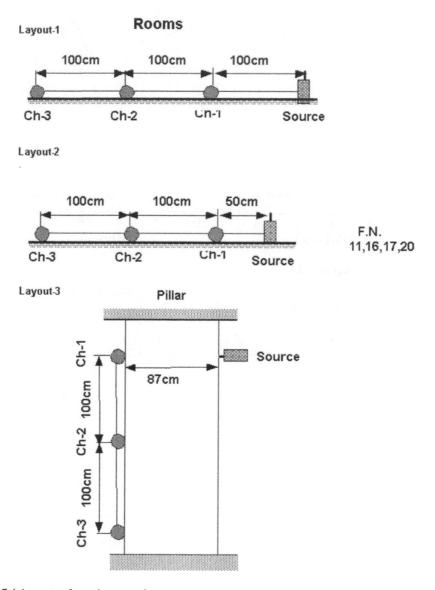

Figure 7.4 Layouts of geophones and source

an accelerometer for triggering (Hamada et al. 2004; Aydan and Geniş, 2004; Aydan et al. 2008). Figure 7.4 shows some views of measurements.

Instrumentation layouts are shown in Figure 7.4. Figure 7.5 shows some of records. The measurements on the floor yielded low elastic wave velocities less than those of intact rocks and pillars, which implies that some low-velocity zones exist beneath the floor as noted from Table 7.1. The pillars, on the other hand, yielded high elastic wave velocities. Furthermore, the wave velocity increases as the elevation increases.

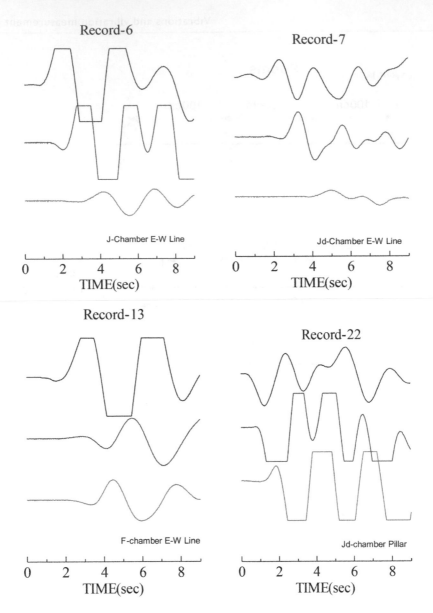

Figure 7.5 Some examples of velocity records

Table 7.1 Elastic wave velocity measurements

Chamber	Direction	Lowest Vp (km s⁻¹)	Highest Vp (km s⁻¹)
Je	N-S	0.862	1.139
Je	E-W	1.200	1.471
J	N-S	1.000	2.090
J	E-W	1.087	1.800
Jd	N-S	1.042	1.139
Jd	E-W	0.980	1.050
I	E-W	1.049	1.220

Chamber	Direction	Lowest Vp (km s⁻¹)	Highest Vp (km s⁻¹)
G	N-S	1.262	1.470
G	E-W	0.971	1.000
F	N-S	0.885	1.471
F	E-W	1.049	2.500
D-1	N-S	1.563	1,754
D-2	N-S	1.389	1.818
D-1	E-W	1.087	1.754
D-2	E-W	1.428	1.563
B	N-S	2.000	2.500
B-1	E-W	2.083	2.501
B-2	E-W	1.370	1.613
Jd-Pillar-1	Vertical	2.549	3.399
Jd-Pillar-2	Vertical	2.195	2.358
Je-Pillar	Vertical	1.611	2.712
F-Pillar1–1	Vertical	1.857	3.064
F-Pillar1–2	Vertical	1.767	2.371
F-Pillar2–1	Vertical	1.652	1.867

Figure 7.6 View of measurements at B7F in Derinkuyu Underground City

7.3.2.2 Measurements at Derinkuyu Underground City

The *in-situ* characterization of the tuff at Derinkuyu Underground City and its variation with depth were also assessed in this study. For this purpose, geotomographic investigations (Figures 7.6–7.7) at its several floors were carried out. The *in-situ* P-wave velocity of rock mass ranges between 0.9 and 1.3 km s⁻¹. However, the variation of V_p with depth was insignificant. Therefore, the rock mass is considered to be fairly uniform.

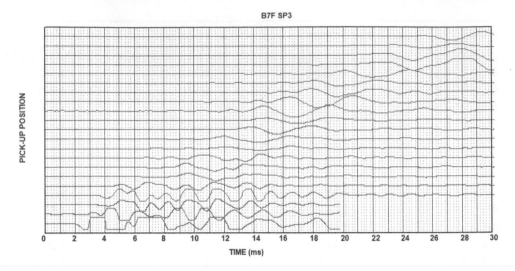

Figure 7.7 Example of vibrations records at B7F

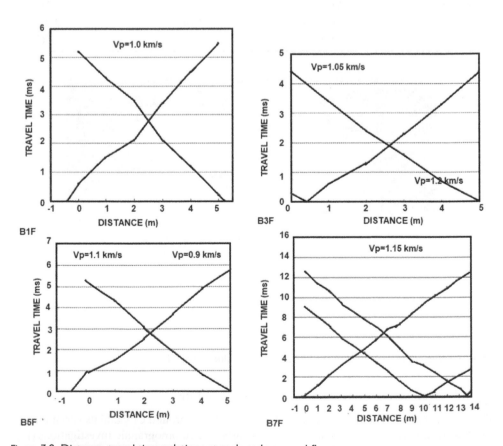

Figure 7.8 Distance–travel time relations at each underground floor.

7.3.2.3 Measurements at a Mitake abandoned lignite mite

Elastic-wave velocity measurements were carried out at an abandoned lignite mine in Mitake town using McSEIS-III (Figure 7.9). Measurements were concerned with the characteristics of lignite pillars that support the overburden rock mass. In addition, some *in-situ* index tests such as Schmidt hammer rebound tests and needle penetration tests were performed at the abandoned mine. The results are summarized in Table 7.2. The -wave velocity of intact lignite samples ranges between 1.57 and 2.26 km s⁻¹.

7.3.2.4 Geotomographic investigations at a fault zone

Inoue and Hokama (2019) carried out some geotomographic investigations in a fault zone in Urasoe City in Okinawa island. This investigation is summarized here.

(A) PROCEDURE

The method is based on P-wave velocity. The vibration source was a hammer, and vibrations were measured using geophones (Figure 7.10). McSeis-SX was used to monitor and process the vibration data. Figure 7.11 illustrates one of the diagrams between distance and

Figure 7.9 In-situ elastic wave velocity measurement

Table 7.2 Results of *in-situ* index tests

Rock	Rebound Number (R)	p-wave velocity (km s⁻¹)	Needle penetration index (N/mm)
Lignite	18–44	1.5	7.0–12.5
Sandstone	14–28	–	20.0–25.0
Mudstone	7–18	–	11.1–20.0

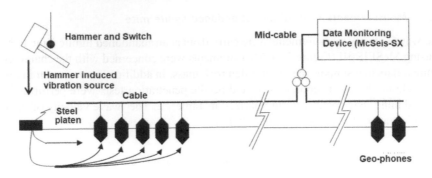

Figure 7.10 Illustration of the measurement technique

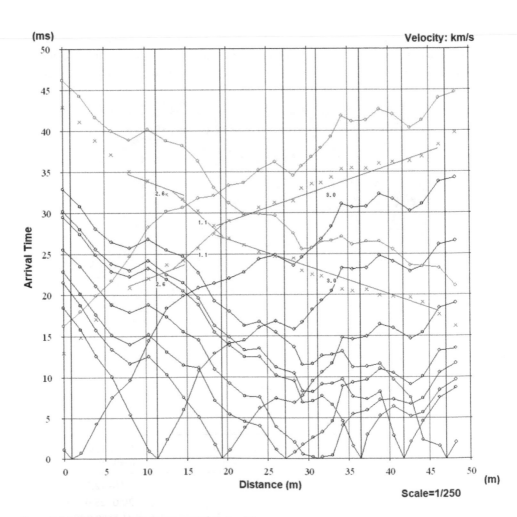

Figure 7.11 Example of distance–arrival time of P-waves

arrival time of P-waves for a given measurement section. Measurements were done along five sections.

The authors utilized the simultaneous iterative reconstruction technique (SIRT) method for geotomographic investigations (Figure 7.12). This method utilizes the least square principle. It is thought to be insensitive to the errors of the measurement data. It can be used to reconstruct high-quality images from even inaccurate data with much noise. Furthermore, it is always convergent. Because of these advantages, SIRT is a good algorithm for the reconstruction of geotomography.

(B) RESULTS

The measurements along five lines were analyzed using the SIRT method. Figure 7.13 shows a 3-D view of the contours of the P-wave velocities along each measurement line. The velocity of elastic P-waves ranges between 0.4 to 3.4 km s^{-1}. The elastic P-wave velocity of intact the Ryukyu limestone samples generally ranges between 3.5 and 4.9 km s^{-1}. When this fact is taken into account, the condition of rock mass varies depending upon the location. In other words, the rock mass condition is highly influenced by the fracturing state due to faulting. Except the weathered surface layers, the wave velocity is expected to decrease in the fracture zones.

Figures 7.14 and 7.15 show the p-wave velocity cons was quite low, in the range of 0.4–1.8 km s^{-1}. The elastic wave velocity of rock mass along the major faults ranged between 0.8 and 1.4 km/s. The elastic wave velocity of the undisturbed zone was more than 3.2 km/s.

7.3.2.5 Effect of impact vibrations at the ground surface and an underground arch structure

One of the major concern was the impacts induced by the landing or takeoff of airplanes on an underground arch structure beneath the runway (Minei et al., 2019). Although it was difficult to do such monitoring using actual planes, some tests on the impact vibrations at the ground surface on arch structure were carried out using a 10 tf truck and 1 tf sandbag dropped from a height of 100–180 cm. the accelerations induced by the passing of the truck over a 10 cm high barrier was quite small and attenuated very rapidly. Therefore, it was decided to use a 1 tf sandbag falling from a height of 100 cm or 180 cm, as

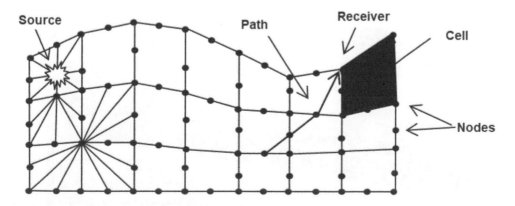

Figure 7.12 Illustration of the concept of SIRT method

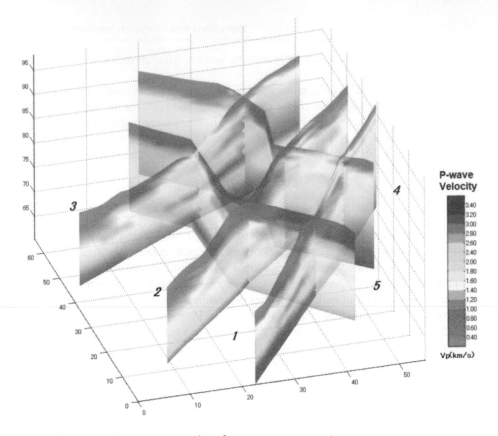

Figure 7.13 P-wave velocity contours along five measurement sections

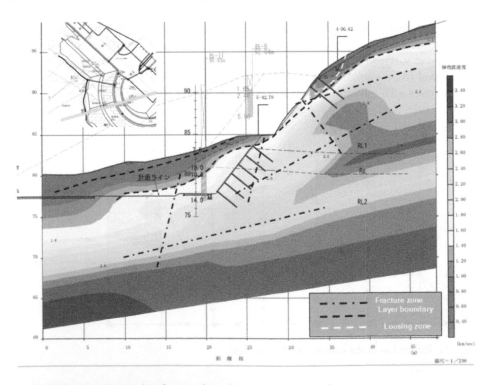

Figure 7.14 P-wave velocity distribution along the rock-cut section 1

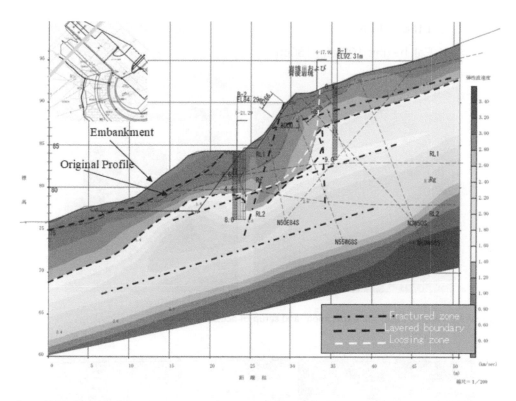

Figure 7.15 P-wave velocity distribution along the rock-cut

shown in Figure 7.16. Besides accelerometers on the ground surface, five accelerometers were installed in the underground arch structure (Figure 7.17). As shown in Figure 7.16, a 1 tf sand bag was dropped at various locations projected on the ground surface, and induced vibrations were measured.

Figure 7.19 shows the attenuation of maximum accelerations measured at various points in relation to the distance from the source point. The attenuation of maximum accelerations are fitted to the following equation:

$$a_{max} = 20000e^{0.8*W*h}\frac{1}{1+r^b} \tag{7.10}$$

where W (tf) is weight of dropped bag, h (m) is the drop height and r (m) is the distance from the drop location. The coefficient b is an empirical value, and it was found that it ranges between 2 (cylindrical) and 3 (spherical). Including the measurements in the underground arch structure, its value is 2.5 and fits the *in-situ* measurements. As noted from Figure 7.19, the vibrations were drastically reduced as a function of the distance from the source area. Therefore, the dynamic impact effects of the airplanes during landing or takeoff would be quite small on the actual underground structure.

1 tf Sand-bag

Fall height: 100 cm, 180cm

Accelerometers

Figure 7.16 View of the *in-situ* experiment and setup of accelerometers

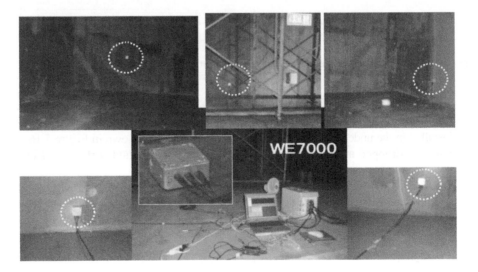

WE7000

Figure 7.17 Locations of accelerometers in the underground arch structure

7.3.2.6 Measurements of vibrations induced by pile construction in Gushikawa bypass bridge next to underground tomb

The assessment of ground conditions is of great importance to assess the response of the ground during construction and in-service of the piles, including the 400-year-old underground tomb, in relation to the construction of pile foundations of the Gushikawa Bypass

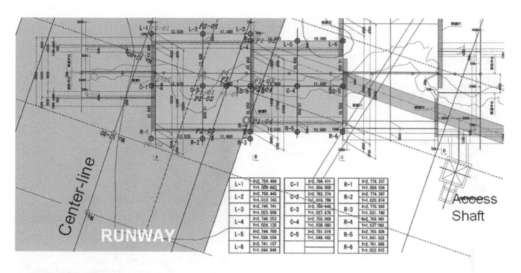

Figure 7.18 Locations of measurement points and sandbag drop

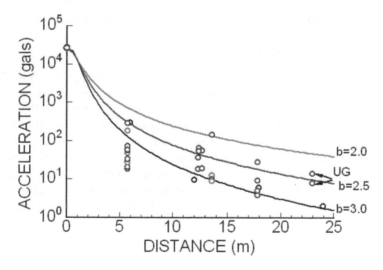

Figure 7.19 Attenuation of maximum accelerations with distance and its comparison with the attenuation relation

Bridge. The authors utilized some dynamic vibration measurements to assess the ground conditions (Tomori et al. 2019). The initial vibration measurements were carried out using five QV3-OAM-EX/W portable accelerometers developed by Aydan *et al.* (2016), which were utilized for different purposes, a vibration measurements system consisting of five TOKYO SOKKI AR-10TF accelerometers, Yokogawa WE7000 measurements station and laptop computer, and TOKYO SOKUSHIN SPC-51 micro tremor device.

As the source of vibration, a 1 tf sandbag was dropped from a height of 1 m above the ground (Figure 7.20(a)). As the bag was torn when it hit the ground, it was decided to use

(a) 1 tonf sand bag   (b) Backhoe bucket strike

Figure 7.20 Views of vibration sources

the backhoe bucket as the vibration source (Figure 7.20(b)). Although it is difficult to adjust the vibration level as desired, it proved to be quite useful as the vibration source. Figure 7.21 shows the layout of accelerometer sensors A1-A5, EX/W1-EX/W5 and SPC51. The main purposes of vibrations were to measure the wave velocity of the ground (Figure 7.21) and the transmission and attenuation of the acoustic emission (AE) signals with distance (Figure 7.22). in particular, the attenuation of AE signals between Sensor AE-1 (top), Sensor AE-3 (above the tomb) and Sensor AE4 (outer sidewall of the tomb) was of great importance. The results indicated that the amplitude of the vibrations was drastically reduced as a function of distance as shown in Figure 7.23. Figure 7.24 shows the attenuation of AE count numbers at AE-1, AE-3 and AE4. The result indicated that AE signals were also drastically reduced with distance.

7.3.2.7 Effect of fault in vibration propagation at Demirbilek open-pit lignite mine

It is well-known that blasting operations cause ground vibrations. The characteristics of the vibrations induced by blasting depend on the blasting geometry, the blasting material and blasting time pattern (Aydan *et al*., 2002, 2014). The results of measurements to be reported in this article were obtained at the Demirbilek open-pit mine in Western Turkey with an emphasis on the effect of fault on the vibration characteristics induced by blasting by using the same blasting geometry, blasting material and blasting time pattern except the location of the blasting hole with respect to the fault strike. Figure 7.25 shows the layout of the blasting experiment. The fault dips to SE with an inclination of 70 degrees, and the north side of the blasting area is very close to the crest of the open-pit mine. Ten blastings were performed.

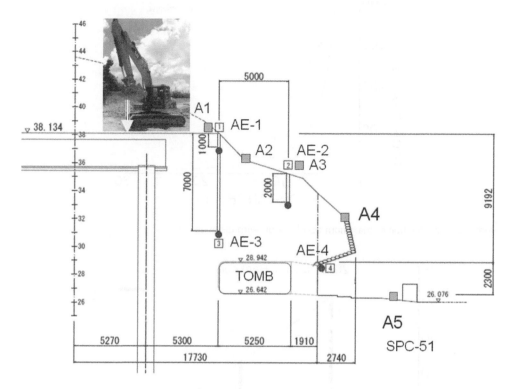

Figure 7.21 Layout of sensors for measuring vibrations

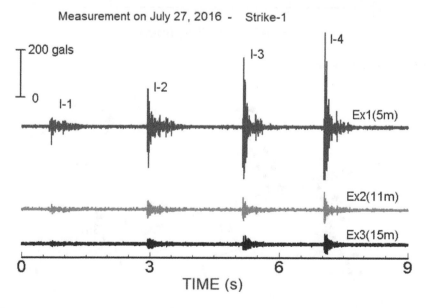

Figure 7.22 Acceleration records induced by the backhoe bucket striking the ground

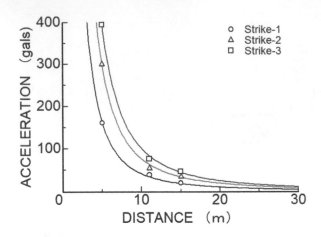

Figure 7.23 Attenuation of maximum acceleration with distance

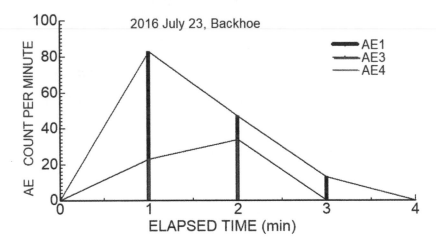

Figure 7.24 Attenuation of AE counts

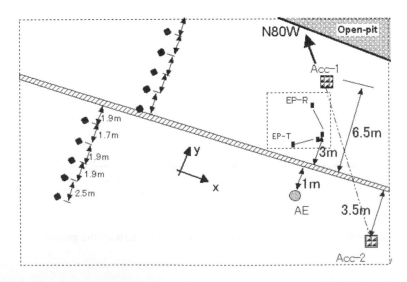

Figure 7.25 Layout of instrumentation for acceleration measurement during blasting at Demirbilek

The blasting operations numbered from DB22 to DB26 involved the blasting on the hanging wall with increasing charge weight while the blasting operations numbered from DB27 to DB31 involved blasting on the footwall of the fault.

Figures 7.26 and 7.27 show the acceleration responses at the observation points set on the footwall and hanging wall of the fault for blasting operations numbered DB26 and DB30. The DB26 and DB30 blasting operations were done on the hanging wall side and footwall side, respectively. The maximum ground acceleration at the footwall side was always less than that on the hanging wall side, even though the blasting was on the footwall side. This may be an indication of the effect of the fault and the free surface on the resulting wave propagations and their characteristics.

Vibrations induced by three different phenomena, namely the fracturing of rocks, faulting and blasting near a fault have some similar characteristics. The maximum accelerations are always higher on the moving side or hanging wall side of the fault. The experimental observations are consistent with maximum ground acceleration measurements on the hangingwall and footwall side of the faults during earthquakes.

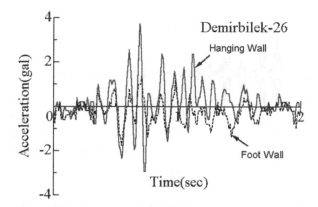

Figure 7.26 Comparison of induced acceleration responses of accelerometers numbered Acc-1 and Acc-2 for direction *x* for blasting operation numbered DB26

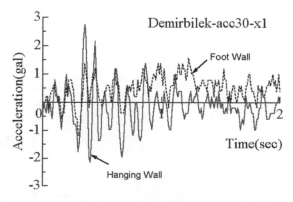

Figure 7.27 Comparison of induced acceleration responses of accelerometers numbered Acc-1 and Acc-2 for direction *x* for blasting operation numbered DB30

7.3.2.8   Ground vibration due to lightning

Vibrations may be caused by lightning in the air as well as in the ground. An interesting record was taken at the Nakagusuku Castle monitoring site at 18:41 on 15 August 2015. Figure 7.28 shows the accelerations records induced by the lightning. The maximum ground acceleration was 346 gals.

7.4   Vibrations by shock waves for nondestructive testing of rock bolts and rock anchors

Vibrations by shock waves for nondestructive testing of rock bolts and rock anchors are briefly presented here. For details, readers are referred to Aydan (2018). Several bar-type tendons (1200 mm long, 36 mm in diameter) and cable-type tendons (1300 mm long and 6 wires with a diameter of 6 mm) with/without artificial corrosion under bonded and unbonded conditions were prepared (Figure 7.29 and Figure 7.30), and the responses of the tendons

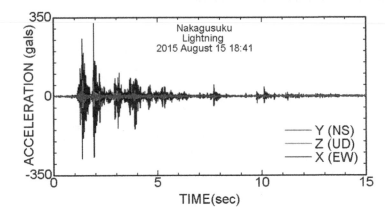

Figure 7.28 Vibrations induced by lightning at Nakagusuku Castle monitoring site

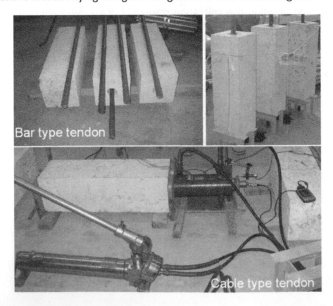

Figure 7.29 Samples with bar-type and cable-type tendons embedded in rock and a typical experimental setup

under single impact waves induced by a impact hammer or special Schmidt hammer–like device (ponchi) were measured. Three different sensors are used, and two of them had a center hole for inducing impact waves on tendons (Figure 7.31). The waves can be recorded as displacement, velocity or acceleration, and the device for monitoring and recording consists of an amplifier and a small handheld-type computer (Figure 7.32). The measurement

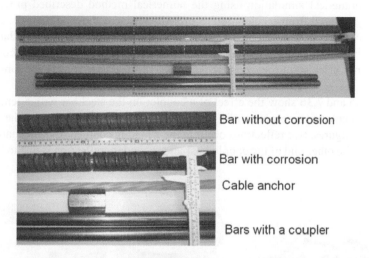

Bar without corrosion

Bar with corrosion

Cable anchor

Bars with a coupler

Figure 7.30 Bar-type and cable-type tendons with/without artificial corrosion used in experiments

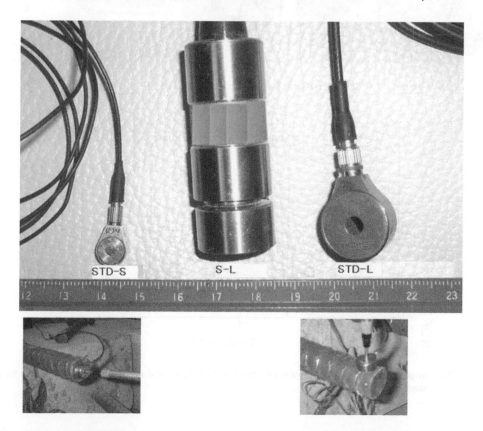

Figure 7.31 Views of sensors

can be done through a single person, and the Fourier spectra of recorded data can be stored in the small handheld computer and visualized in the site of measurement. In the following, only the results obtained are given without any reference to the sensor or hammer unless it is mentioned.

Figure 7.33 shows the wave responses of a 1200 mm steel bar induced by an impact hammer and its numerical simulation using the numerical method described previously. The wave velocity inferred from acceleration records directly was about 5520 m s^{-1}. The Fourier spectra of wave records induced by the impact hammer and a special Schmidt hammer–like device (ponchi) are shown in Figure 7.34. The frequency content interval is about 2300 Hz and the inferred velocity of the steel bar was 5520 m s^{-1}. These results are consistent with each other.

Figures 7.35 and 7.36 show the effect of a coupler on the steel bar with a length of 1000 mm and the acceleration response of a single 1000 mm bar also shown in the figures. As noted from the figures, two reflections occur in the 2 m long coupled bar. The main reflection is due to from the other end of the coupled bar and secondary reflection is due to the coupler.

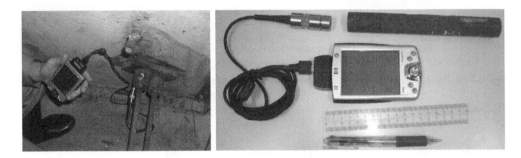

Figure 7.32 Views of PC-pocket-type sampling and recording device

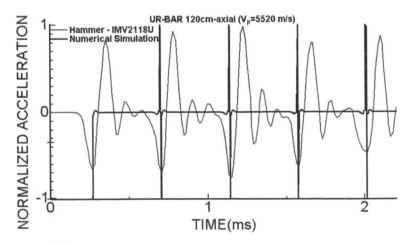

Figure 7.33 Impact hammer–induced wave response and its numerical simulation for a 1200 mm long steel bar

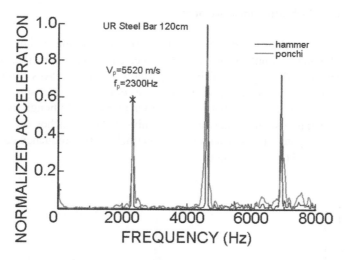

Figure 7.34 Normalized Fourier spectra of recorded acceleration records induced by impact hammer and a special Schmidt hammer–like device

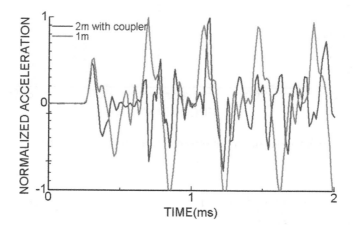

Figure 7.35 Impact hammer–induced wave responses of single and two 1 m long bars connected to each other with a coupler

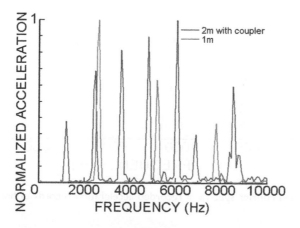

Figure 7.36 Normalized Fourier spectra of recorded acceleration records of single and two 1 m long bars connected to each other with a coupler

Figure 7.37 shows the Fourier spectra of 1200 mm bar with and without an artificially induced area reduction to simulate corrosion. As expected from the numerical analysis high-frequency content would be generated by the partially reflected waves from the artificial corrosion zone. This feature is clearly observed in the computed Fourier spectra. Figure 7.38 shows the effect of bonding of the bar to the surrounding rock. Although the frequency of the bonded bar is slightly smaller than that of the unbonded rock anchor, the Fourier spectra for the first mode is quite close to each other. Nevertheless, the frequency content starts to change after the second or higher modes.

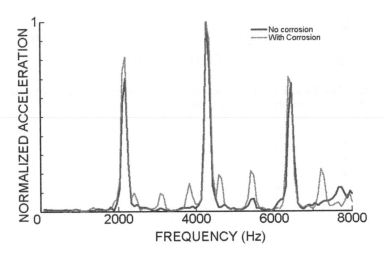

Figure 7.37 Normalized Fourier spectra of recorded acceleration records for bars with or without corrosion

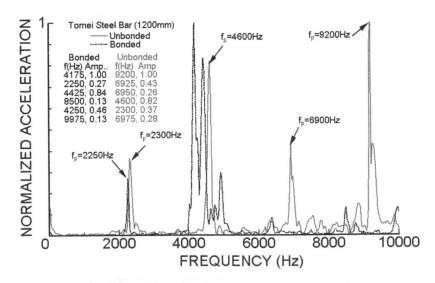

Figure 7.38 Normalized Fourier spectra of recorded acceleration records of bonded and unbonded bars

References

Aydan, Ö. (2003) Actual observations and numerical simulations of surface fault ruptures and their effects engineering structures. *The Eight U.S.-Japan Workshop on Earthquake Resistant Design of Lifeline Facilities and Countermeasures against Liquefaction.* Technical Report, MCEER-03–0003. pp. 227–237.

Aydan, Ö. (2018) *Rock Reinforcement and Rock Support.* CRC Press, Taylor and Francis Group, 486p, London, ISRM Book Series, No. 6.

Aydan, Ö., Bilgin, H.A. & Aldas, U. G. (2002) The dynamic response of structures induced by blasting. *International Workshop on Wave Propagation, Moving Load and Vibration Reduction.* Okayama, Baklema, 3–10.

Aydan, Ö., Daido, M., Tano, H., Tokashiki, N. & Ohkubo, K. (2005a). A real-time multi-parameter monitoring system for assessing the stability of tunnels during excavation. ITA Conference, Istanbul. pp. 1253–1259.

Aydan, Ö. & Geniş, M. (2004) Surrounding rock properties and openings stability of rock tomb of Amenhotep III (Egypt). *ISRM Regional Rock Mechanics Symposium, Sivas.* pp. 191–202.

Aydan, Ö., Geniş, M. & Bilgin, H.A. (2014) The effect of blasting on the stability of benches and their responses at Demirbilek open-pit mine. *Environmental Geotechnics. ICE,* 1(4), 240–248.

Aydan, Ö., Sakamoto, A., Yamada, N., Sugiura, K. & Kawamoto, T. (2005b) The characteristics of soft rocks and their effects on the long term stability of abandoned room and pillar lignite mines. Post Mining 2005, Nancy.

Aydan, Ö., Tano, H., Geniş, M., Sakamoto, I. & Hamada, M. (2008) Environmental and rock mechanics investigations for the restoration of the tomb of Amenophis III. *Japan-Egypt Joint Symposium New Horizons in Geotechnical and Geoenvironmental Engineering, Tanta, Egypt.* pp. 151–7.2.

Aydan, Ö., Tano, H., Ideura, H., Asano, A., Takaoka, H., Soya, M. & Imazu, M. (2016). Monitoring of the dynamic response of the surrounding rock mass at the excavation face of Tarutoge Tunnel, Japan. EUROCK2016, Ürgüp, 1261–1266.

Aydan, Ö., Tano, H., Imazu, M., Ideura, H. & Soya, M. (2007) The dynamic response of the Taru-Toge tunnel during blasting. *ITA WTC 2007: Congress and 42st General Assembly, San Francisco, USA.*

Aydan, Ö., Ulusay, R., Hasgür, Z. & Hamada, M. (1999) The behavior of structures built on active fault zones in view of actual examples from the 1999 Kocaeli and Chi-Chi earthquakes. *ITU International Conference on Kocaeli Earthquake, Istanbul.* pp. 131–142.

Hamada, M., Aydan, Ö. & Tano, H. (2004) Rock mechanical investigation: Environmental and rock mechanical investigations for the conservation project in the royal tomb of amenophis III. *Conservation of the Wall Paintings in the Royal Tomb of Amenophis III.* First and Second Phases Report, UNESCO and Institute of Egyptology, Waseda University. pp. 83–138.

Haruyama H and Narita, A. (2008), On the characteristics of the modified open TBM used in the excavation of the Kuriko evacuation tunnel, *Tohoku Technological Research Meeting,* Sendai, JSCE, VI-5.

Inoue, H. & Hokama, K. (2019) Assessment of rock mass conditions of Ryukyu Limestone formation for a rock-cut in Urasoe Fault Zone (Okinawa) by elastic wave velocity tomography. *Proceedings of 2019 Rock Dynamics Summit in Okinawa.* pp. 744–749.

Minei, H., Nagado, Y., Ooshiro, Y., Aydan, Ö., Tokashiki, N. & Geniş, M. (2019) An integrated study on the large-scale arch structure for protection of karstic caves at New Ishigaki Airport. *Proceedings of 2019 Rock Dynamics Summit in Okinawa.* pp. 390–395.

Tomori, T., Yogi, K., Aydan, Ö. & Tokashiki, N. (2019) An integrated study on the response of unsupported underground cavity to the nearby construction of piles of Gushikawa By-Pass Bridge. *Proceedings of 2019 Rock Dynamics Summit in Okinawa.* pp. 408–413.

Chapter 8

Degradation of rocks and its effect on rock structures

It is well-known that rocks surrounding rock engineering structures or constituting rock mass are prone to degradation when they are subjected atmospheric conditions and/or gas/fluids percolating through rocks and rock discontinuities. Degradation results from the alteration of minerals, the weakening of particle bonds and/or solution of particles. Figure 8.1 shows several examples of degradation situations of rocks. Properties of intact rock can be drastically changed due to degradation processes. In this chapter, major fundamental processes are explained, and the effects of degradation processes on the properties of rocks are described.

Figure 8.1 Examples of degradation situations of some rocks

Table 8.1 Degradation of some common minerals

Original Minerals	Weathering Process	Resulting Minerals
Minerals containing Fe, Mg: olivine, pyroxene, amphibole	H_2CO_3 (= H_2O + CO_2) alteration, oxidation	Clay minerals, Fe-oxides
Feldspars	H_2CO_3 alteration	Clay minerals
Biotite (micas)	Hydrolysis	Clay minerals
Calcite, gypsum	Solution by water	None

8.1 Degradation of major common rock-forming minerals by chemical processes

Table 8.1 summarizes some common resulting minerals due to degradation (weathering, oxidation, alteration processes). This subsection presents some of these processes. Feldspars are one of the common rock-forming minerals. Oxidation is the ionic reaction with some components in the original mineral. For example, the oxidation of iron (Fe) results in the hematite mineral:

$$4\ Fe + 3\ O2 \text{-->} 2\ Fe2\ O3 \text{ (hematite)}$$

Hydrolysis is a chemical reaction of minerals with water containing dissolved CO_2. For example, orthoclase transforms into kaolin as a result of the chemical reaction given by:

$$2\ K\ Al\ Si_3O6 + 6\ H_2O + CO_2 \text{ ---> } Al_2\ Si_2\ O5\ (OH)_4 + 4\ H_2SiO_4 + K_2CO_3$$
$$\text{(orthoclase) (kaolinite)}$$

The resulting minerals from the chemical processes are generally clay minerals. It is generally known that dark color minerals are much more vulnerable to degradation compared with whitish or glassy color minerals.

While plagioclase is more resistant to weathering, orthoclase is much more vulnerable to weathering when they are exposed to water or acid environments. It is well-known that muscovite is resistant to weathering. However, it may be easily disintegrated due to their platy structure in a freezing–thawing environment. On the other hand, biotite is quite vulnerable to weathering particularly in tropical regions.

Calcite and evaporates can be easily solved in water containing dissolved CO_2. This process is expressed in the following form:

$$H_2O + CO_2 + CaCO_3 \text{--> } Ca{+}{+} + 2\ HCO_3$$

8.2 Degradation by physical/mechanical processes

8.2.1 Freezing–thawing process and its effects on rocks

Freezing–thawing cycles are an important process in the degradation of rocks with time. This process repeats itself endlessly if atmospheric air temperature drops below zero during the cold seasons. There is a great concern in many countries to assess the behavior of

(a) Toyohama Tunnel   (b) Zelve antique semi-underground settlement

Figure 8.2 Examples of freezing–thawing–induced failures in nature

engineering structures subjected to freezing–thawing cycles as this process may result in failures and fatalities. The collapse of tuffaceous rock mass above the portal of the Toyohama Tunnel in Hokkaido island of Japan killed more than 10 people traveling through the tunnel at the time of the collapse (Figure 8.2(a), Ishikawa and Fujii, 1997). Particularly rocks whose mechanical properties are influenced by water content are quite vulnerable to degradation and disintegration when they are subjected to freezing and thawing cycles.

The Alps are the highest mountains in Europe, and their peaks are more than 4000 m high. The temperatures are generally below 0 degree when the altitude is greater than 2500 m. Temperature measurement in rock 4 m deep from the surface at the north face of the Sphinx Observatory of the Jungfraujoch was recorded to be fluctuating between +0.8°C and −9°C. Although the rock temperature would depend upon the depth from the surface, the observation data implies that the fluctuation of rock temperature may range between 8 and 10 °C.

The measured air temperature difference descending from Gornergrat (3130 m) to Zermatt (1620 m) was about 15°C, while the elevation difference between two stations is about 1510 m. The limit altitude of trees in the vicinity of the Jungfraujoch and Gornergrat observatories is about 2500 m. The degradation of rocks above the tree limit altitude was quite severe, and the surface of mountains were covered with debris of disintegrated rock blocks as seen in Figure 8.3. The disintegration of rocks due to freezing and thawing process was particularly severe in schistose rocks, particularly green schist and micaschist as seen in Figure 8.3(c).

One of the interesting observations was done near the Zermatt station of Zermatt-Rothorn Cable Car Route. Rock bolts were used to support the rock blocks in the vicinity of this station. The freezing and thawing process caused the disintegration of rock, which had fallen off. Nevertheless, the pieces of rocks underneath the rock bolt plate were still in place as seen in Figure 8.4. The rock was green-schist.

Many antique and modern underground and semiunderground openings are excavated in tuffs in the Cappadocia region. The erosion and subsequent partial or total collapse of these structures take place from time to time, and the preservation of these antique underground or semiunderground remains as the assets of past civilizations is of paramount importance. The failures of these structures generally occur during thawing or after heavy rains as a result of the degradation due to freezing–thawing and decrease of strength properties of tuffs with saturation (Aydan and Ulusay, 2003, 2013).

(a) Disintegration of rock nearby Blauherd Station (Zermatt-Rothorn Route)

(b) Disintegration of rock nearby Rotendoden Station (Zermatt-Gornergrat Route)

(c) Disintegration of green schist and micaschist

Figure 8.3 Effect of freezing–thawing in the vicinity of Gornergrat Observatory

Figure 8.4 Disintegration of rocks in the vicinity of rock bolted rock surface in Zermatt

(a)   Freezing–thawing experiments

The general procedure used for freezing–thawing experiments involves freezing rocks under saturated conditions at −20–30 °C and thawing in an oven under the temperature of +20–100 °C. However, these conditions do not exist in nature, and the degradation of material properties such as uniaxial compressive strength drastically decreases. Therefore, the validity and applicability of current freezing–thawing procedures used on rocks are quite questionable. Furthermore, the fully saturated condition may not always be observed in nature.

Rocks are tuff obtained from Oya in the Tochigi Prefecture and Asuwayama in the Fukui Prefecture of Japan and Derinkuyu and Zelve in the Cappadocia Region of Turkey. Table 8.2 gives physico-mechanical properties of tuffs used in the experiments. Two sets of samples were prepared. One set of samples were directly subjected to temperature cycles under dry conditions, while the samples of the other set were kept immersed in water in their containers, as seen in Figure 8.3(b). Samples were taken out from the climatic chamber at certain cycles to measure their physico-mechanical properties using nondestructive testing procedures such as P-wave velocity measurement, a needle-penetration device and weight change. The strain–stress responses of samples within the elastic range at the selected cycles were also measured, and their deformation modulus were determined.

An environmental chamber LHU-113, whose temperature can be varied between − 20 °C and +85 °C produced by ESPEC, was used (Figure 8.5). The temperature cycles can be

Table 8.2 Geomechanical properties of tuffs (dry)

Property	Derinkuyu	Zelve	Oya	Asuwayama
Unit weight (kN m⁻³)	13.7–15.9	10.9–15.1	13.5–15.7	20.1–22.0
UCS (MPa)	7.9–10.3	0.8–3.6	8.7–24.8	28.6–37.7
Tensile strength (MPa)	0.5–0.9	0.26–0.46	0.8–1.5	2.1–3.8
Elastic modulus (GPa)	1.5–3.6	0.5–2.2	0.9–3.0	3.2–4.5
Poisson ratio	0.2–0.26	0.27–0.31	0.25–0.3	0.15–0.21
Elastic wave velocity (km s⁻¹)	1.5–2.2	1.1–1.8	1.6–1.8	2.8–3.2

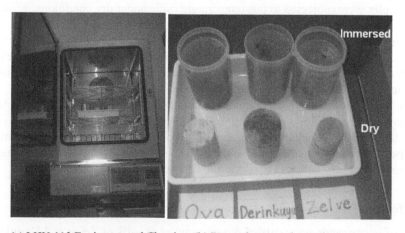

(a) LHU-113 Environmental Chamber (b) Dry and saturated samples

Figure 8.5 Views of the environmental chamber and samples

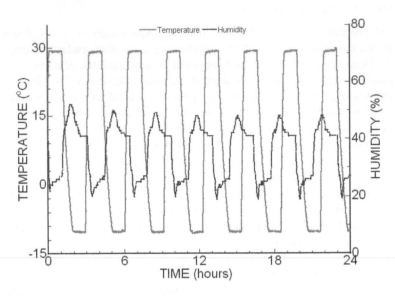

Figure 8.6 Monitored temperature and humidity in the environmental chamber

programmed as desired. As yearly temperature variations were generally between −10 ℃ and 30 ℃ in the areas of rock sampling, the temperature of the chamber was varied between −10 ℃ and 30 ℃, and the total duration of each cycle was 3 h. The temperature and humidity of the chamber were monitored continuously as shown in Figure 8.6.

Temperature measurements in Derinkuyu and Zelve (Aydan *et al.*, 2014; Ulusay *et al.*, 2013) indicated that temperature may decrease below the freezing point during the period of time starting from November to March, as seen in Figure 8.7. The area receives rainfall and snow during this period also. The effect of cyclic the freezing–thawing process on tuff samples of Derinkuyu and Zelve has been recently investigated by the authors. During these tests, samples were subjected to a temperature cycle ranging between −10℃ and 20℃. Some of the samples were subjected to this temperature cycle under dry conditions, while the rest were subjected under saturated conditions. The experiments clearly indicated that the effect of the temperature cycle has no big influence on the properties of Cappadocia tuffs while they may destroy the bonding structure of rocks under saturated conditions. The effect of freezing–thawing on tuffs with larger water content absorption capacity is much higher than that on tuffs with less water absorption capacity.

Figure 8.8(a) shows the views of dry and saturated tuff samples of Zelve subjected to the same temperature and humidity environment. While the saturated sample was heavily damaged by the freezing–thawing process, there was no visible damage to the dry sample. The same type of experiments carried out at the University of the Ryukyus in 2015 yielded very similar results for very soft rocks and mortar samples (Figure 8.8(b)).

(b)   Effects of freezing–thawing on physico-mechanical properties

A series of experiments were carried out on samples of tuffs of Zelve and Derinkuyu subjected to freezing–thawing cycles under dry and saturated conditions with high temperature variations, and changes in weight, elastic wave velocity, NPI and UCS were determined.

Derinkuyu   Zelve

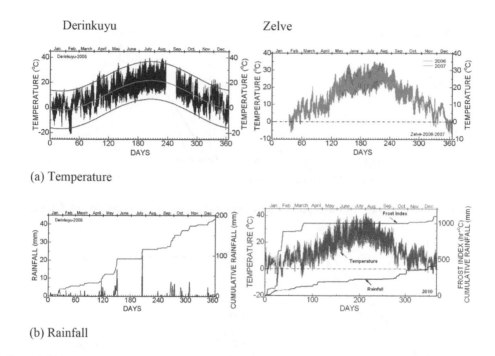

(a) Temperature

(b) Rainfall

Figure 8.7 Variations of temperature and rainfall in Derinkuyu and Zelve

(a) Zelve tuff samples   (b) Tuff, granite, mortar and porphryite samples

Figure 8.8 Views of dry and saturated samples subjected to the same temperature environment

Figures 8.9 and 8.10 show the change in weight of dry and saturated samples, and stress–strain relationship in dry samples, respectively. Change in weight of the saturated samples indicated that occurrence of spalling is directly related with freezing–thawing cycles in nature. Except for mechanical effects resulting in volumetric change during the freezing–thawing test, samples in the dry state were not affected by freezing–thawing cycles.

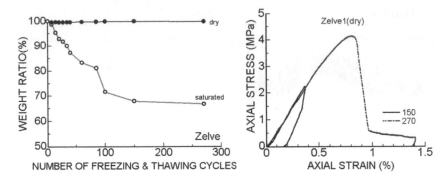

Figure 8.9 Effect of freezing–thawing cycle on (a) weight and (b) stress–strain relation of Zelve tuff

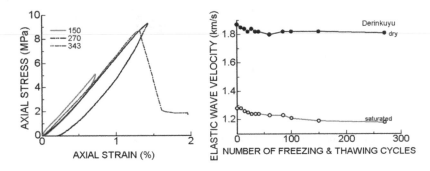

Figure 8.10 Effect of freezing–thawing cycles on uniaxial stress–strain responses and elastic wave velocity of Derinkuyu tuff

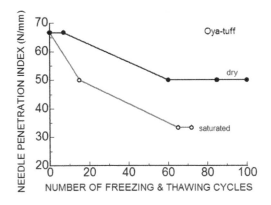

Figure 8.11 Variation of NPI with the cycle number of thawing–freezing for Oya tuff

The decrease in UCS of the Zelve tuff is about 40% at the end of the 40th cycle of the freezing–thawing tests. Except for mechanical effect resulting in volumetric change during the freezing–thawing test, samples in the dry state were not affected.

The value of the needle penetration index (NPI) is expected to decrease as the cycle number of thawing–freezing increases. Figure 8.11 shows the variation of NPI with the cycle number of thawing–freezing for Oya tuff. As noted from the figure, the value of NPI drastically decreases with the increasing cycle number of thawing–freezing for saturated samples of Oya tuff.

8.2.3   Cyclic saturation and drying (slaking)

(a)   Cyclic saturation and drying process on soft rocks

It is well-known that some soft rocks may absorb or desorb water, resulting in volumetric changes, which may lead to their disintegration. An experimental procedure is described to measure the water content migration characteristics and associated volumetric variations. An experimental device illustrated in Figure 8.12 was developed (Aydan *et al.*, 2006). The experimental setup consists of an automatic scale, an electric current inductor, electrodes, isolators, laser displacement transducer, voltmeter, rock sample and laptop computers to monitor and to store the measurement parameters and temperature-humidity unit consisting of sensors and logger. Rock samples collected from the Tono Underground Mine in Central Japan were first fully soaked with water for a certain period of time. Then they are put on the automatic scale and dried. During the drying process, the weight, height and voltage changes of the sample were continuously measured. The temperature and humidity changes of the drying environment were also continuously monitored.

Figure 8.13(a) shows temperature, humidity, shrinkage strain, weight change and electrical resistivity variations on both fine-grain and coarse-grain sandstone samples under laboratory conditions. While the weight change (water content) of coarse-grain sandstone was slightly large than that of fine-grain sandstone, there was a remarkable difference between the shrinkage strains of samples. The shrinkage strain of fine-grain sandstone was more than twice that of coarse-grain sandstone. The electrical resistivity of samples increases as the samples lose their water content. The relation between water content and electrical

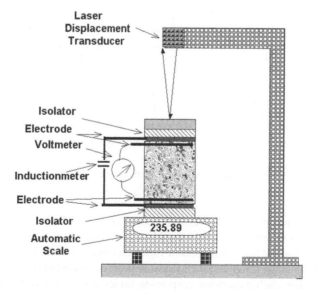

Figure 8.12 Illustration of experimental setup

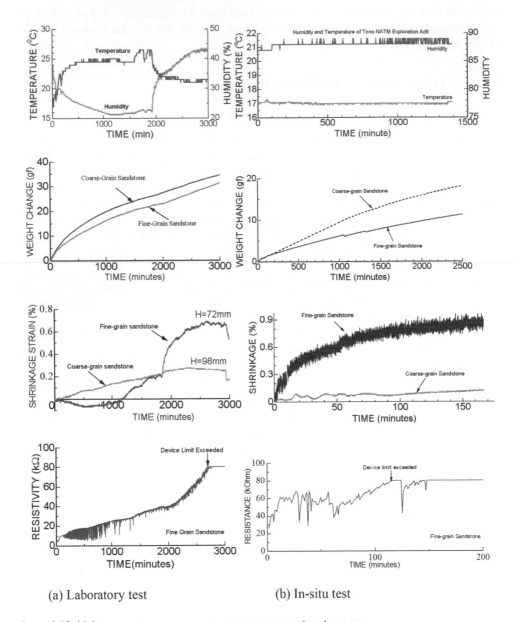

(a) Laboratory test   (b) In-situ test

Figure 8.13 Multiparameter responses on water content migration tests

resistivity is studied and presented elsewhere (Kano *et al.*, 2004). It is considered that if the electrical resistivity of surrounding rock could be measured continuously *in-situ*, it may be quite useful for evaluating the water content variations and associated volumetric variations.

The same experimental setup and the same samples were used in the NATM adit in order to observe the same multiparameter responses under *in-situ* conditions. The measured responses are shown in Figure 8.13(b).

(a) Fine grain sandstone   (b) Coarse grain sandstone

Figure 8.14 Views of samples after the completion of the experiments

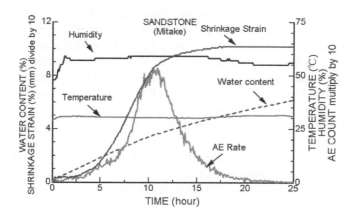

Figure 8.15 Multiparameter responses of a sandstone sample from Mitake

The temperature and humidity in the adit were almost constant. Even though the weight change (water content) of coarse-grain sandstone was twice that of fine-grain sandstone, the shrinkage strain of fine-grain sandstone sample, was almost 4 times that of the coarse-grain sandstone sample under *in-situ* conditions. These experimental results clearly indicated that fine-grain sandstone was prone to high volumetric strains in relation to water content variations. Tangential and axial cracks appeared on the outer surface of fine-grain sandstone samples while such cracks were not observed on coarse-grain sandstone samples (Figure 8.14). This experimental observation could be directly associated with the magnitude of shrinkage strains.

Because of some technical problems, acoustic emission responses could not be measured. However, similar experiments were carried out on a sandstone sample from Mitake, which is near the Tono mine and belongs to the same geologic formation. The results are shown in Figure 8.15. As noted from the figure, there is a distinct acoustic emission response when the sample loses its water content. The largest acoustic emission activity corresponds to the largest shrinkage strain and water content variation rate. These acoustic emission activities are generally associated with cracking in samples. Therefore, the acoustic emissions in actual boreholes could not be directly related to cracking in the surrounding rock.

(b)   Effect of cyclic saturation and drying on mechanical properties of rocks

The cyclic saturation and drying tests were carried out in accordance with the procedures given by International Society for Rock Mechanics and Rock Engineering (ISRM (2007)) and American Society for Testing Methods (ASTM (2000)). The weight loss and uniaxial compressive strength (UCS) of the tuff samples from Avanos were determined at different cycles. Due to the fact that the samples subjected to durability tests would be broken during uniaxial compression tests at certain cycles, various specimens prepared from the blocks were used in these tests. The UCS was determined on dry samples following 2, 4, 9, 11 and 14 test cycles by applying this method. As seen from Figure 8.16, after wetting–drying tests, some change in weight is observed, and the weight loss at the end of 14 test cycles is about 4.5% (Fig. 8.16). Figure 8.16 suggests that the reduction in UCS between the 2nd and 14th cycles is about 1.05 MPa. This change in UCS means the total loss in strength is about 40%. In freezing–thawing tests, the specimens were subjected to a total of eight test cycles. The weight loss at the end of eight test cycles is about 6% (Fig. 8.16). Figure 8.16 indicates that the reduction in UCS between the 1st and 11th cycles is about 1.05 MPa. This change in UCS means the loss in strength is about 40%. Comparison of the wetting–drying and freezing–thawing test results indicates that freezing–thawing cycles are more effective on decrease in weight loss and UCS of the tuff studied. Based on the 4-cycle slake durability tests, Id values from the 1st to 4th cycles are between 81.5–98%, 69.3–94%, 56.7–88.3% and 40.7–77.1%, respectively. Particularly after the 2nd cycle, a considerable disintegration of the samples was observed.

Based on the wetting–drying tests, the UCS of the Kavak tuff decreased by 4% and 22% at the end of 22nd and 40th cycles, respectively. The same tests on the Zelve tuff suggest that a 36.3% decrease in the UCS occurs at the end of 5th cycle.

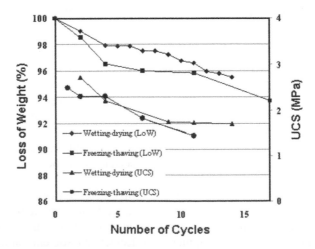

Figure 8.16 Variation of loss of weight and UCS with number of cycles

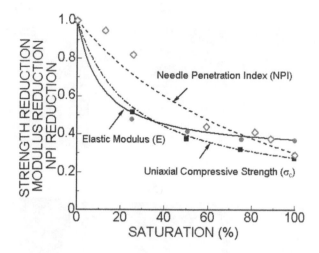

Figure 8.17 Variation of normalized elastic modulus and uniaxial compressive strength and needle penetration index (NPI) in relation to saturation

The mechanical properties of soft rocks prone to water absorption and desorption may depend upon the degree of saturation (Aydan and Ulusay, 2003, 2013). These property changes may also greatly affect their responses of rock engineering structures. Figure 8.17 shows an example of changes of the mechanical and index properties of Oya tuff.

8.3 Hydrothermal alteration

The alteration process is due to percolating hydrothermal fluids in rock mass, and it may act on rock mass in a positive or negative way. The positive action of the alteration may heal existing rock discontinuities by rewelding through the deposition of ferro-oxides, calcite or siliceous filling material. On the other hand, the negative action of the alteration would cause the weakening of the bonding of particles of rocks and producing clayey materials similar to the chemical degradation. As a result of alteration, rocks may be transformed into kaolin or chlorite minerals. As the intact rock is one of the most important parameters influencing the mechanical response of rock masses, the negative action of hydrothermal alteration may account for the degradation of intact rock. While the weathering of rocks is mostly observed near ground surface up to a depth of 40 m and their effects disappear with depth, alteration may be observed at greater depths. Figure 8.18 shows different stages of alteration of rhyolite of Okumino.

8.4 Degradation due to surface or underground water flow

Surface and underground water flow over the surface of rock surfaces or through rock mass may also cause the degradation of rock mass. Figure 8.19 shows some examples of degradation due to surface or underground water flow in the Zelve valley of the Cappadocia region in Turkey and Ishigaki island. It is observed that heavy rains and rapid surface water flow

Figure 8.18 Rhyolite of Okumino subjected to hydrothermal alteration at different stages of degradation: (1) fresh, (2) stained, (3) slight alteration, (4) moderate alteration, (5) heavy alteration, (6) decomposed

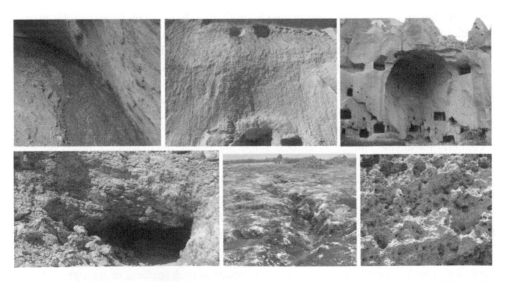

Figure 8.19 Some examples of degradation due to surface or underground water flow in Zelve valley of Cappadocia region in Turkey and Ishigaki island

induces the degradation of rock mass due to both the saturation of rocks and surface erosion and the washing out of filling material of rock discontinuities in soft rocks such as tuffs. The surface flow or percolation of rock mass in limestone and evaporates may dissolve rock, resulting in the widening of pores and fractures, which may also lead to large-scale cavities.

8.5   Biodegradation

As previously described, the findings have demonstrated that the degradation is ultimately formed by a combination of physical and chemical processes, and the climate or lithology are among the controlling factors affecting the rate of rock weathering. There are some new considerations for the degradation of rocks in geoscience and geoengineering (Ehrlich and Newman, 2009). This degradation is assumed to be related to bacterial activities, which may be another very important cause of degradation of rocks. In this subsection, some biodegradation studies undertaken on tuff samples in Cappadocia Region are described.

(a)   Sampling locations

Four tuff samples (size: approximately 1000 cm³) from the 5th floor of Derinkuyu Underground City, Zelve semiunderground (cliff) settlement and Uçhisar Fairy Chimney settlement were collected (Figure 8.20). The Derinkuyu site has been monitored by Aydan and his group (Aydan *et al.*, 1999a, 1999b, 2007a, 2007b; Aydan and Ulusay, 2003, 2013) since 1996. Multiparameter monitoring program has been implemented in this site, and the sample collected is the underground 5th floor. Lighting at this site is due to only electricity. However, the room is about 6 m from the main ventilation shaft.

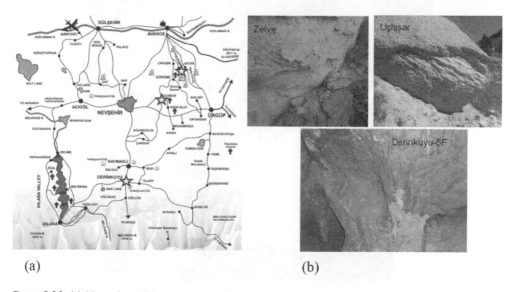

(a)   (b)

Figure 8.20 (a) Map of and (b) views of sampling locations

The sample collected in the Zelve site is at the toe of semi-underground settlement where the toe erosion is severe. Multiparameter monitoring has been implemented in this area since 2004. The third sample is collected from Uçhisar where huge fairy chimneys are found. The sample was obtained from the toe of a fairy chimney suffering the heavy spalling problem.

(b)   Microscopic investigations

The bacteria in rocks, so-called endolithic bacteria, actively bore into the host rock by solubilizing cementing mineral grains in their attempt to gain access to nutrients and energy (Konhauser, 2007). To date, their contributions in various rock types, such as limestone, dolomite, sandstone, granite, basalt, tuff and the like, have been well documented (e.g. Büdel *et al.*, 2004; Hoppert *et al.*, 2004; Hall *et al.*, 2008). In the investigation, four samples as shown in Figure 8.21 were selected, and microscopic conditions were observed by a biological microscope (Matsubara and Aydan, 2016). In the study, 1 g of rock sample was mixed with 9 g of sterile distilled water, and little mixed water samples were observed.

Figure 8.22 shows the microscopic images in the rock samples. As shown in the figure, numerous microorganisms were recognized in all rock samples. Qualitatively, it is the sample from the Derinkuyu-5F that has the largest number of microorganisms in comparison to other samples. On the other hand, the number of microorganisms in the sample from the Zelve is the smallest.

Although the number of microorganisms may depend on the time and period of sampling, it is understood that microorganisms would be more active at the Derinkuyu-5F. This

Figure 8.21 Views of samples obtained from Zelve valley, Fairy Chimney at Uçhisar and the fifth floor of Derinkuyu Underground City

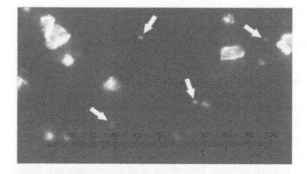

(a) Zelve Valley

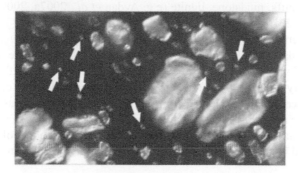

(b) Uçhisar Fairy Chimney settlement

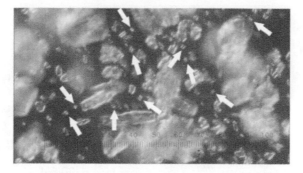

(c) 5th floor of Derinkuyu Underground City

Figure 8.22 Microscopic conditions of inner rock samples (The white arrows indicate microorganisms; scale bar: 100 = 1 mm).

difference may depend on environmental conditions. It is noted that (1) the outcrops were dried almost constantly with sunlight in summer but were wet almost constantly in winter at the Zelve site; (2) the Uçhisar site is in the shade, and the outcrops were wet almost constantly; (3) the 5th floor of Derinkuyu site is always dark, and constant temperature and high humidity are continuous. Interestingly, this site is always dark, so that the chances that they are photosynthetic bacteria such as cyanobacteria are virtually nil. Therefore, they would be chemolithoautotroph or chemoheterotroph. Although the identification of the bacteria and other detailed analyses are an issue to be addressed in the future, endolithic bacteria may possibly induce the rock weathering in Cappadocia Region.

8.6   Degradation rate measurements

The degradation of rock due to different causes, such as the cyclic drying–wetting, freezing–thawing, winds and rainfall, surface or underground water flow, results in erosion, slabbing, spalling and raveling (Figure 8.23). Due to these effects, tuffs, particularly those close to ground surface at the toe regions of semiunderground rock structures at valley bottoms, are prone to degradation. Furthermore, limestone and evaporates are dissolved by sea waves, winds, river flow or percolating rainwater. There are some studies on the degradation rate of soft rocks such as tuffs, mudstone, lignite (e.g. Aydan *et al.*, 2007a, 2007b, 2008a, 2008b; Aydan and Ulusay, 2016).

   The studies undertaken on tuffs of the Cappadocia region by Aydan *et al.* (2007a, 2007b, 2008a, 2008b), Aydan and Ulusay (2013 and Kasmer *et al.* (2008) are described here as the preservation of antique underground cities and semiunderground settlements, as well as fairy chimneys, very much depends upon the erosion or degradation rate of tuffs. The annual erosion rate has been either computed from existing structures or measured *in-situ* (Figure 8.24). There are many historical structures with well-known dates of construction. If the dates of structures are known, it is possible to evaluate the average rate of degradation. The technique proposed by Aydan *et al.* (2007, 2008a, 2008b) adopted herein and more data are

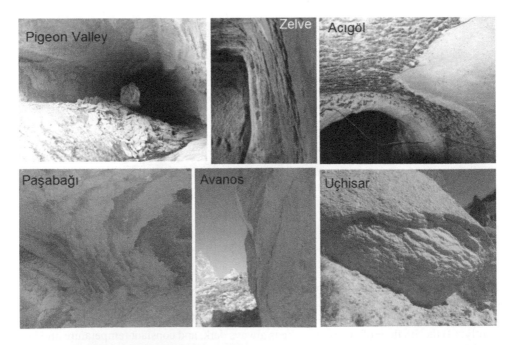

Figure 8.23 Views of spalling and slabbing problems

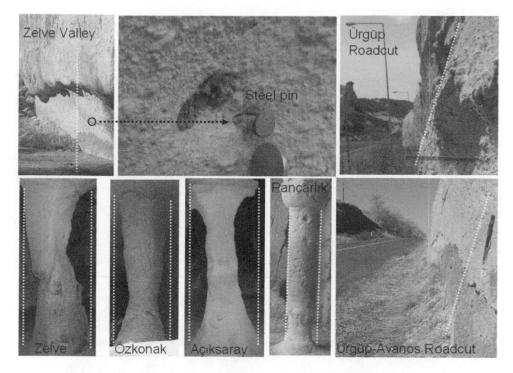

Figure 8.24 Some actual examples for computation of degradation rates

presented. Daily temperature changes in the Cappadocia region are very high, and freezing generally occurs between November and March.

The amount of erosion was also studied by Kasmer *et al.* (2008) in the 1st valley in ZOAM at two selected locations between 2006 and 2011 using nails inserted in rock mass in different locations. Location 1 was the toe of a steep natural valley slope, while Location 2 was a huge tuff block at the bottom of the valley using stainless nails and a sensitive device. The measurements indicated that there is an increasing initial trend in the amount of erosion at both locations particularly in the spring season. Particularly in the winter season, the toe of the slope becomes wet due to capillarity and the stream flowing in the valley. In the spring season, the rock tends to dry. As a result of these repeated drying–wetting cycles, the amount of erosion showed an increase when compared to that in the hot summer season. The amount of erosion throughout the five-year period was between 1.07 and 6.21 mm and by assuming a homogeneous erosion, the annual average erosion is determined between 0.21 and 1.24 mm. By considering that the valley has been used for settlement for more than 1000 years and this monitoring covers a five-year period, the amount of erosion due to natural factors seems to be important.

Table 8.3 gives estimated degradation rates for several locations. The erosion rate in Cappadocia generally ranges between 0.04 and 0.5 mm/yr (Aydan *et al.*, 2007a, 2007b, 2008a, 2008b; Aydan and Ulusay, 2013). However, Kasmer *et al.* (2008) reported that the annual average erosion could be up to 1.6 mm in the Zelve valley as given in Table 8.3. As noted from the table, the erosion or degradation rate of tuffs with larger amounts of clay minerals is

Table 8.3 Degradation rate

Location	Depth (mm)	Time (years)	Rate (mm a⁻¹)
Ürgüp	600–1200	1500	0.24–0.48
Avanos	280–750	1500	0.19–0.5
Derinkuyu	60–135	1500	0.04–0.09
Ihlara	300–750	1500	0.2–0.5
Açıksaray	150–600	1500	0.1–0.4
Pancarlık	150–300	1500	0.1–0.2
Göreme	75–225	1500	0.05–0.15
Özkonak	300–450	1500	0.2–0.3
Selime	150–300	1500	0.1–0.2
Zelve	150–2400	1500	0.1–1.6
Ürgüp Roadcut	20–50	50	0.4–1.0
Ürgüp-Avanos Roadcut	8–15	30	0.3–0.5

Figure 8.25 Views of differential erosion and weathering

much higher than that of tuffs with less clay mineral content. The degradation is particularly high at the toe of the cliffs below the semiunderground openings, roads and valleys.

The resistance of the tuff layers in association with different eruption episodes of volcanoes of the region differs, and differential weathering and erosion take place. Particularly, the erosion rates are higher at the toe of rock layers, which results in overhanging configurations (Figure 8.25).

8.7   Needle penetration tests for measuring degradation degree

Aydan *et al.* (2014) proposed a method to evaluate the degree of degradation (weathering) of soft rocks utilizing the needle penetration index (NPI). Aydan *et al.* (2014) defined a parameter (W) for quantitative evaluation of the degradation state by the following equation:

$$W = 100\beta \left(\frac{1 - \dfrac{NPI_w}{NPI_f}}{\beta\left(1 - \dfrac{NPI_w}{NPI_f}\right) + \dfrac{NPI_w}{NPI_f}} \right) \tag{8.1a}$$

or

$$\frac{NPI_w}{NPI_f} = 1 - \frac{W}{W + \beta(100 - W)} \tag{8.1b}$$

where NPI_w and NPI_f are values of NPI for the completely decomposed and fresh conditions, respectively. The degradation state parameter (W) has a value of 0 for the fresh state and 100 for the completely decomposed state. Coefficient β in Equation (8.1) ranges between 0.2 and 0.6.

A series of needle penetration index experiments were carried out on the samples shown in Figure 8.26; the results are given in Table 8.4. Aydan *et al.* (2006, 2008a, 2008b, 2014) also reported NPI values for fresh tuff samples at the respective places. The needle penetration index and visual weathering state results indicate that the weathering state was quite high in the Derinkuyu sample while it was light in the Zelve samples.

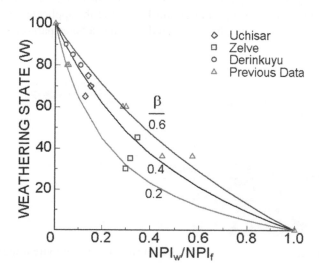

Figure 8.26 Relation between NPI and weathering state

Table 8.4 NPI values and weathering state of tuff samples

Location	NPI (fresh)	NPI (degradated)	Degradation State (%)
Derinkuyu	33–46	1.5–5.0	85–95
Uçhisar	14–25	2.0–3.2	65–75
Zelve	20–33	7–10	30–45

8.8   Utilization of infrared imaging technique for degradation evaluation

Compared to the passive-type infrared thermographic imaging technique, active-type infra-red thermographic imaging technique should be also useful in evaluating the soundness of various structures for maintenance purposes (Aydan, 2019). In this respect, the utilization of artificial heat sources such as heaters, coolers, sunlight radiation, dynamic excitation, wind may be useful.

Figure 8.27a shows the visible and infrared thermographic images of Nakagusuku Castle remain where castle walls were reconstructed utilizing original blocks and newly replaced blocks, which are made of Ryukyu limestone. As the original blocks are partly weathered compared with the newly replaced limestone blocks, their heat absorption characteristics under the sunlight heating are different from each other. This observational fact clearly indicates that the active infrared thermographic imaging technique could be quite useful particularly in the evaluation of the soundness of rocks as well as rock engineering structures in geoengineering for maintenance purposes.

An active infrared thermographic imaging attempt was done on weathered sandstone, locally known as Niibi stone. The main purpose was to see if the weathered zone could be distinguished from the nonweathered part. Figure 8.27b shows both visible and infrared thermographic images of the partially weathered Niibi sandstone. Infrared thermographic images correspond to the images just after and 120 s after the application of the heat shock. This experiment clearly indicates that weathered and unweathered parts can be distinguished

(a) Visible image   (b) Infrared thermographic image

Figure 8.27a Visible and infrared thermographic images of a reconstructed wall of Nakagusuku Castle

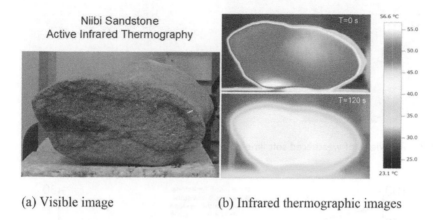

(a) Visible image   (b) Infrared thermographic images

Figure 8.27b Visible and infrared thermographic images of weathered Niibi sandstone

and that the rate of cooling, which could be determined from continuous monitoring, can yield the variation of their thermo-physical characteristics. Therefore, the active infrared thermographic imaging technique may also be another efficient tool in characterizing the weathering of rocks and its thermo-physical state.

8.9   Degradation assessment of rocks by color measurement technique

Most degradation assessment procedures in rock mechanics is based on visual inspections and some index tests. When rock degrades, its color undergoes variations. Mostly the surface become whitish, yellowish or reddish due to the alteration of some ferrous minerals contained in the rock matrix. Particularly, the variation of the rock color could be a measure to assess its weathering state. For this purpose, a chroma meter developed by Minolta CR-800 was used. There are various modes of color measurements. Kano *et al.* (2004) reported that L-a-b mode is the most suitable mode for characterizing the degradation state of rocks. For this purpose, an orthogonal coordinate system is defined. The axis L is a measure of surface darkness. It is value +100 for white color and −100 for black color. The axis a is a measure of variation between red (+100) and green (−100), while the axis b is a measure between yellow (+100) and blue (−100). Figure 8.28 shows two views of weathered soft limestone observed outside the tomb. The weathered soft limestone gives the impression of rock of a thinly layered yellow to brown material.

Figure 8.29 shows the variation of parameters of L, a, b for fresh and weathered soft limestone and hard limestone. Although the variations of hard limestone are very small, very distinct variations occur for soft limestone.

Figure 8.30 shows the variation of parameters of L, a, b from the entrance to the deepest level (Jee) of the tomb. Although some fluctuations are observed, the state of rock mass throughout the tomb is remains almost the same. In other words, the weathering of rock mass in the tomb is almost negligible for 3400 years compared to that of rock subjected to the harsh weather conditions outside.

Figure 8.28 Two views of weathered soft limestone near the entrance of the tomb

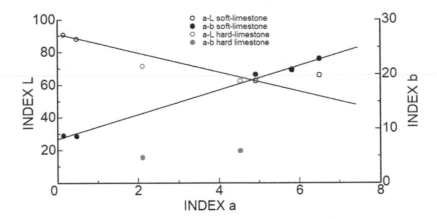

Figure 8.29 Variation of parameters *L*, *a*, *b* for weathered and nonweathered soft and hard limestone

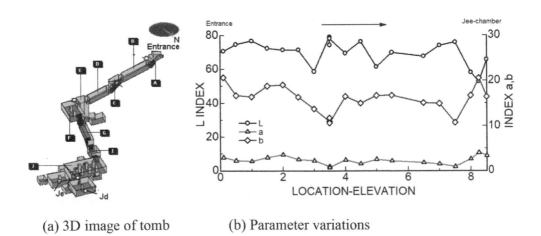

(a) 3D image of tomb   (b) Parameter variations

Figure 8.30 Variation of parameters *L*, *a*, *b* of rock surface from the entrance to Jee chamber of the tomb

8.10   Effect of degradation process on the stability of rock structures

8.10.1   Antique rock structures

(a)   Observations

BENDING AND TOPPLING OF BLOCKS

This problem particularly occurs at locations where the pillars left in the rock-hewn openings fail and/or when the toe of slopes is eroded by water or wind effect (Figure 8.31). As a result of changes in the stress state due to toe erosion, thermal loading due to freezing–thawing and property changes due to cyclic wetting and drying process, spalling and slabbing in cliffs and underground openings may occur. These phenomena result in the formation of slabs with different thicknesses on natural slopes or the walls of underground openings and consequently cause a change in geometry of openings to trigger failures (Figure 8.31)

Figure 8.31 Some partial failures, spalling, slabbing and collapses in Cappadocia region

(b)   Analytical solutions

The degradation process may be approximated using some functional forms. For the simplicity, we assume that the degradation is a linear function of time and is given in the following form:

$$\Delta L = a \cdot t \tag{8.2}$$

where constant a is called the degradation rate.

One of the main parameters influencing the long-term stability of rock structures is the creep strength characteristic of rocks. The creep strength (σ_{cr}) in terms of short-term strength (σ_s) can be represented in the following form (Aydan and Nawrocki, 1998). $A_w(t)$ is assumed to be time dependent to account for the effect of erosion.

$$\sigma_{cr} = \sigma_s \left(1 - b \ln \left(\frac{t}{\tau} \right) \right) \tag{8.3}$$

where b, t and τ are the empirical constant, time and short-term test duration, respectively. The value of b generally ranges between 0.0186 and 0.0583 for the Cappadocian tuffs (Aydan and Ulusay, 2013; Ito *et al.*, 2008).

The tensile or uniaxial compression strength of the rock mass is estimated from the procedure suggested by Aydan *et al.* (2014). In this procedure, property φ_m, such as UCS or tensile strength of the rock mass, is obtained from the following formula by using the RMQR value of the respective property of intact rock φ_i, and the value of β can be taken as 6 on the basis of experimental data from construction sites in Japan (Aydan *et al.*, 2014):

$$\frac{\varphi_m}{\varphi_i} = \alpha_0 - (\alpha_0 - \alpha_{100}) \frac{\text{RMQR}}{\text{RMQR} + \beta (100 - \text{RMQR})} \tag{8.4}$$

STABILITY OF OPENINGS NEXT TO CLIFFS AND FAIRY CHIMNEYS

Based on the observations of the authors at several locations in Cappadocia, most of the instabilities occur as a result of the collapse of toes of openings next to cliffs (Figure 8.31). In such failures, the erosion of the toe by natural agents and decrease in long-term strength of the rock are the main causative factors. In addition, instability problems were also observed in openings in fairy chimneys (Figure 8.31). Herein, the stability of openings next to cliffs and in fairy chimneys through the utilization of creep tests is evaluated by using simplified approaches as described by Aydan *et al.* (2007).

In the model illustrated in Figure 8.32, it is assumed that the pillar or wall at the valley-side carries half of the burden over the opening. Due to its conical shape, a fairy chimney is considered an axially symmetrical rock structure including a circular opening at its center. Although the real stress distributions in these openings are slightly different from those in these methods recommended in this study, it is considered that the approaches used will be helpful in assessing the conditions of the instabilities investigated. The time-dependent safety factor (SF) of the wall next to the cliff and in fairy chimney is written as follows:

$$SF = \frac{\sigma_{cr}(t)}{\gamma H} \frac{A_w(t)}{A_t} \tag{8.5}$$

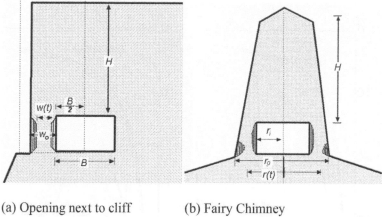

(a) Opening next to cliff   (b) Fairy Chimney

Figure 8.32 Simplified mechanical models for opening next to cliff and fairy chimney

where γ, σ_{cr}, H, A_t and $A_w(t)$ are unit weight, creep strength, overburden height, total area supported and wall area. The area ratios for the continuous wall next to the cliff and in the cylindrical fairy chimney shown in Figure 8.32(a) specifically take the following forms, respectively:

$$\frac{A_t}{A_w(t)} = \frac{w/2+t_o}{t^*(t)} \tag{8.6a}$$

$$\frac{A_t}{A_w(t)} = \frac{r_o^2}{r^2(t)-r_i^2} \tag{8.6b}$$

The erosion rate in Cappadocia generally ranges between 0.04 and 0.5 mm a^{-1} (Aydan *et al.*, 2007a, 2007b, 2008a, 2008b, 2008c, 2008d; Aydan and Ulusay, 2013). The reduction of supporting area (A_w) due to erosion/degradation may be counted as proposed by Aydan *et al.* (2006). For example, the wall thickness of the openings next to cliffs and fairy chimneys may be given in the following form:

$$w(t) = w_o - \eta t \tag{8.7a}$$

$$r(t) = r_o - \eta t \tag{8.7b}$$

Equation 8.3 can be adopted for the creep strength versus failure time function, together with or without the degradation model to evaluate the safety of openings. Figure 8.33 shows the safety factor of the openings next to cliffs and in fairy chimneys as a function of time and overburden for strength properties of tuffs without degradation. It is clear from Figure 8.33 that an increase in overburden heightens the probability of failure of pillars in openings next to cliffs, and the collapse of fairy chimneys increases. Comparison of Figures 8.33(a) and 8.33(b) also indicates that openings next to cliffs are more likely to fail in the long term as compared with those of fairy chimneys. The observations by the authors in the Zelve, Göreme and Ihlara valleys confirm this result.

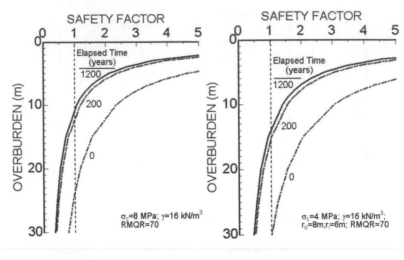

(a) Opening next to cliff   (b) Fairy Chimney

Figure 8.33 Safety factor variation with depth

(c.1)   (c.2)

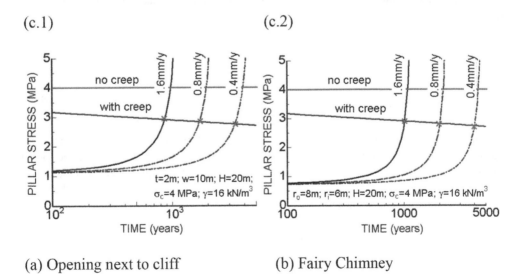

(a) Opening next to cliff   (b) Fairy Chimney

Figure 8.34 Effect of erosion rate and decrease in strength on pillar stability for openings next to cliffs and fairy chimneys, respectively

By considering this erosion range, the calculations carried out for an opening with a width of 10 m and an overburden height of 20 m are shown in Figure 8.34(c1). This figure suggests that the stability of openings next to cliffs is affected by erosion. Similarly, an analysis was also carried out for a fairy chimney with a base diameter of 8 m using the same erosion range and the results shown in Figure 8.34(c2) indicate that the collapse potential of fairy chimneys will increase after 1000 years.

ROOF FAILURE

The analytical models are based on bending and arching theories developed by the authors to incorporate not only the gravitational load of the roof but also the effects of discontinuities, creep and degradation of rock mass (i.e. Aydan, 1989, 2008; Aydan *et al.*, 2005a, 2005b, 2006, 2007a, 2007b; Aydan and Tokashiki, 2011). Aydan and Tokashiki (2011) recently considered also the effect of load from topsoil as well as the concentrated loads resulting from superstructures or vehicles (e.g. airplanes) in developing an analytical method based on bending theory (Figure 8.35). The limit of roof span (L) normalized by roof rock layer thickness (h_r) under its deadweight and topsoil with thickness of (h_s) can be obtained in the following form (Aydan and Tokashiki, 2011):

$$\frac{L}{h_r} = \sqrt{\beta \frac{\sigma_t(t)}{\left(\gamma_r h_r(t) + \gamma_s h_s\right)}} \tag{8.8}$$

where $\sigma_t(t)$ is the tensile strength of roof layer. The value of β would take the values of 1/3, 2/3 and 2 for cantilever, simple and built-in beams. The effect of the degradation of the roof layer may be also estimated from the following relation:

$$h_r(t) = h_{ro} - \eta t \tag{8.9}$$

The arching theory for assessing the stability of roofs of the shallow underground openings is generally based on three-hinged beams. As discussed by Aydan (1989), the rotation of

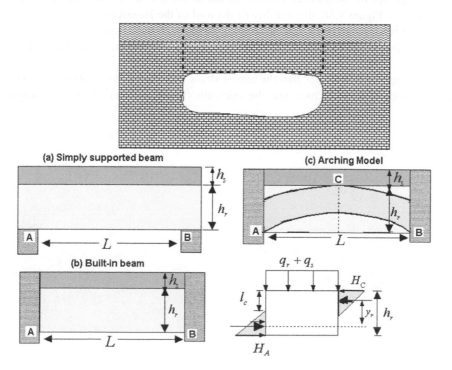

Figure 8.35 Simplified mechanical models for estimating roof stability

rock layer, in which the arch is formed, should be initiated for the arch action to take place. The failure modes of an arch are various, and they are crushing at the crown and/or abutments, vertical or horizontal sliding at abutments and sliding along an existent discontinuity within the arch. The detailed derivations for each failure mode are presented in the doctorate thesis of Aydan (1989) and also in the article by Kawamoto *et al.* (1991).

The classical arching theory assumes that the failure mode of the arch takes place by crushing at the crown or abutments and the final form for the limit of roof span (L), normalized by rock roof layer thickness (h_r) under its own weight and topsoil with thickness of (h_s), as follows (i.e. Aydan and Tokashiki, 2011):

$$\frac{L}{h_r} = \sqrt{\beta \frac{\sigma_c(t)}{\left(\gamma_r h_r(t) + \gamma_s h_s\right)}} \qquad (8.10)$$

where σ_c is the compressive strength of the roof layer. The value of β would take the following values 4/3 when crack length (ℓ_c) is zero. However, the maximum resistance of the arch would be attained when the value of β is 3/2 after some manipulations concerning the crack length in the arch. The effect of concentrated loads can be taken into account if necessary.

The tensile and compressive strengths of rock mass can be obtained using the formulas suggested by Aydan *et al.* (2014), Aydan and Kawamoto (2000) and Tokashiki and Aydan (2010) in terms of those of intact rock and the RMQR values of rock masses (Aydan *et al.*, 2014). When Equations (8.8) and (8.9) are used for stability assessment, the effects of creep and degradation rate of rock mass in these equations can be taken into account using the method suggested by Aydan *et al.* (2005a, 2006b, 2006, 2007a, 2007b, 2008a, 2008b)

As shown in Figure 8.36, the stability of the roof of the largest room in the Avanos Congress Center is first analyzed using the bending theory with built-in conditions (Ulusay *et al.*, 2013). The span of the room is 12 m with a 5 m thick solid roof and 3 m thick soil deposits at the top.

The bending stress indicated that the tensile stress will exceed the tensile strength of solid rock at the top of the solid roof near the sidewalls (Figure 8.37(a)). The next computation

Figure 8.36 Plan of the Congress Center

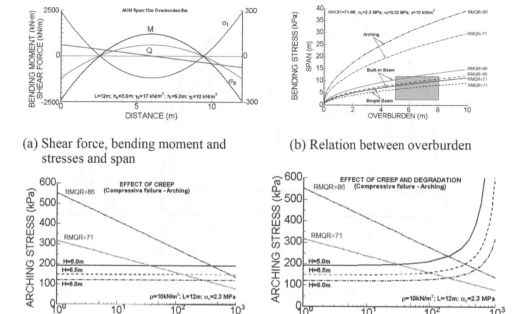

(a) Shear force, bending moment and
    stresses and span

(b) Relation between overburden

(c) Long-term stability for creep effect only

(d) Long-term stability for creep and
    degradation

Figure 8.37 Computed results using various analytical models for the roof stability of the Congress
Center

was carried out for three different conditions, namely simple-beam, built-in beam and arching (Figure 8.37(b)). The simple beam and built-in beam conditions clearly implied that
some cracking should occur, and this conclusion was in accordance with our observations.
The stability analyses for arching action indicated that the opening should be stable in the
short term even though it may be cracked. The next computation was carried out for the
long-term response with the consideration of the long-term strength of the tuffs of Cappadocia, including those of the Congress Center. The pure creep analysis with the consideration
of the arching model implied that the openings should remain stable for longer than 30
years under the most unfavorable conditions (Figure 8.37(c)). However, if the degradation
resulting from the cyclic freezing–thawing and wetting–drying taken into account, the stable
duration becomes shorter (Figure 8.37(d)). The degradation rate is based on the observations
of the tuff of Cappadocia (Aydan *et al.*, 2008a, 2008b).

The actual failure mode of semiunderground openings observed in Frig valley was analyzed by Aydan and Kumsar (2016). Figure 8.38 illustrates the physical model and idealized
mechanical model of the failure mode. The tensile stresses develop above the opening as the
erosion depth increases. The cracks propagate downward, and the rock body acts like dead
load on the floor slab of the opening. Once the erosion depth reaches a critical depth, the
whole mass topples. In other words, the final stage of the failure implies a flexural toppling

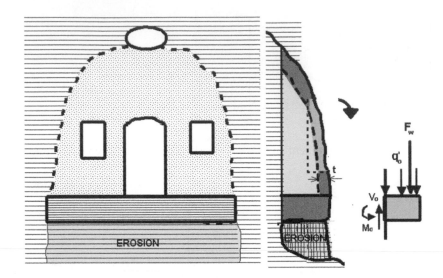

Figure 8.38 Illustration of the mechanical model for assessing the stability of semi-underground openings

failure. This mode of failure of underground openings, slopes and cliffs has been previously studied in detail by Aydan and Kawamoto (1992). The maximum flexural tensile stress can be shown to be:

$$\sigma_{ft} = 6\gamma_r \left(\frac{h}{2} + H\frac{t}{d}\left(1 - \frac{t}{2d}\right)\right)\left(\frac{d}{h}\right)^2 \tag{8.11}$$

where γ_r, h, H, t and d are unit weight of rock, thickness of the floor slab, height of unstable body, thickness of overhanging wall and erosion depth. Although the shearing failure is less likely, the maximum shear stress according to the bending theory may be given in the following form:

$$\tau_{max} = \frac{3}{2}\gamma_r \left(h + H\left(\frac{t}{d}\right)\right)\left(\frac{d}{h}\right) \tag{8.12}$$

The failure would occur whenever the tensile or shear strength of rock mass is exceeded.

This approach is applied to the typical situations observed in Frig valley. The most important item is the evaluation of the tensile and cohesion of rock masses. The approach proposed by Aydan et al. (2014) is adopted in this study. For the RMQR values of rock mass given in Table 8.5, the normalized tensile strength or cohesion of rock mass would be 0.23–0.26 times that of intact rock. Furthermore, the tensile strength of rock mass would decrease with time due to sustained creep load (Aydan and Ulusay, 2013) and the long-term strength would obey Equation (8.3).

First a static stress analyses is carried out, and the critical erosion depth is computed for various rock mass tensile strength ratios, as shown in Figure 8.39. Parameters assumed in

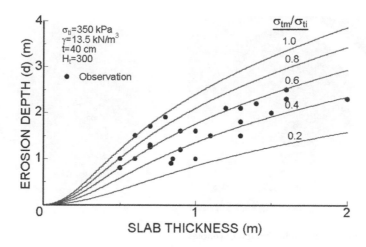

Figure 8.39 Critical erosion depth as a function of slab thickness

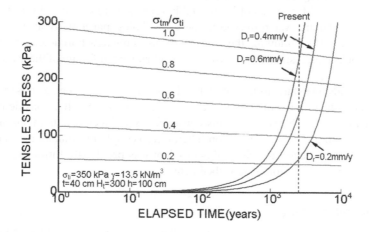

Figure 8.40 Estimation of collapse time of semiunderground openings

the computations are also given in the figure. The results imply that the rock mass strength ratio could range between 0.4 and 1.0.

Next computation is carried out to evaluate the failure time of semiunderground openings due to toe erosion and creep of rock mass. Again the rock mass strength ratio and erosion rates (D_r) were varied as a function of time. The results are shown in Figure 8.40. It is interesting to note that major collapses may start to occur after 1270 years.

8.10.2   Coupled analyses for failure in boreholes due to cyclic wetting and drying

Tono mine is one of the uranium mines of Japan, and it has been the subject of extensive research regarding nuclear waste disposal in sedimentary rocks. The top formation of the mine consists of sedimentary rocks ranging from sandstone to siltstone. An exploration adit

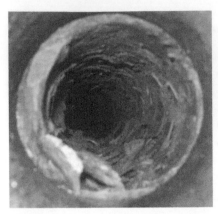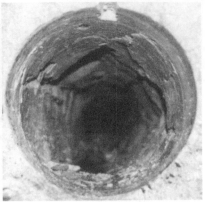

Figure 8.41 Views of yielding horizontal boreholes

was excavated by using the NATM, and this adit was therefore named the NATM adit. At an extension of this adit, a series of horizontal boreholes were excavated for measuring *in-situ* stresses. The failure was observed in these boreholes, particularly at their crown and sidewalls (Figure 8.41). Following these observations, some new boreholes were drilled into sedimentary rocks to investigate the causes and mechanisms of failure of these nonsupported boreholes.

Fine-grain sandstone starts to fracture while losing its water content, as observed in laboratory tests. The situation is similar to the reverse swelling problem. It is considered that rock shrinks as it loses its water content. This consequently induces results in shrinkage strain leading to the fracturing of rock in tension. Therefore, a coupled formulation of the problems was considered to be necessary. Since this problem was previously formulated by the first author, its summarized version is described here.

(a)   Mechanical modeling

The water content variation in rock may be modeled as a diffusion problem. Thus the governing equation is written as:

$$\frac{d\theta}{dt} = -\nabla \cdot \mathbf{q} + Q \tag{8.13}$$

where θ, q, Q and t are water content, water content flux, water content source and time, respectively. If water content migration obeys Fick's law, the relation between flux q and water content is written in the following form:

$$\mathbf{q} = -k\nabla\theta \tag{8.14}$$

where k is water diffusion coefficient. If some water content is carried out by the groundwater seepage or airflow in open space, this may be taken into account through the material derivative operator in Equation (8.13). However, it would be necessary to describe or evaluate the seepage velocity or airflow.

If the stress variations occur at slow rates, the equation of motion without the inertial term may be used in incremental form as given here:

$$\nabla \cdot \dot{\sigma} = 0 \tag{8.15}$$

The simplest constitutive law for rock between stress and strain fields would be a linear law, in which the properties of rocks may be related to the water content in the following form:

$$\dot{\sigma} = \mathbf{D}(\theta)\dot{\varepsilon}_e \tag{8.16}$$

The volumetric strain variations associated with shrinkage (inversely swelling) may be related to the strain field in the following form:

$$\dot{\varepsilon}_e = \dot{\varepsilon} - \dot{\varepsilon}_s \tag{8.17}$$

(b) Finite element modeling

The finite element form of the water content migration field takes the following form after some manipulations of Equations (8.13) and (8.14) through the usual finite element procedures:

$$[M]\{\dot{\theta}\} + [H]\{\theta\} = \{Q\} \tag{8.18}$$

where

$$[M] = \int [\mathrm{N}]^{\mathrm{T}} [\mathrm{N}] dV, [H] = k \int [B]^{\mathrm{T}} [B] dV, \{Q\} = \int [\bar{N}]^{\mathrm{T}} \{q_n\} d\Gamma$$

Similarly, the finite element form of the incremental equation of motion given by Equation (8.19) is obtained as follows:

$$[K]\{\dot{U}\} = \{\dot{F}\} \tag{8.19}$$

where

$$[K] = \int_V [B]^{\mathrm{T}} [D][B] dV, \{\dot{F}\} = \int_V [B]^{\mathrm{T}} [D]\{\dot{\varepsilon}_s\} dv + \int_s [\bar{N}]^{\mathrm{T}} \{\dot{t}\} ds$$

(c) Analyses and discussions

The first group analyses were concerned with the simulations of displacement, strain and stress field around a circular borehole in a hydrostatic stress field. Specifically, the effects of the sandstone type and borehole diameter were analyzed. Figure 8.42 shows the computed results for displacement, water content and stress fields for fine- and coarse-grain sandstones for a borehole with a diameter of 200 mm at the overburden of the adit. Since the water migration characteristics of both fine- and coarse-grain sandstones were the same, the resulting water content migration distributions with time were the same. However, displacement, strain and stress fields were entirely different for each sandstone type. Since the volumetric variation of fine-grain sandstone as a function of water content is much larger than that of coarse-grain sandstone, the shrinkage of the borehole in fine-grain sandstone is greater than that in coarse-grain sandstone. Consequently, the radial stress in the close vicinity of the borehole wall becomes tensile in fine-grain sandstone. This, in turn, implies that there would be fractures parallel to the borehole wall if the tensile strength of rock were exceeded. Furthermore, such fractures would only occur in the vicinity of boreholes in fine-grain sandstone, as observationally noted *in-situ*.

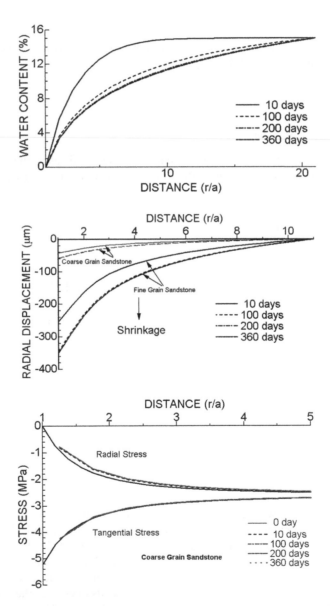

Figure 8.42 Variations of computed water content, displacement, stress fields for fine- and coarse-grain sandstones

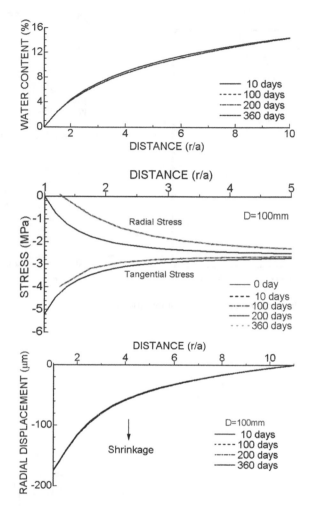

Figure 8.43 Variations of computed water content, displacement, stress fields of a borehole with a diameter of 100 mm for fine-grain sandstone

Next, the borehole diameter was changed from 200 mm to 100 mm, and the rock was assumed to be fine-grain sandstone. The computed results shown in Figure 8.43 indicated that the water content variation reached the steady state rapidly. However, the computed strain and stress fields are the same as those of the borehole with a diameter of 200 mm except for the magnitude of radial displacement.

The final computational example was concerned with a circular borehole under two dimensional initial *in-situ* stress fields. It is observed that the bottom of the borehole was wet or covered with water *in-situ*. In order to take into account this observation in computations, the boundary conditions for the water content migration field and displacement field were assumed as illustrated in Figure 8.44. The other properties were the same as those used in axisymmetric simulations. The computed displacement field and associated yielding zone are shown in Figure 8.45.

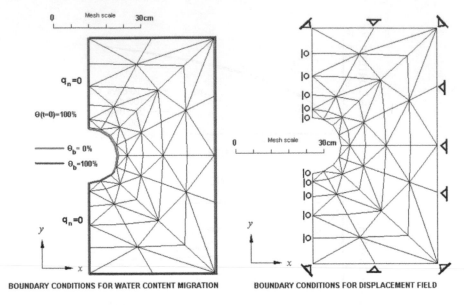

Figure 8.44 Assumed boundary conditions in computations

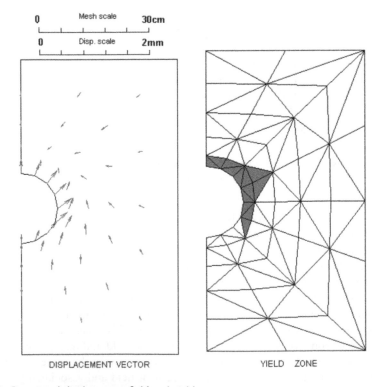

Figure 8.45 Computed displacement field and yield zone

As noted from the figure, the bottom of the borehole heaves, and the crown of the borehole shrinks upward. In other words, the upper part of the borehole expands outward due to water content loss. The displacement and stress fields of the surrounding rock are entirely different at the lower and upper parts of the borehole. As a result, yielding occurs only in the upper part of the borehole. This computational result is in accordance with actual observations. The yielding zone is not depleted in this computation. However, if the yielding zone were depleted in the computation region, the process would repeat itself after each depletion of the yielded zone.

References

Aydan, Ö. (1989) The stabilisation of rock engineering structures by rockbolts. Doctorate Thesis, Nagoya University, Faculty of Engineering.

Aydan, Ö. (2001) Modelling and analysis of fully coupled hydro-thermo-diffusion phenomena. In *Proceedings of International Symposium on Clay Science for Engineering*, Balkema, IS-SHIZUOKA. pp. 353–360.

Aydan, Ö. (2003) The moisture migration characteristics of clay-bearing geo-materials and the variations of their physical and mechanical properties with water content. *2nd Asian Conference on Saturated Soils*, UNSAT-ASIA.

Aydan, Ö. (2008). New directions of rock mechanics and rock engineering: Geomechanics and Geoengineering. 5th Asian Rock Mechanics Symposium (ARMS5), Tehran, 3–21.

Aydan, Ö. (2012) The inference of physico-mechanical properties of soft rocks and the evaluation of the effect of water content and weathering on their mechanical properties from needle penetration tests. ARMA 12–639, *46th US Rock Mechanics/Geomech. Symp.*, Paper No. 639, 10p (on CD).

Aydan Ö. (2019) Infrared thermographic imaging in geoengineering and geoscience. In: LaMoreaux J. (eds) *Environmental Geology*, 413–438. Encyclopedia of Sustainability Science and Technology Series. Springer, New York.

Aydan, Ö., Daido, M., Tano, H., Nakama, S. & Matsui, H. (2006) The failure mechanism of around horizontal boreholes excavated in sedimentary rock. *50th US Rock mechanics Symposium*, Paper No. 06–130 (on CD).

Aydan, Ö. & Kawamoto, T. (1992) The stability of slopes and underground openings against flexural toppling and their stabilisation. *Rock Mechanics & Rock Engineering*, 25(3), 143–165.

Aydan, Ö. & Kawamoto, T. (2000) The assessment of mechanical properties of rock masses through RMR rock classification system. GeoEng2000, UW0926, Melbourne.

Aydan, Ö. & Kumsar, H. (2016) A geoengineering evaluation of antique underground rock settlements in Frig (Phryrgian) Valley in the Afyon-Kütahya region of Turkey. *EUROCK2016, Ürgüp*. pp. 853–858.

Aydan, Ö. & Nawrocki, P. (1998) Rate-dependent deformability and strength characteristics of rocks. *International Symposium on the Geotechnics of Hard Soils-Soft Rocks,* Napoli, 1, 403–411.

Aydan, Ö., Sakamoto, A., Yamada, N., Sugiura, K. & Kawamoto, T. (2005a) The characteristics of soft rocks and their effects on the long term stability of abandoned room and pillar lignite mines. Post Mining 2005, Nancy.

Aydan, Ö., Sakamoto, A., Yamada, N., Sugiura, K. & Kawamoto, T., (2005b): A real time monitoring system for the assessment of stability and performance of abandoned room and pillar lignite mines. Post Mining 2005, Nancy.

Aydan, Ö. & Tokashiki, N. (2011) A comparative study on the applicability of analytical stability assessment methods with numerical methods for shallow natural underground openings. *The 13th International Conference of the International Association for Computer Methods and Advances in Geomechanics, Melbourne, Australia*. pp. 964–969.

Aydan, Ö. & Ulusay, R. (2003) Geotechnical and Geoenvironmental characteristics of man-made underground structures in Cappadocia, Turkey. *Engineering Geology*, 69, 245–272.

Aydan, Ö. & Ulusay, R. (2013) Geomechanical evaluation of Derinkuyu antique underground city and its implications in geoengineering. *Rock Mechanics and Rock Engineering, Vienna*, 46, 738–754.

Aydan, Ö. & Ulusay, R. (2016) Rock engineering evaluation of antique rock structures in Cappadocia Region of Turkey. *EUROCK2016, Ürgüp*. pp. 829–834.

Aydan, Ö., Tano, H., Watanabe, H., Ulusay, R. & Tuncay, E. (2007a) A rock mechanics evaluation of antique and modern rock structures in Cappadocia Region of Turkey. *Symposium on the Geology of Cappadocia, Nigde*. pp. 13–23.

Aydan, Ö., Tano, H. & Geniş, M. (2007b) Assessment of long-term stability of an abandoned room and pillar underground lignite mine. *Rock Mechanics Journal of Turkey*, (16), 1–22.

Aydan, Ö., Ulusay, R., Tano, H. & Yüzer, E. (2008a). Studies on Derinkuyu underground city and its implications in geo-engineering. *First Collaborative Symposium of Turkey-Japan Civil Engineers, 5 June, 2008, İTÜ, İstanbul*, Proceedings. pp. 75–92.

Aydan, Ö., Tano, H., Ulusay, R. & Jeong, G.C. (2008b). Deterioration of historical structures in Cappadocia (Turkey) and in Thebes (Egypt) in soft rocks and possible remedial measures. *Proceedings of the International Symposium of Conservation Science for Cultural Heritage 2008*, National Research Institute of Cultural Heritage, Korea. pp. 55–65.

Aydan, Ö., Ulusay, R., Tano, H. & Yüzer, E. (2008c) Studies on Derinkuyu underground city and its implications in geo-engineering. *Procs. of the First Collab. Symp. of Turkish-Japan Civil Engineers, İTÜ, İstanbul*. pp. 75–92.

Aydan, Ö., Tano, H., Geniş, M., Sakamoto, I. & Hamada, M. (2008d) Environmental and rock mechanics investigations for the restoration of the tomb of Amenophis III. *Japan-Egypt Joint Symposium New Horizons in Geotechnical and Geoenvironmental Engineering, Tanta, Egypt*. pp. 151–162.

Aydan, Ö., Ulusay, R. & Tokashiki, N. (2014) A new Rock Mass Quality Rating System: Rock Mass Quality Rating (RMQR) and its application to the estimation of geomechanical characteristics of rock masses. *Rock Mechanics & Rock Engineering*, 47, 1255–1276.

Büdel, B., Weber, B., Kühl, M., Pfanz, H., Sültemeyer, D. & Wessels, D. (2004) Reshaping of sandstone surfaces by cryptoendolithic cyanobacteria: bioalkalization causes chemical weathering in arid landscapes. *Geobiology*, 2, 261–268.

Ehrlich, H.L. & Newman, D.K. (2009) *Geomicrobiology*. CRC Press, London.

Hall, K, Guglielmin, M. & Strini, A. (2008) Weathering of granite in Antarctica: I. Light penetration into rock and implications for rock weathering and endolithic communities. *Earth Surface Processes and Landforms*, 33, 295–307.

Hoppert, M., Flies, C., Pohl, W., Günzl, B. & Schneider, J. (2004) Colonization strategies of lithobiotic microorganisms on carbonate rocks. *Environmental Geology*, 46, 421–428.

Ishijima, Y. & Fujii, Y. (1997) A study on the mechanism of slope failure at Toyohama tunnel, Feb. 10, 1996. *International Journal of Rock Mechanics and Mining Sciences*, 34(3–4), 87.e1–87.e12.

Ito, T., Aydan, Ö., Ulusay, R. & Kasmer, Ö. (2008) Creep characteristics of tuff in the vicinity of Zelve antique settlement in Cappadocia Region of Turkey. *Proceedings of the International Symposium 2008; 5th Asian Rock Mechanics Symposium, Tehran, 24–26 November 2008, Namaye Penhan*. pp. 337–344.

Kano, K., Doi, T., Daido, M. & Aydan, Ö. (2004) The development of electrical resistivity technique for real-time monitoring and measuring water-migration and its characteristics of soft rocks. *Proceedings of 4th Asia Rock Mechanics Symposium, Kyoto*. pp. 851–854.

Kasmer, Ö., Ulusay, R. & Aydan, Ö. (2008) Preliminary assessments of the factors affecting the stability of antique underground openings at the Zelve Open Air Museum (Cappadocia, Turkey). *2nd European Conf. of Int. Assoc. for Engineering Geology, 15–19 September 2008, Madrid*.

Kawamoto, T., Aydan, Ö. & Tsuchiyama, S. (1991) A consideration on the local instability of large underground openings. *Int. Conf., GEOMECHANICS'91, Hradec*. pp. 33–41.

Konhauser, K.O. (2007) *Introduction to Geomicrobioplogy*. Wiley-Blackwell, Oxford, pp. 192–234.

Matsubara, H. & Aydan, Ö. (2016) The effect of biological degradation of tuffs of Cappadocia, Turkey. *EUROCK2016, Ürgüp*. pp. 871–876.

Tokashiki, N. & Aydan, Ö., 2010. The stability assessment of overhanging Ryukyu limestone cliffs with an emphasis on the evaluation of tensile strength of rock mass. *Journal of Geotechnical Engineering JSCE*, 66(2), 397–406.

Ulusay, R., Akagi, I., Ito, T., Seiki, T, Yüzer, E. & Aydan, Ö. (1999) Long term mechanical characteristics of Cappadocia tuff. *Proceedings of the 9th International Congress on Rock Mechanics*, Paris. pp. 687–690.

Ulusay, R., Aydan, Ö., Geniş, M. & Tano, H. (2013) Assessment of stability conditions of an underground congress centre in soft tuffs through an integrated rock engineering methods (Cappadocia, Turkey). *Rock Mechanics and Rock Engineering*, 46(6), 1303–1321.

Chapter 9

Monitoring of rock engineering structures

9.1 Deformation measurements

The monitoring of ground movements resulting from excavation and from creep-degradation of surrounding rock in rock engineering structures may be required for stability assessments and environmental safety.

The monitoring of ground movements may be accomplished by using direct and indirect techniques. Direct techniques may involve inclinometers, extensometers of the mechanical or fiberoptic type, geodetic measurements or GPS technique. Indirect methods may involve aerial photogrammetric methods or the InSAR method.

The most important aspect in monitoring is the long-term reliability and repeatability of measurements because the measurements must be carried out over a long period of time, i.e. years. When ground movements are accelerated, the measurement intervals may be required to be shortened to time units such as hours or minutes, and the combination of several techniques may be necessary. A brief outline of the major techniques is presented in the following paragraphs.

9.1.1 Direct measuring techniques

(a) Inclinometers

An inclinometer basically measures the deviation of inclination of each segment of the casing (Figure 9.1). The ground movements are converted to displacements on the basis of measured deviations and the location of each casing segment. The device consists of inclinometer casing, traversing probe and readout unit. The casing is permanently installed in a borehole. Important features of casing include the diameter of the casing, the coupling mechanism, groove dimensions and straightness, and the strength of the casing.

The traversing inclinometer probe is the standard device for surveying the casing. Recently, some of the traversing probe is equipped with fiber-optic sensors. The traversing probe obtains a complete profile because it is drawn from the bottom to the top of the casing. The first survey establishes the initial profile of the casing. Subsequent surveys reveal changes in the profile of the casing, if movement has occurred.

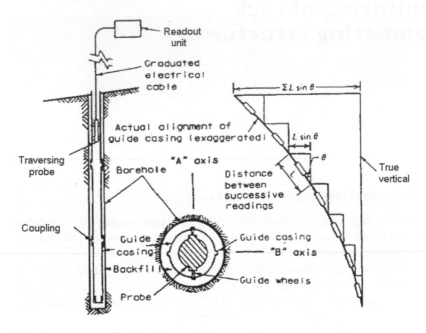

Figure 9.1 Illustration of ground movement measurement by an inclinometer

(b) Extensometers

An extensometer monitors changes in the distance between one or more anchors fixed to the ground and a reference head at the borehole collar (Figure 9.2). Components of an extensometer include anchors, rods or wires with protective tubing, and a reference head. The anchors, with wires or rods attached, are installed in a borehole. Borehole extensometers have anchored benchmarks.

A borehole is drilled to a depth at which the strata are stable. It is then lined with a steel casing with slip-joints to prevent crumpling as subsidence occurs. An inner pipe rests on a concrete plug at the bottom of the borehole and extends to the top. This inner pipe then transfers the stable elevation below to the surface. The wires/rods span the distance between the anchors and the reference head, which is installed at the borehole collar.

Measurements are obtained at the reference head with a sensor or a micrometer, either of which measures the distance between the top (near) end of the anchor and a reference surface. A change in the distance with respect to initial measurement indicates that movement has occurred. Movement may be referenced to a borehole anchor that is installed in stable ground or to the reference head, which can be surveyed. In recent years, fiberoptic extensometers have been developed and applied in the practical monitoring of ground movements.

(c) Optical leveling technique

Ground movements may be systematically measured using optical levelling devices (Figure 9.3). For this purpose, a network of levelling points is established, and measurements are performed at certain time intervals. The repeatability of the measurements obtained from

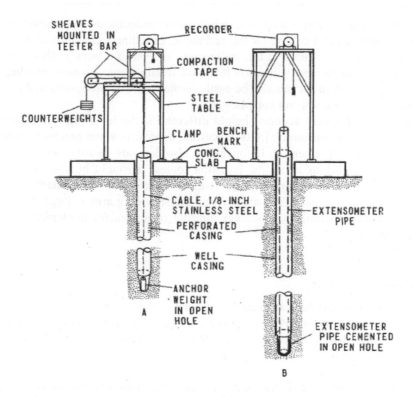

Figure 9.2 Illustration of ground movement measurement by extensometers

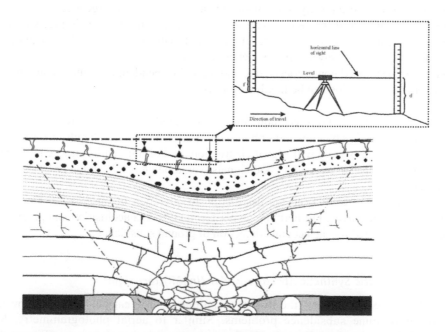

Figure 9.3 Illustration of ground movement measurement by levelling technique

different surveys and different loops and stability of the monuments with respect to its local environment are the main criteria for this technique. Levelling technique can provide very precise height differences up to a few millimetres in level of accuracy. This technique is relatively flexibly implemented in areas that have usually dense housing, building and/or vegetation. The benchmarks can also be easily located. Its data processing and analysis of the measured results are also not complicated.

Although it yields very accurate height differences, the levelling technique is relatively slow and time-consuming in its execution, especially when precise levelling procedures are being implemented. Its operation also depends on time, weather and also environmental conditions along the levelling routes. It should be also noted that the monitored points should generally be associated with a certain benchmark located on a stable zone outside the subsiding area. When the subsiding area is large, then connection to the stable benchmarks is another limiting constraint for implementation of the levelling technique.

GPS LEVELING TECHNIQUE

GPS technique has also been applied to subsidence measurements in recent years (Figure 9.4). First, a network of GPS stations is established. The GPS surveys are carried out at stations using dual-frequency geodetic-type GPS receivers. There are several advantages of using GPS technique:

- GPS provides the three-dimensional displacement vector with two horizontal and one vertical components so that the subsidence is obtained three-dimensionally.
- GPS provides the displacement vectors in a unique coordinate reference system, so that it can be used to effectively monitor subsidence in a relatively large area.
- GPS can yield the displacement vectors with a several millimetres in precision level, which is relatively consistent in the temporal and spatial domains, so that it can be used to detect even relatively small subsidence signals.
- GPS can be utilized continuously, day and night, independent of weather conditions, so that its field operation can be flexibly optimized.

However, surveys by the GPS technique may show slightly worse standard deviations due to the signal obstruction by trees and/or buildings around the station. Therefore, the obtained precision level of the survey with signal obstructions and/or multipaths may be in the order of 1–3 cm, while the surveys with good signals would be generally in the order of several millimeters.

The problem may result from the destruction or alteration of observation monuments inside the urbanized areas. Furthermore, expertise in GPS data acquisition and precise data processing is required for the accurate detection of ground subsidence.

(d) Interferometric Synthetic Aperture Radar (InSAR) technique

Although it is relatively new, the InSAR technique has a great potential for long-term monitoring of mine subsidence problems. Similar to aerial photogrammetry, SAR images of the same area taken at certain time intervals are necessary (Figure 9.5).

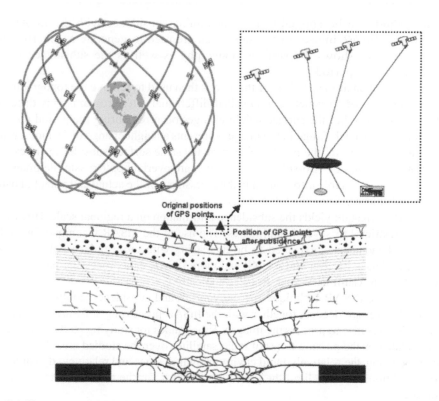

Figure 9.4 Illustration of ground movement measurement by GPS technique

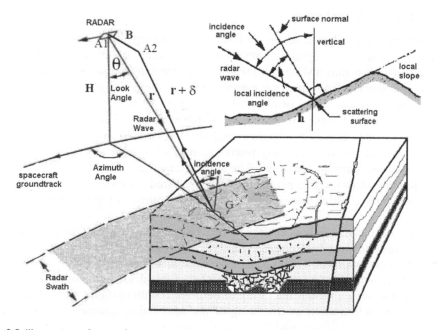

Figure 9.5 Illustration of ground movement measurement by InSAR technique

Several methods can be applied to monitor mining subsidence with the InSAR technique. Several InSAR images of the same area are used to generate an interferogram. This interferogram shows the surface topography as well as the subsidence that has occurred during the period.

A digital elevation model (DEM) of the area is then used to remove the topography from the interferogram and to generate a so-called differential interferogram. This differential interferogram shows that the surface height changes occurred during the period due to the mining subsidence. Each fringe in the image represents a height change of 28 mm. From the differential interferogram, the points affected by the same surface subsidence are extracted and contoured as lines of "iso-subsidence," which are overlaid onto a satellite image. This combination gives valuable information in the assessment of the actual and potential damage caused by subsidence.

The InSAR technique yields the subsidence information on a regional scale. However, its accuracy is restricted to several centimetres. The use of the InSAR technique for studying subsidence phenomena is expected to be increasing with more radar satellites in space (e.g. ERS, Radarsat, Envisat and ALOS) and more InSAR data processing packages available (e.g. Atlantis, Gamma, Vexcel, Roi-Pac and Doris).

In order to use INSAR technique for studying ground subsidence, multitemporal radar images of the area are needed, together with the InSAR processing software and hardware, as well as the expertise to process the images. Time frames for studying land subsidence will also be dictated by the passing times of radar satellites over the studied area. In the context of data processing, the relatively rapid environmental changes and relatively dynamic atmospheric conditions can also limit the potential of InSAR for the accurate detection of land subsidence.

(e) Laser scanning technique

The basic principle is based on the emission of a light signal (Laser) by a transmitter and receiving the return signal by a receiver. The scanner uses different techniques for distance calculation that distinguish the type of instrument in the receiving phase. The distance is computed from the time elapsed between the emission of the laser and the reception of the return signal or phase shift based on when the computation is carried out by comparing the phases of the output and return signals. The laser scanner devices operate by rotating a pulsed laser light at high speed and measuring reflected pulses with a sensor. The scanner automatically rotates around its vertical axis, and an oscillating mirror moves the beam up and down. The scanner calculates the distance of a measured point together with its angular parameters. The measured points constitute a set of points called cloud points, which are used to quantify the geometry of the structure or surface in 3-D.

As an application of this concept, the authors have tried to evaluate the performance of a tunnel in the Okinawa Prefecture. Figure 9.6 shows a digital image of the tunnel during the construction phase. This type of evaluation would provide a quick evaluation of the state of the tunnel and possible locations where some degradation of support systems may occur, and some unusual fracturing or deformed configurations of the liners resulting from large deformation or fracturing of the surrounding ground may be assessed.

Figure 9.6 Laser scanner and a digital image of a tunnel under construction obtained from laser scanning

Furthermore, this technique could be also utilized for the maintenance and long-term deformation monitoring of rock engineering structures such as tunnels, slopes, underground powerhouses.

9.2   Acoustic emission techniques

Acoustic emission (AE) signals are generated by the sudden release of elastically stored energy during rock mass fracturing caused by excavation, hydrofracturing, slippage and other engineering operations. The piezoelectric sensors receive the vibrations and transform them into electrical signals. The AE systems consist of sensors, data transmission cable, data acquisition system, and data processing unit. The data transmission may be through cable/wires, optical fiber, or wireless. The power source may be an ordinary electric power supply or battery-operated power. The AE signals can be stored directly on the system, and they may be used to evaluate the location and type of event. However, this type of monitoring requires extensive data storage. On the other hand, if the system is based on pulse counting, the system can be quite compact. Tano *et al.* (2005) developed a very compact acoustic emission system consisting of AE sensor, amplifier and pulse-counting logger. The system operates using batteries (Figure 9.7).

The protection of instrumentation devices during a blasting operation is extremely difficult. In previous studies, the devices were installed in larger holes at the sidewall and covered by some protection sheaths (Aydan *et al.*, 2005a). In this study, the instruments were put in an aluminum box attached to a rock bolt head and protected by a semicircular steel cover fixed to the tunnel surface by bolts. Figure 9.8 illustrates the installation of the instruments in close proximity to the tunnel face. The instruments installed at the crown were about 1.5 m from the tunnel face while the instruments installed at the mid-height of the sidewall were about 2.5 m from the tunnel face during 1–3 September 2015 and 2 m during 5–7 October 2015 monitoring. AE counts were recorded for 1 s intervals, and the recording was started and stopped by a remotely operated switch device.

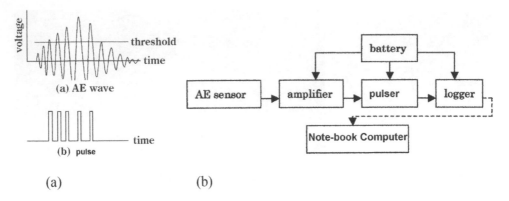

Figure 9.7 (a) Principle of AE counting, (b) block diagram of AE counting system

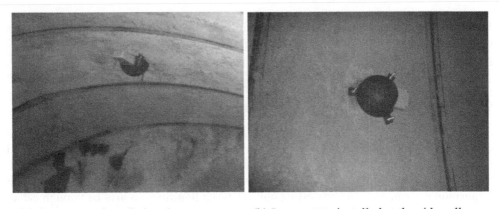

(a) Instruments installed at the crown   (b) Instruments installed at the sidewall

Figure 9.8 Views of the instruments installed at the crown and sidewall of the tunnel

9.3   Multiparameter monitoring

The mechanical properties of soft sedimentary rocks are influenced by the variation of water content. The compressive strength and elastic modulus of soft rocks generally decrease with the increase of water content, or vice versa. The effect of degradation of rock by the cycles of wetting and drying cause the flaking of rock near the surface and it falls under the effect of gravity. This process repeats itself endlessly, which subsequently causes the reduction of the supporting area of the pillars in the long term. Therefore, it is of first interest to monitor the environmental parameters such as temperature, humidity, air pressure within the abandoned mines and water content, temperature variations in surrounding rocks in association with the degradation process.

It is also known that, when rocks subjected to high stresses start to fail, the stored mechanical energy in rock tends to transform itself into different forms of energy according to the energy conservation law. Some of these transformations involve the variations of electrical potential, magnetic field, heat release and kinetic energy in the surrounding rock mass. Recent experimental studies on various types of rocks by Aydan and his co-workers (Aydan *et al.*, 2001, 2002, 2003), including some from abandoned mines indicated distinct variations of multiparameters previously mentioned during deformation and fracturing processes. Therefore, they may be used in the assessment of the short-term stability of abandoned mines.

The real-time multiparameter measurement system involves electric potential (EP) variations, acoustic emissions (AE), temperature and water content of rocks (RT), temperature, humidity and air pressure within the abandoned lignite mines above the groundwater table and outside (Aydan *et al.*, 2005b) (Figure 9.9). The rocks around abandoned mines above the groundwater level are much more prone to degradation due to environmental variations in time (Aydan *et al.*, 2005a). The temperature, humidity and air pressure sensors are installed at a certain spacing from the mine entrance to monitor the spatial distribution and variations with time. Temperature and water content sensors are also installed at several depths in some selected pillars and monitored with time.

Geoelectric potential monitoring devices and electrodes are set up within the abandoned mine. It is generally desirable to install the devices near geological discontinuities such as faults and fracture zones, which are much more sensitive to stress changes within the surrounding rock mass. When rock mass is damp or saturated, the electrical resistance of

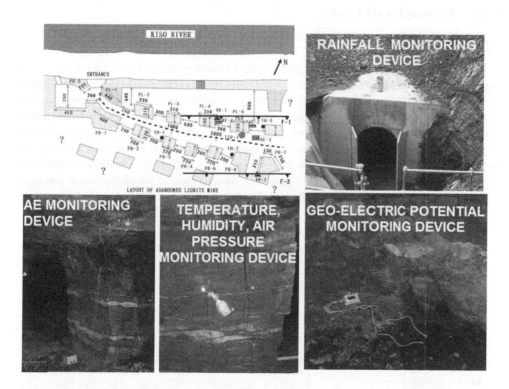

Figure 9.9 Application of multiparameter system for monitoring degradation and long-term response of surrounding rock mass at an abandoned lignite mine

Source: *From Aydan et al. (2005b)*

ground is in the order of kiloOhms. Therefore, it would be sufficient to use devices, whose impedance is in the order of megaohms. Otherwise, it would be necessary to use devices having impedance in the order of gigaohms. Therefore, the measured electric potential variations by the devices are directly related to those of the surrounding rock mass. The amplitude and orientation of geoelectric potential variations are used to infer the likely location and magnitude of sources of instability.

The AE system, limited only to counting AE events (Tano *et al.*, 2005; Aydan *et al.*, 2005b), has proved to be useful for the long-term monitoring of rock fracturing in abandoned lignite mines. Pulse signals, which correspond to AE waves exceeding a threshold, are discriminated through a pulsar and recorded onto a pulse counter (logger) as AE rate counts. Such a limited specification of the rate counting reduces the system cost so two AE systems can be used as one set. One of the AE systems is called an active unit, while the other one is called a dummy unit. The active unit is directly attached to the rock burst, while the dummy AE sensor is not in contact with the rock burst. If signals are counted on both systems, the count of the active unit is deleted from the measured data. This active-dummy counting system increases the reliability of the AE monitoring and checks the noise condition in field.

9.4   Applications of monitoring system

9.4.1   Amenophis III tomb

(a)   Instrumentation

In the tomb of Amenophis III, the wall between the J-chamber and the Jd-chamber is in a critical condition concerning the overall stability of the tomb in the vicinity of J-chamber and adjacent chambers such as Je, Jc and Jd (Aydan *et al.*, 2008). Acoustic emission monitoring units consisting of acoustic emission sensors, together with their amplifiers and loggers, were installed at the wall between J-chamber and Jd-chamber, at the pillar of Je-chamber and at pillar 3 of J-chamber (Figure 9.10). In addition, a displacement gap gauge was installed at the same wall together with amplifier and logger units. Figure 9.11 shows some views of the installation locations of acoustic emission measurement units. Figure 9.12 shows a view of the displacement gap gauge installed at the crack crossing the north wall of J-chamber adjacent to Jd-chamber.

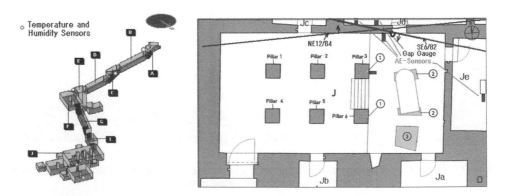

Figure 9.10 Locations of climatic and AE sensors and displacement gap gauge

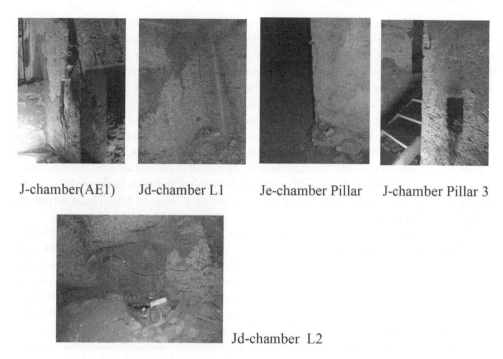

J-chamber(AE1)   Jd-chamber L1   Je-chamber Pillar   J-chamber Pillar 3

Jd-chamber L2

Figure 9.11 Views of installation locations of acoustic emission sensors

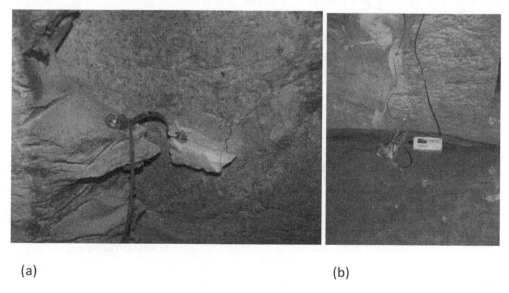

(a)   (b)

Figure 9.12 Views of (a) displacement gap gauge, (b) logger unit

(b)   Temperature and humidity measurements

Temperature and humidity variations from the entrance down to the Jd-chamber were monitored at an interval of 1 h (Figure 9.13). The results of measurements were downloaded on 16 February 16, 2004. Out of six instruments, we were able to download data from the four instruments installed at the entrance, F, Jd and Je chambers. Nevertheless, the results are sufficient to have a general idea of the temperature and humidity measurements for almost a period of one year throughout the tomb. Figure 9.10 shows the variations of temperature and humidity between 20 March 2003 and 16 February 2004. As noted from the figure, temperature variations of Chambers Je and Jd remain almost constant throughout the measurement period, and its value is about 28°C, except for the period of human activity in the tomb. The human activity period is associated with the cleaning of the wall art of the tomb, and it roughly starts in November and ends in the middle of May. During these periods, the temperatures fluctuate within a range of 5°C. These periods are easily noticed in the figure.

The mean temperature of Chamber F is about 1°C below that of Chambers Je and Jd. Since the location of Chamber F has a higher elevation and is directly connected to the entrance, the fluctuation of the temperature of this chamber is larger. The fluctuation range is about 7°C during the human activity period, and thereafter it remains almost constant. The largest temperature variations were observed at the entrance of the tomb. The highest temperature is about 40°C in June (note that the entrance of the tomb is situated at the northern side of the west valley), and the lowest temperature is about 10°C during the period between December and February. Furthermore, the daily fluctuation is about 10–12°C during the human activity period, and it is about 5–7°C when there is no human activity in the tomb.

The mean humidity in Chambers Je and Jd is about 20–22% when there is no human activiy in the tomb. However, the daily fluctuations become very large and range within 10–40%. The humidity of Chamber F is higher as it is directly connected to the entrance. The humidity variation at the entrance of the tomb ranges between 10% and 65% during the period of measurement. During summer, the humidity is lower, and it becomes larger during

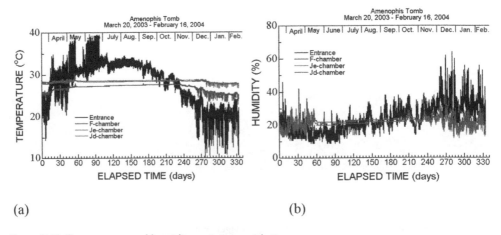

(a)   (b)

Figure 9.13 Temperature and humidity variations with time

the winter period. In other words, June is driest month of the year, while December is the most humid month of the year.

(c) Monitoring results

Acoustic emissions observations were done at four locations. Since the human activities within the tomb may cause some acoustic emission events, the acoustic emission counts occurring during the daytime are not taken into account and special emphasis was given to those occurring after work hours (from 19 p.m. till 6 a.m. of the next day). If this criterion is applied to the results of measurements, there was almost no acoustic emission activity for AE instruments numbered AE1, AE2 and AE4 while AE instrument numbered AE3 showed a distinct acoustic emission activity during the period of measurements as shown in Figure 9.14(a). After a relatively calm period between April and May, acoustic emission activity started in June and continued in a linear fashion. After a calm period in August, a very sharp acoustic emission activity started at the beginning of October. Another sharp acoustic emission activity was observed at the beginning of December when the cleaning operation in the tomb started. These acoustic emission activities clearly indicate that localized cracking has been taking place in the wall between Chambers J and Jd.

The data from the displacement gap gauge installed at the relatively continuous open fracture on the north wall of Chamber J is available only for the period between 15 May and 3 November 2003. Figure 9.14(b) shows the response of the displacement gap gauge during the aforementioned period. The maximum range of the displacement is within 0.05 mm. Soon after the installation of the gap gauge, the displacement of the fracture has the mode of opening. It seems that the opening displacement of the fracture resembles the temperature variations at the entrance of the tomb even though the temperature of the chamber in which the gauge is installed remains almost constant. Although it is difficult to make a quantitative statement about the effect of the outside temperature, it would be natural to expect that the deformation behavior of the tomb would be influenced by the atmospheric temperature variations and crustal straining due to the motion of the Earth. However, a sudden variation in the displacement response occurred in 11 October 2003. After a certain closure, the crack started to open up. At this moment, it is very difficult

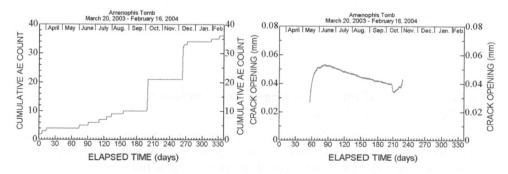

Figure 9.14 (a) Acoustic emission response of the sensor AE3 with time, (b) displacement response of the fracture with time

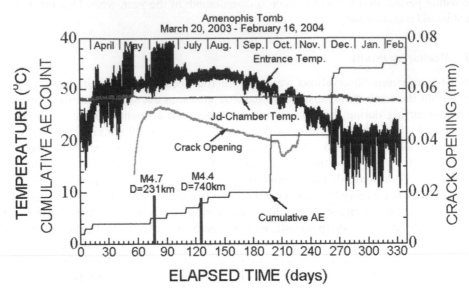

Figure 9.15 Comparison of parameters measured by *in-situ* monitoring systems

to make further comment. Nevertheless, the sudden changes of displacement variation should be associated with some new crack occurrence in the north wall of the main chamber.

All relevant data is replotted in Figure 9.15 in order to discuss the implications of the measurement results from *in-situ* monitoring systems. It seems that the acoustic emission activity started at the end of May 2003 has some relation to the seismic activity along the Nile river to the north of the tomb. If the outside temperature increases, the whole hill should expand. Consequently the inward closure of the tomb should occur, which subsequently causes the opening of the fracture. On the other hand, if the outside temperature decreases, the reverse response should occur, provided that the surrounding rock behaves linear elastically. The sudden increase of AE activity and subsequent variation of displacement response simply implies that some new fracturing took place, and this further led to the opening of the fracture. In other words, there is a high possibility of further propagation of the existing fracture in the north wall of the main chamber of the tomb.

9.4.2 Application of laser surveying to an abandoned lignite mine in Mitake

The laser surveying technique has been greatly improved in recent years. This method has been now applied to various rock engineering projects involving slopes, underground excavations and mining-related structures, although it was quite time-consuming to do the laser surveying at initial stages. The author utilized this technique in an abandoned mine,

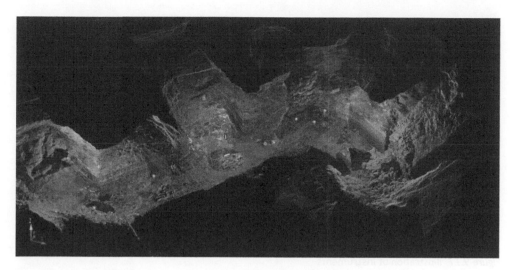

Figure 9.16 Laser surveying result of an abandoned mine at Mitake town

exploiting the room and pillar technique at Mitake town in Japan. Figure 9.16 shows one example of the laser surveying used in the abandoned mine with highly complex geometry of within 20 minutes. The processed results can be easily visualized immediately after surveying. This method may be quite useful in surveying various structures. In addition, it may be used for monitoring the response of rock engineering structures in the short and long term.

9.4.3 Application of acoustic emission measurements at underground tomb

A monitoring program was undertaken to check that there was no negative effect on the nearby underground tomb of the construction of piles. The monitoring involved a multiparameter monitoring system consisting of acoustic emission (AE) sensors installed at four locations, displacement measurements between layers L2–L3, inclination at the top of the excavation and at the tomb entrance; vibration due to machinery or earthquakes was monitored as illustrated in Figure 9.17. Some of results are presented in this section.

Figure 9.18 shows an example of acoustic emissions recorded at five AE sensors. The accelerometer at the ground surface was also used to check the vibrations caused by the machinery operations. In addition, major construction operations and their timing were also recorded. Most of the acoustic emissions were due to construction activities during the daytime. However, some special attention was paid to the acoustic emissions after 9:00 p.m. and before 8:00 a.m. during a typical workday. The results indicated that there were no acoustic emissions 9:00 p.m. and before 8:00 a.m and that they were due to the construction-induced vibrations during the working hours.

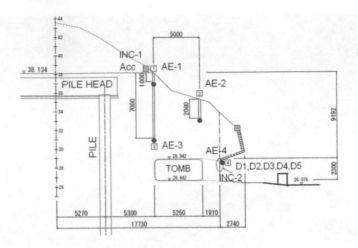

Figure 9.17 Illustration of instruments

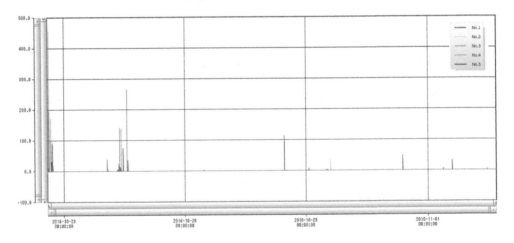

Figure 9.18 AE counts observed during 21 October–2 December 2016

9.4.4 Applications of climatic monitoring in Taru-Toge Tunnel

Three locations, which were about 1.5–2.5, 31 and 76 m from the tunnel face, were chosen for temperature, humidity and air pressure variations (Figures 9.19 and 9.20). The air pressure measurements can be also used for investigating the blasting pressure wave propagation in the tunnel besides identifying the exact blasting times.

Blasting operations also cause carbon dioxide (CO_2) emissions. A CO_2 monitoring device produced by T&D was used for this purpose and set just in front of the mobile ventilation equipment. The CO_2 measurement was done during 5–7 October 2015 monitoring (Figure 9.20).

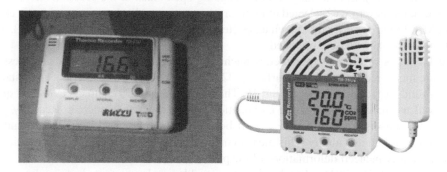

Figure 9.19 Temperature, humidity and air pressure and CO_2 devices

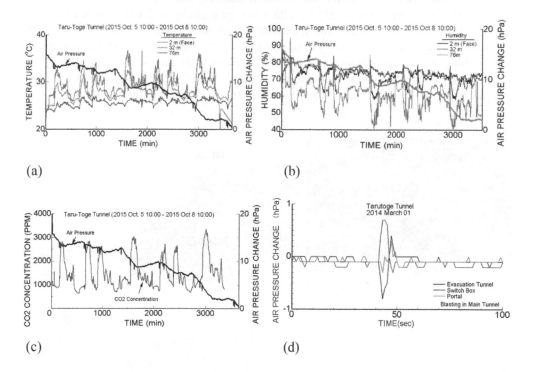

(a)

(b)

(c)

(d)

Figure 9.20 Measurement results

9.4.5 Applications of multiparameter monitoring system in underground powerhouse

The multiparameter system fundamentally covers all measurable quantities such as displacement, acoustic emissions, electric potential or electrical resistivity, water level changes, climatic parameters such as temperature, humidity and CO_2 and temperature changes of rock

(Aydan *et al.*, 2019). In this study, a thoroughgoing open crack at the access tunnel next to the powerhouse cavern was selected to monitor its movement (displacement), acoustic emissions (AE), together with climatic changes (temperature and humidity) (Figure 9.21). The unit is fundamentally battery operated.

The locations of climatic parameters such as temperature, humidity and air pressure (THP) are measured at the ground surface (132 m elevation). Two CO_2 sensors are installed at the ground surface and Underground 1F. The monitoring of CO_2 is to check the air quality as well as the condition for the carbonation environment for concrete. Recently, two temperature sensors are installed in a short borehole to monitor rock temperature and air temperature around its vicinity. This measurement is expected to yield some information of cyclic thermally induced deformations of the powerhouse.

Aydan *et al.* (2016a, 2016b) has developed a portable accelerometer, which can be used in four different modes. For strong motion observations during earthquakes, every accelerometer should have a triggering level to start and stop recording for a given time interval and sampling rate and to store in a digital format. The minimum sampling rate is 1 Hz. The device is called QV3-OAM-XXX, and it has the ability with a storage capacity of 2 GB. The power of the accelerometer can be an internal battery, external battery, solar energy or ordinary 100–240 V electricity. In the case of solar energy or ordinary electricity, the power is stored in an external battery through an adapter from a power supply such as solar panels or electric outlet. The system adopted at the powerhouse is designed to utilize the ordinary electricity (100 V) or two external batteries. Currently, the system utilizes electricity available in the powerhouse.

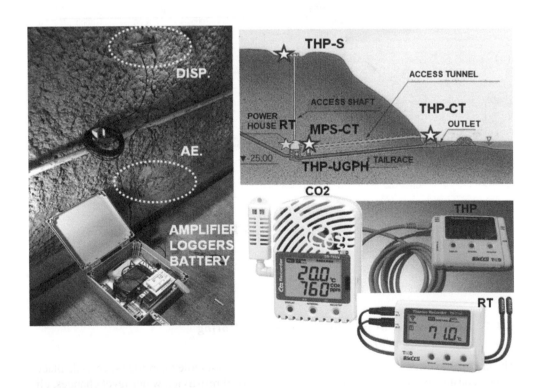

Figure 9.21 Views and locations of installed multiparameter monitoring system

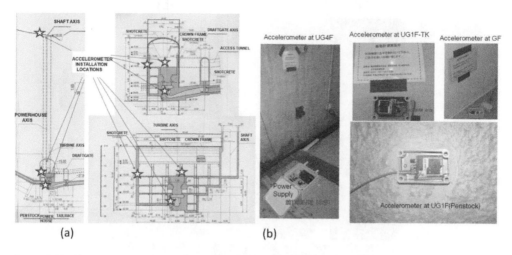

Figure 9.22 (a) Installation locations of accelerometers, (b) views of installed accelerometers

Four accelerometers were installed at the power station site. Three accelerometers were installed in the powerhouse as illustrated in Figure 9.22. One accelerometer is at the Underground 4F, which is at the bottom of the powerhouse, two accelerometers at Underground 1F, which is the midlevel of the powerhouse. One of the accelerometers is fixed to the sidewall at the penstock side, and the other accelerometer is fixed to the middle of the end wall of the cavern. The fourth accelerometer is installed at the surface. Figure 9.22 shows some views of the installed accelerometers. The main purposes of the installation is to observe the seismic response of the cavern during earthquakes and to evaluate the ground motion amplifications as pointed out by previous pioneering researchers (Nasu, 1931; Komada and Hayashi, 1980).

9.4.6 Applications of multiparameter monitoring system at Nakagusuku Castle

A multiparameter monitoring system was also initiated by the authors at Nakagusuku Castle. The system at the castle was actually installed about three years before the one installed at Katsuren Castle, which is probably the first attempt regarding masonry structures in the world. The monitoring was initiated in December 2013, and it is ongoing. During the period of measurements, some earthquakes occurred and a long-term, creep-like separation of a huge crack in Ryukyu Limestone layer extending to the Shimajiri formation layer has been taking place. Figure 9.23 shows the installation location.

An earthquake with a moment magnitude of 6.5 occurred at 5:10 a.m. on 13 March 2014 (JST) in the East China Sea at a depth of 120 km on the western side of Okinawa island. Another earthquake occurred at 11:27 a.m. on the same day near Kumejima island. Although the magnitude of the earthquake was intermediate and far from the location, some permanent displacement occurred, as seen in Figure 9.24.

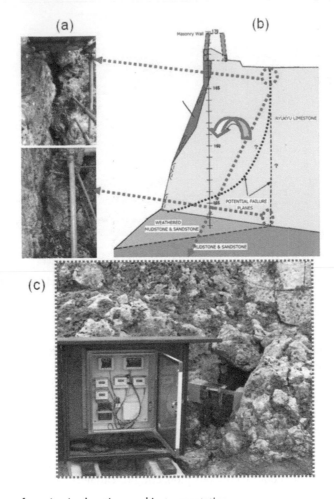

Figure 9.23 Views of monitoring locations and instrumentation

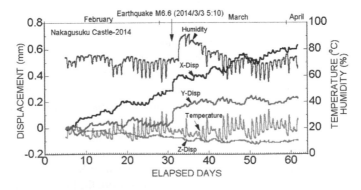

Figure 9.24 Monitoring results during February to March 2014

9.5   Principles and applications of drone technology

9.5.1   Drones

Drones are essentially unmanned aerial vehicles (UAV) that are equipped with high-quality cameras, which can take photos at exact intervals, and gyroscopes, an inertial measurement unit (IMU) and controllers to fly smoothly. For a drone to fly correctly, the inertial measurement unit (IMU), gyro stabilization and flight controller technology are all essential. Drones generally use three- and six-axis gyro stabilization technology to provide navigational information to the flight controller. An inertial measurement unit detects the current rate of acceleration using one or more accelerometers. A magnetometer may be used to assist the IMU on drones against orientation drift.

Drones may be equipped with a number of sensors such as distance sensors (ultrasonic, laser, Lidar) or chemical sensors for digital mapping or other purposes. As Lidar, which is an acronym for laser interferometry detection and ranging, can penetrate forest canopy and are widely used for topographical mapping.

Aerial photogrammetry is used in topographical mapping using digital or digitized aerial photos of area with known control points. Aerial photographs were taken from a camera mounted on the bottom of an airplane and later were digitized. These days, digital photographs are used together with a record of height and position using GPS and/or other positioning sensors. The plane flies over the area to take overlapping photographs (generally 60% overlapping) over the entire area of interest (Figure 9.25). When it is used for mapping and measuring the displacements of structures following the earthquakes, three-dimensional coordinates of the common points on pre- and post-earthquake photographs were determined. Hamada and Wakamatsu (1998) used this technique to determine the liquefaction induced displacements. This technique is now utilized together with images from drones. However, the fundamental principles remain the same.

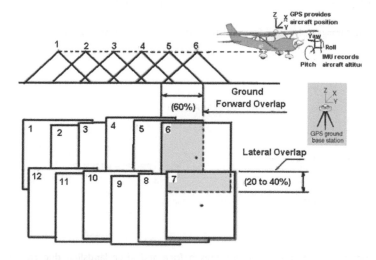

Figure 9.25 Illustration of concept of aerial photogrammetry

9.5.2   Applications to slope and cliffs

*(a)   Slope stability problems and landslide caused by
the 2016 Kumamoto earthquakes*

Kumamoto Prefecture suffered by two successive earthquakes occurring on 14 and 16 April 2016 (Aydan *et al.*, 2018a). These two earthquakes were associated with well-known faults in the region. While the first earthquake on 14 April had a moment magnitude of 6.1–6.2 (Mj 6.5), the strong motions at Mashiki town were more than 1500 cm s^{-2}. The second earthquake with a moment magnitude of 7.0 (Mj 7.3) occurred on April 16, resulting in surface ruptures due to faulting and induced strong motions over a large area. The second earthquake was particularly destructive and caused widespread damage to rock engineering structures, including the built environment. The causes of the damage were high ground motions and permanent straining, which is one of the well-known characteristics of intraplate earthquakes associated with surface faulting.

The drone was used the first time to estimate the geometry of the landslide body and the volume of landslide body. Figure 9.26 shows the digital image of the landslide, while Figure 9.27 compares the longitudinal profiles of the landslide area before and after the earthquake. As noted from the figures, it is very easy to evaluate the geometry of the slope failures for evaluating the landslide body.

(a) Front view   (b) Top view

Figure 9.26 Digital images obtained from drones using the aerial photogrammetry technique

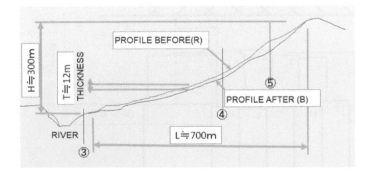

Figure 9.27 Comparison of longitudinal profiles before and after landslide due to the Kumamoto earthquake

(b)   Application of drone surveying to cliffs and steep slopes

The investigation of the possibility of the failures of slopes and cliffs or the back-analysis of the failed cliffs and slopes requires the exact geometry of the topography. Figure 9.28 shows the applications in the shore of the Gushikawa in Itoman City in the south of Okinawa island and the shore at the southern part of the Miyako island. As noted from the figures, it is quite easy to evaluate digitally the geometry of slopes and cliffs. In particular, the evaluation of cliffs is quite cumbersome due to the overhanging rock mass with toe erosion.

9.5.3 Applications to sinkholes

The evaluation of the geometry of sinkholes is an extremely difficult and dangerous task due to the unstable configuration and unseen cracks. A sinkhole recently occurred in a Ryukyu limestone quarry in Kumejima island during the quarrying. The excavator fell into the sinkhole together with its operator. Luckily no one was hurt. The evaluation of the size and geometry of the sinkhole was necessary. The drone utilizing the aerial photogrammetry technique was applied at this site, and the results are shown in Figures 9.29 and 9.30. It is interesting to note that the overhanging part of the sinkholes can be accurately evaluated.

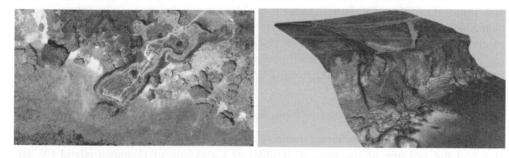

Figure 9.28 (a) Digital image of the cliffs in the vicinity of Gushikawa Castle remains in Itoman City, (b) digital image of the cliffs in the southern shore of Miyako island

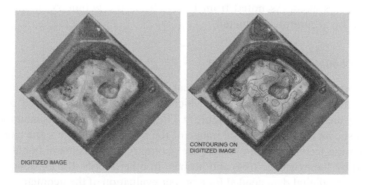

Figure 9.29 Evaluation of the sinkhole geometry

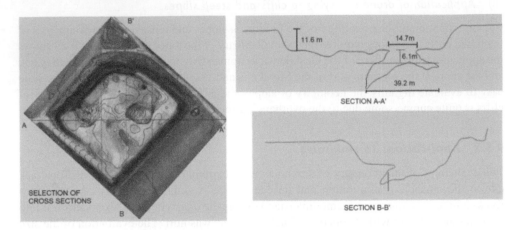

Figure 9.30 Selection of cross sections and tracing

9.5.4 Applications to tsunami boulders

There are many tsunami boulders in the major islands of the Ryukyu archipelago (Aydan and Tokashiki, 2019). The largest tsunami boulder is probably the one in Shimoji island near Shimoji Airport. The quantification of the geometry and position of these boulders are of great importance in assessing the past major earthquake and tsunami events in a given region. Both drones based on the aerial photogrammetry technique and the laser scanning technique were used to evaluate the tsunami boulders in Okinawa island and Shimoji island.

(a) Kasakanja tsunami boulder in Okinawa island

A drone based on the aerial photogrammetry technique was utilized to evaluate the geometry and the position of the tsunami boulder at Kasakanja of Okinawa island (Aydan 2018; Aydan and Tokashiki, 2019). Figure 9.31 shows a view of the drone in operation near the tsunami boulder in Kasakanja. Figure 9.32 shows the topography of the investigated area together with projections on a chosen cross section and in plan. The skill of the operator is also important when the investigations are carried out in areas where overhanging cliffs exist. As noted from Figure 9.32, the geometry of the overhanging cliffs can also be accurately evaluated.

(b) Tsunami boulder in Shimoji island

Shimoji Airport has a 4 km long runway near this tsunami boulder. As drones could not fly near the airports due to restrictions, which are automatically imposed on drones, the laser scanning technique was used. Figure 9.33(a) shows the actual tsunami boulder, and Figure 9.33b shows the laser-scanned image of the boulder from the same angle. Although the laser scanning technique can evaluate the geometry of the tsunami boulder, it is somewhat affected by the existence tress and bushes. In other words, the existence of trees and bushes disturb the digital data needed for a proper evaluation of the geometry of the tsunami boulders.

Figure 9.31 View of the drone operation near the tsunami boulder at Kasakanja

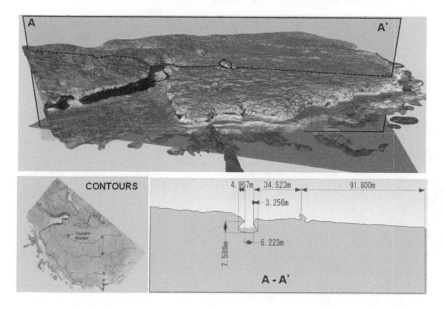

Figure 9.32 Processed digital topography of the Kasakanja tsunami boulder and its close vicinity

(a)   (b)

Figure 9.33 (a) View of the tsunami boulder, (b) digital laser scanned image of the tsunami boulder in Shimoji island

9.5.5 Applications to masonry castles

Many historical masonry structures are in the Okinawa Prefecture, Japan. The northeast corner of the Katsuren Castle collapsed during the 2010 earthquake off Okinawa island (Figure 9.34). Therefore, there is a great concern about the long-term performance and stability of masonry structures during earthquakes in the Okinawa Prefecture.

In Katsuren Castle and Nakagusuku Castle, some long-term monitoring and strong motion observations are implemented (Figures 9.35–9.37). The drone-based aerial photogrammetry technique was used to observe the current state of Katsuren Castle and Nakagusuku Castle with a particular attention to locations where continuous measurements were undertaken (Aydan *et al.*, 2016b). These measurements are going to be repeated and compared with those from continuous monitoring results. The repetitions of the measurements using the aerial photogrammetry technique are expected to provide the overall behavior of the castles in the long term three-dimensionally. These types of drone monitoring are also among the first to utilize the drone technology in the world.

Figure 9.34 Collapse and damage to the retaining walls of Katsuren Castle during the 2010 off-Okinawa island earthquake

Source: After Aydan *et al.* (2016b, 2018b)

Figure 9.35 3-D digital image of Katsuren Castle

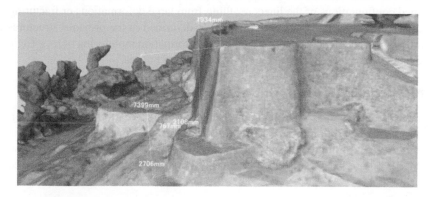

Figure 9.36 3-D digital image of Katsuren Castle at its northeast corner, where continuous monitoring is implemented

Figure 9.37 3-D digitized image of Nakagusuku Castle

9.6   Applications to maintenance monitoring

Japan has established regulations to carry out compulsory checks on the long-term performance of infrastructures every five years. For this purpose, the authorities or public and private companies and establishments owning the structures have been implementing various techniques to evaluate the state of the structures every five years. Needless to say, such evaluations should be such that they are independent of the techniques employed. The techniques vary from very simple procedures to the very sophisticated. In this respect, the utilization of the drone-based and/or laser scanning techniques could be of great use. As an application of this concept, the authors have tried to evaluate the performance of a tunnel in Okinawa Prefecture. Figure 9.38 shows a digital image of the tunnel during the construction phase. As the tunnels have concrete liners at the final stage of construction, it would be quite practical to evaluate the configuration of the tunnel in a 3-D digital form and check its geometrical changes every five years. This type of evaluation would provide a quick evaluation of the state of the tunnel and possible locations where some degradation of support systems may occur, and some unusual fracturing or deformed configurations of the liners resulting from large deformations or fracturing of the surrounding ground may be assessed. The concept described in the previous structures could be also utilized for the maintenance and long-term deformation monitoring of rock engineering structures such as tunnels, slopes and underground power houses.

9.7   Monitoring faulting-induced deformations

The permanent deformation of ground may be induced due to earthquake faulting, or creeping faults may also be monitored by the utilization of the drone-based and/or laser scanning techniques in a similar fashion to those described for previous structures (Aydan 2012; Aydan 2017). Figure 9.39 shows an application of an application of the drone-based aerial

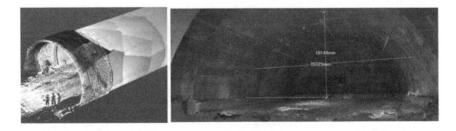

Figure 9.38 Digital images of tunnels obtained from laser scanners

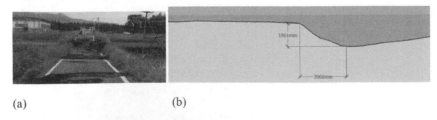

(a)   (b)

Figure 9.39 (a) View of the evaluation of ground deformations induced by earthquake faulting at a site in the 2016 Kumamoto faulting, (b) measured subsidence

photogrammetry technique at a site during the Kumamoto earthquake. It is expected that the utilization of the drone-based and/or laser scanning techniques would be quite useful for the evaluations of the deformation of the ground, as well as structures induced by earthquake faulting or permanent movements resulting from ground liquefaction or other causes would be quite effective in years to come (Aydan *et al.*, 2018a; Aydan 2016). These achievements may also lead to better evaluations of the effects of earthquake faulting on structures, as well as that of permanent ground movements.

References

Aydan, Ö. (2012) Ground motions and deformations associated with earthquake faulting and their effects on the safety of engineering structures. In: Meyers, R. (ed.) *Encyclopedia of Sustainability Science and Technology*. Springer, New York, 3233–3253.

Aydan, Ö. (2016) *Issues on Rock Dynamics and Future Directions*. Keynote. ARMS2016, Bali, 20p, on USB.

Aydan, Ö. (2019) Some thoughts on the risk of natural disasters in Ryukyu Archipelago. *International Journal of Environmental Science and Development*, 9(10), 282–289.

Aydan, Ö. & Tokashiki, N. (2019) Tsunami Boulders and Their Implications on the Mega Earthquake Potential along Ryukyu Archipelago, Japan. *Bulletin of Engineering Geology and Environment*. DOI: 10.1007/s10064-09-1378-3.

Aydan, Ö., Minato, T. & Fukue, M. (2001) An experimental study on the electrical potential of geomaterials during deformation and its implications in Geomechanics. *38th US Rock Mechanics Symposium*, Washington, Vol. 2, 1199–1206.

Aydan, Ö., Ito, T., Akagi, T., Watanabe, H. & Tano, H. (2002). An experimental study on the electrical potential of geomaterials during fracturing and sliding. *Korea-Japan Joint Symposium on Rock Engineering*, Seoul, Korea, July, 211–218.

Aydan, Ö., Tokashiki, N., Ito, T., Akagi, T., Ulusay, R. & Bilgin, H.A. (2003) An experimental study on the electrical potential of non-piezoelectric geomaterials during fracturing and sliding. *9th ISRM Congress*, South Africa, 73–78.

Aydan, Ö., Daido, M., Tano, H., Tokashiki, N. & Ohkubo, K. (2005a) A real-time multi-parameter monitoring system for assessing the stability of tunnels during excavation. *ITA Conference, Istanbul*. pp. 1253–1259.

Aydan, Ö., Sakamoto, A., Yamada, N., Sugiura, K. & Kawamoto, T. (2005b) A real time monitoring system for the assessment of stability and performance of abandoned room and pillar lignite mines. Post Mining 2005, Nancy.

Aydan, Ö., Tano, H., Ulusay, R. & Jeong, G.C. (2008) Deterioration of historical structures in Cappadocia (Turkey) and in Thebes (Egypt) in soft rocks and possible remedial measures. *2008 International Symposium on Conservation Science for Cultural Heritage, Seoul*. pp. 37–41.

Aydan, Ö., Tano, H., Imazu, M., Ideura, H. & Soya, M. (2016a) The dynamic response of the Taru-Toge tunnel during blasting. *ITA WTC 2016 Congress and 42st General Assembly, San Francisco, USA*.

Aydan, Ö., Tokashiki, N. & Tomiyama, J. (2016b) Development and application of multi-parameter monitoring system for historical masonry structures. *44th Japan Rock Mechanics Symposium*. Tokyo, pp. 56–61.

Aydan, Ö., Tomiyama, J., Matsubara, H., Tokashiki, N. & Iwata, N. (2018a) Damage to rock engineering structures induced by the 2016 Kumamoto earthquakes. *The 3rd Int. Symp on Rock Dynamics, RocDyn3, Trondheim*, 6p, on CD.

Aydan, Ö., Tokashiki, N., Tomiyama, J., Morita, T., Kashiwayanagi, M., Tobase, T. & Nishimoto, Y. (2019). A study on the dynamic and multi-parameter responses of Yanbaru Underground Powerhouse. *Proceedings of 2019 Rock Dynamics Summit in Okinawa*, 7–11 May 2019, Okinawa, Japan, ISRM (Editors: Aydan, Ö., Ito, T., Seiki T., Kamemura, K., Iwata, N.), pp. 414–419.

Aydan, Ö., Nasiry, N.Z., Ohta, Y. & Ulusay, R. (2018b) Effects of earthquake faulting on civil engineering structures. *Journal of Earthquake and Tsunami*, 12(4), 1841007 (25 pages).

Hamada, M. & Wakamatsu, K. (1998) A study on ground displacement caused by soil liquefaction. *Geotechnical Journal JSCE*, 596(III-43), 99–208.

Tano, H., Abe, T. & Aydan, Ö. (2005) The development of an *in-situ* AE monitoring system and its application to rock engineering with particular emphasis on tunneling. *ITA Conference, Istanbul*. pp. 1245–1252.

Chapter 10

Earthquake science and earthquake engineering

10.1 Introduction

Earthquakes are known to be one of the natural disasters resulting in huge losses of human lives as well as of properties experienced in the 1999 Kocaeli, Düzce, Chi-chi, and 1995 Kobe earthquakes. It is well-known that ground motion characteristics, deformation and surface breaks of earthquakes depend on the causative faults. While many large earthquakes occur along the subduction zones, which are far from the land, and their effects appear as severe shaking, the large in-land earthquakes may occur just beneath or near urban and industrial zones as observed in the recent great earthquakes. Earthquakes are due to the temporary instability of Earth's crust resulting from stress state changes. While the accumulation of stress takes a long time, from seconds to thousands of years, which is called the stick phase, the stress release occurs in a few seconds to 500–600 s, and it is called the slip phase. This chapter addresses the scientific and engineering aspects of earthquakes.

10.2 Earthquake occurrence mechanics

10.2.1 Uniaxial compression experiments in relation to earthquakes

Tests on instrumented samples of various rocks, such as Ryukyu limestone, tuff, granite, porphryte, andesite, sandstone and the like, were performed (e.g. Aydan *et al.*, 2007, 2011; Ohta, 2011; Ohta and Aydan, 2004, 2010). Two examples (Fuji-TV No.1 and Mitake Sandstone MS2). Fuji-TV No.1 is a prismatic granite sample ($100 \times 100 \times 200$ mm). The acceleration responses start to develop when the applied stress exceeds the peak strength and it attains the largest value just before the residual state is achieved, as seen in Figure 10.1(a). This pattern was observed in all experiments. Another important aspect is that the acceleration of the upper plate is much larger than that of the lower plate. This is also a common feature in all experiments. In other words, the amplitude of accelerations of the mobile part of the loading system is higher than that of the stationary part.

Mitake Sandstone MS2 sample (height: 93 mm; diameter: 45 mm) is a soft rock, and an accelerometer was attached to the sample at the mid-height. Figure 10.1(b) shows the axial stress and acceleration response as a function of time. The failure of this rock sample is ductile, and the maximum acceleration is much less than that for the granite sample. Nevertheless, the maximum acceleration occurs just before the residual state, which is very similar to that observed during the fracturing of hard brittle granite sample.

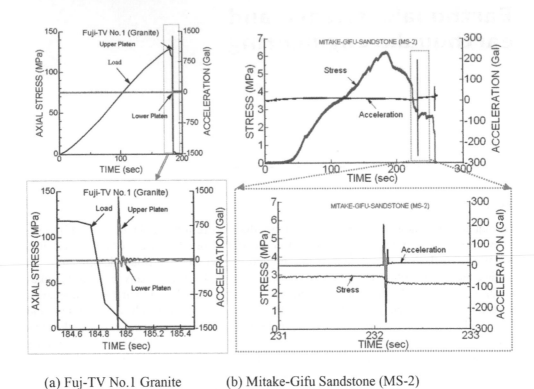

(a) Fuj-TV No.1 Granite   (b) Mitake-Gifu Sandstone (MS-2)

Figure 10.1 Acceleration and axial responses of a granite sample denoted: (a) Fuji-TV No.1, (b) Mitake-Gifu sandstone (MS-2)

Fundamentally, the observed acceleration responses during the fracturing of various rocks are similar to one another except their absolute values. The most striking feature is the chaotic acceleration response during the initiation and propagation of the macroscopic fracture of the sample. This chaotic response is very remarkable for the radial acceleration component in particular, and probably this phase is associated with the small fragment detachments before the final burst of rock samples. The small fragments result from splitting cracks aligned along the direction of loading before they coalesce into a large shear band. Furthermore, the audible sounds of fracturing are emitted from the rock during this phase.

As reported by Aydan (2003a) and his coworkers (Aydan *et al.*, 2007, 2011), work done (according to the definition in continuum mechanics) on tested samples and maximum acceleration (Aydan, 2003a) increases proportionally to the maximum acceleration, and it is always higher on the mobile part compared with that of the stationary side of the loading system. This result should probably have very important implications in many disciplines of geoscience and earthquake engineering for inferring and understanding ground motions.

10.2.2   Stick-slip phenomenon for simple mechanical explanation of earthquakes and some experiments

Brace and Byerlee (1910) and Byerlee (1970) were first to suggest the stick-slip phenomenon as a possible explanation of mechanics of earthquakes. Nevertheless, the stick-slip phenomenon is well-known in the field of tribology (e.g. Bowden and Leben, 1939, Jaeger and Cook, 1979).

10.2.2.1   Simple theory of stick-slip phenomenon

In this model, the basal plate is assumed to be moving with a constant velocity v_m, and the overriding block is assumed to be elastically supported by the surrounding medium, as illustrated in Figure 10.2. The basic concept of modeling assumes that the relative motion between the basal plate and overriding block is divergent and follows the formulation by Bowden and Leben (1939). Let as assume that the motion of the plate can be modeled as a stick-slip phenomenon. The governing equation of the motion of the overriding block may be written.

During the stick phase, the following holds:

$$\dot{x} = v_s, \quad F_s = k \cdot x \tag{10.1}$$

where v_s is belt velocity, and k is the stiffness of the system. The initiation of slip is given as (Figure 10.3):

$$F_y = \mu_s N \tag{10.2}$$

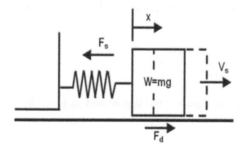

Figure 10.2 Mechanical modeling of stick-slip phenomenon

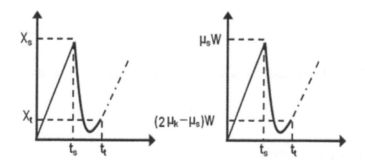

Figure 10.3 Frictional forces during a stick-slip cycle

where μ_s is static friction coefficient, N is normal force. For the block shown in Figure 10.4, it is equal to block weight W, and it is related to the mass m and gravitational acceleration g through mg. During the slip phase, the force equilibrium yields:

$$-kx + \mu_k W = m\frac{d^2x}{dt^2} \tag{10.3}$$

where μ_k is dynamic friction angle. The solution of the preceding equation can be obtained as:

$$x = A_1 \cos\Omega t + A_2 \sin\Omega t + \mu_k \frac{W}{k} \tag{10.4}$$

If initial conditions ($t = t_s$, $x = x_s$ and $\dot{x} = v_s$) are introduced in Equation (10.4), the integration constants are obtained as follows:

$$x = \frac{W}{k}(\mu_s - \mu_k)\cos\Omega(t - t_s) + \frac{v_s}{\Omega}\sin\Omega(t - t_s) + \mu_k\frac{W}{k}$$

$$\dot{x} = -\frac{W}{k}(\mu_s - \mu_k)\Omega\sin\Omega(t - t_s) + v_s\cos\Omega(t - t_s) \tag{10.5}$$

$$\ddot{x} = -\frac{W}{k}(\mu_s - \mu_k)\Omega^2\cos\Omega(t - t_s) - v_s\Omega\sin\Omega(t - t_s)$$

where $\Omega = \sqrt{k/m}$ and $x_s = \mu_s\frac{W}{k}$.

At $t = t_t$, velocity becomes equal to belt velocity, which is given as $\dot{x} = v_s$. This yields the slip period as:

$$t_t = \frac{2}{\Omega}\left(\pi - \tan^{-1}\left[\frac{(\mu_s - \mu_k)W\Omega}{k \cdot v_s}\right]\right) + t_s \tag{10.6}$$

where $x_{s=} v_s .t_s$. The rise time, which is the slip period, is given by:

$$t_r = t_t - t_s \tag{10.7}$$

Rise time can be specifically obtained from Equations (10.10) and (10.5) as:

$$t_r = \frac{2}{\Omega}\left(\pi - \tan^{-1}\left[\frac{(\mu_s - \mu_k)W\Omega}{k \cdot v_s}\right]\right) \tag{10.8}$$

If belt velocity can be omitted, that is, $v_s \approx 0$, the rise time reduces (t_p) to the following form:

$$t_r = \pi\sqrt{\frac{m}{k}} \tag{10.9}$$

The amount of slip is obtained as:

$$x_r = |x_t - x_s| = 2\frac{W}{k}(\mu_s - \mu_k) \tag{10.10}$$

The force drop during slip is given by:

$$F_d = 2(\mu_s - \mu_k)W \qquad (10.11)$$

It should be noted that this formulation does not consider the damping associated with slip velocity. If the damping resistance is linear, the governing equation (10.4) will take the following form:

$$-kx - \eta\dot{x} + \mu_k W = m\frac{d^2x}{dt^2} \qquad (10.12)$$

10.2.2.2   Device of stick-slip tests

Figure 10.4 shows a view of the experimental device. The experimental device consists of an endless conveyor belt and a fixed frame. The inclination of the conveyor belt can be varied so that tangential and normal forces can be easily imposed on the sample as desired. To study the actual frictional resistance of the interfaces of the rock blocks, the lower block is stuck to a rubber belt while the upper block is attached to the fixed frame through a spring as illustrated in Figure 10.4(a). Some experiments were conducted using the rock samples of granite with planes having different surface morphologies. The base blocks were 200–400 mm long, 100–200 mm wide and 40–100 mm thick. The upper block was 100–200 mm long, 100 mm wide and 50–100 mm high.

When the upper block moves together on the base block at a constant velocity (stick phase), the spring is stretched at a constant velocity. The shear force increases to some critical value, and then a sudden slip occurs with an associated spring force drop. Because the instability sliding of the upper block occurs periodically, the upper block slips violently over the base block. Normal loads can also be easily increased in experiments.

To measure the frictional force acting on the upper block, the load cell (KYOWA LUR-A-200NSA1) is installed between the spring and fixed frame. During experiments, the displacement of the block is measured through a laser displacement transducer produced by KEYENCE and a contact type displacement transducer with a measuring range of 70 mm,

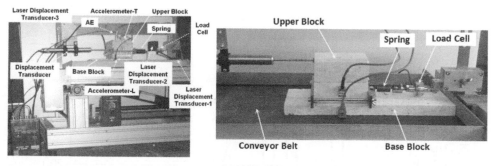

(a) Overall view   (b) Detailed view

Figure 10.4 Stick-slip experimental setup

while the acceleration responses parallel and perpendicular to the belt movement are measured by a three-component accelerometer (TOKYO SOKKI) attached to the upper block.
The measured displacement, acceleration and force are recorded onto laptop computers.

10.2.2.3   Stick-slip experiment

Many stick-slip experiments were performed on various natural rock blocks as well as on
other types of blocks made of foam, plastic, wood and aluminum (e.g. Aydan, 2003a; Aydan,
2019; Aydan *et al.*, 2019; Ohta and Aydan, 2010). Here experimental results on discontinuities in granite are quoted as an example. Three different combinations of the surface roughness conditions of granite blocks were investigated while keeping the system stiffness, upper
block weight and base velocity constant. These combinations are rough-to-rough (tension
joint), rough-to-smooth (saw-cut), smooth-to-smooth interfaces. The measured response of
a discontinuity with a rough-to-rough combination is shown in Figure 10.5. As noted from
the figure, the velocity of the upper block starts to change before the slippage.

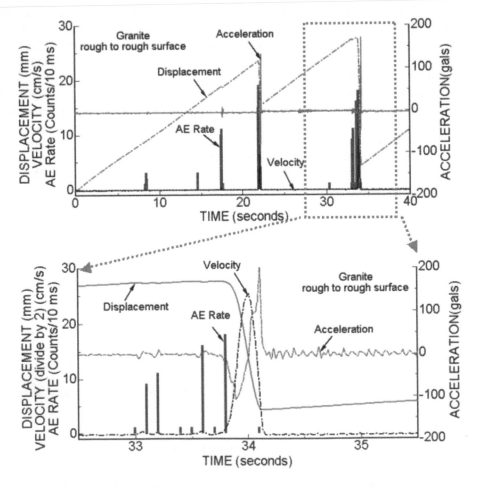

Figure 10.5 The response of a discontinuity surface during a stick-slip experiment

Figures 10.6 and 10.7 show a series of stick-slip experiments on discontinuities in granite and Ryukyu limestone, together with interpretations of peak and residual friction angles. The peak (static) friction angle can be evaluated from the T/N response while the residual (kinetic) friction angle is obtained from the theoretical relation (10.11). Some tilting experiments were carried out on the same discontinuity planes (Aydan *et al.*, 2019). The peak (static) friction angles for both the discontinuity plane obtained from tilting tests and stick-slip experiments were very close to each other. The residual or kinetic friction angles for a rough discontinuity plane of granite are also very close to each other. Similarly, the residual (kinetic) friction angles of saw-cut discontinuity plane of Ryukyu limestone obtained from stick-slip experiments are very close to those obtained from tilting experiments. Nevertheless, the kinetic or residual friction angle is generally lower than those obtained from the tilting experiments. As expected from the theoretical formulas derived in the previous subsection, the phenomenon would be periodic. If the peak and residual friction angles are the

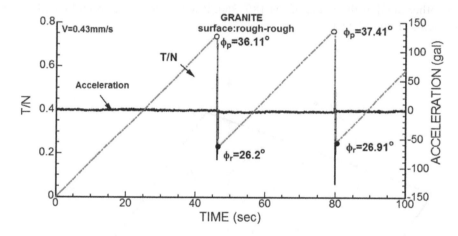

Figure 10.6 Stick-slip response of rough discontinuity plane of granite

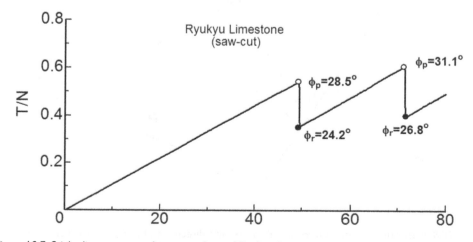

Figure 10.7 Stick-slip response of saw-cut plane of Ryukyu limestone

same, the slip would be continuous with a given velocity, which may be the fundamental explanation of fault-creep observed in some segments of the North Anatolian Fault in Turkey and the San Andreas Fault in the United States.

10.3   Causes of earthquakes

The main cause of earthquakes is the stress changes in the Earth's crust and its temporary mechanical instability. Let's consider the average stress changes on a fault plane in view of results of stick-slip experiments. The stress state and geometrical parameters of the faults can be illustrated as shown in Figure 10.8. Vectors n, s and b are normal, sliding and neutral vectors with respect to the fault plane, respectively. Neutral vector b is perpendicular to the plane defined by normal and sliding vectors n, s. Parameters p, d and i stand for dip (plunge), dip direction and sliding direction on the fault plane. When the value of i is 0–180, it corresponds to the faults with a normal component. On the other hand, if the value of i is 180–360, it will correspond to faults with reverse components (Figure 10.9).

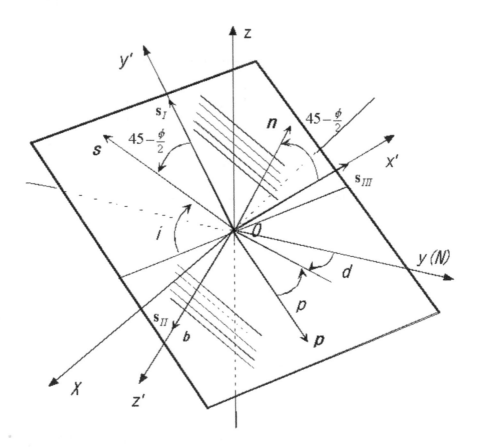

Figure 10.8 Illustration of notation for a fault plane with directions of principal stresses, slip, normal and neutral vectors

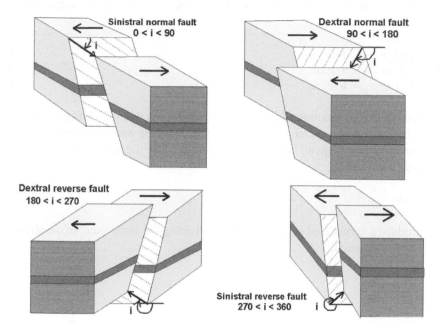

Figure 10.9 Illustration of the definition of striation angle on a fault surface

The normal and sliding vectors in terms of their components can be specifically written:

$$\mathbf{n} = \left\{ n_x \quad n_y \quad n_z \right\}, \ \mathbf{s} = \left\{ s_x \quad s_y \quad s_z \right\} \tag{10.13}$$

where $n_x = \sin p \sin d$, $n_y = \sin p \cos d$, $n_z = \cos p$, $s_x = -\cos p \sin d \sin i + \cos d \cos i$, $s_y = -\cos p \cos d \sin i - \sin d \cos i$, $s_z = \cos p \sin i$.

As the neutral vector b is perpendicular to the plane of normal and sliding vectors, it can be mathematically expressed as given here:

$$\mathbf{b} = \mathbf{s} \times \mathbf{n} \tag{10.14}$$

The traction vector t acting in the fault plane can be related to stress tensor σ and the normal n of the fault plane through the Cauchy equation (e.g. Mase, 1970; Eringen, 1980):

$$\mathbf{t} = \sigma \cdot \mathbf{n}$$

The stress tensor can be written in matrix form as given here:

$$\sigma = \begin{bmatrix} \sigma_{xx} & \sigma_{xy} & \sigma_{xz} \\ \sigma_{xy} & \sigma_{yy} & \sigma_{yz} \\ \sigma_{xz} & \sigma_{yz} & \sigma_{zz} \end{bmatrix} \tag{10.15}$$

Furthermore, the following relations can be written for the normal, shear and neutral vectors:

$$\sigma_N = \mathbf{n} \cdot \mathbf{t} \text{ or } \sigma_N = \mathbf{n} \cdot \sigma \cdot \mathbf{n} \tag{10.16a}$$
$$\sigma_S = \mathbf{s} \cdot \mathbf{t} \text{ or } \sigma_S = \mathbf{s} \cdot \sigma \cdot \mathbf{n} \tag{10.16b}$$
$$\sigma_B = \mathbf{b} \cdot \mathbf{t} \text{ or } \sigma_B = \mathbf{b} \cdot \sigma \cdot \mathbf{n} \tag{10.16c}$$

Aydan (1995) has both theoretically and numerically shown that the vertical component σ_{zz} of the stress tensor can be taken as a quantity obtained by the multiplication of depth h and unit weight γ of rock, as given here, by taking into account the sphericity of the Earth and gravitational acceleration:

$$\sigma_{zz} = \gamma h \tag{10.17}$$

Let us introduce a normalized stress obtained by dividing the component of the stress tensor by its vertical component:

$$\mathbf{N} = \begin{vmatrix} N_{xx} & N_{xy} & N_{xz} \\ N_{xy} & N_{yy} & N_{yz} \\ N_{xz} & N_{yz} & N_{zz} \end{vmatrix} \tag{10.18}$$

where N_{zz} is equal to 1. Let us consider a coordinate system $ox'y'z'$ whose axes are aligned with the principal stress components as shown in Figure 10.8:

$$\sigma' = \begin{vmatrix} \sigma_I & 0 & 0 \\ 0 & \sigma_{II} & 0 \\ 0 & 0 & \sigma_{III} \end{vmatrix} \tag{10.19}$$

Similarly, the normalized principal stress tensor by the vertical stress can be written as:

$$\mathbf{N}' = \begin{vmatrix} N_I & 0 & 0 \\ 0 & N_{II} & 0 \\ 0 & 0 & N_{III} \end{vmatrix} \tag{10.20}$$

The shearing of rock takes place along the direction of the maximum shear stress according to the least work principle of the mechanics. Therefore, the shear stress on the plane of normal and sliding vectors must be nil, implying that the direction of the neutral vector must coincide with that the intermediate principal stress. Thus this can be mathematically expressed as:

$$\sigma_B \equiv \sigma_{II} \tag{10.21}$$

and

$$\mathbf{b} \equiv \mathbf{s}_{II} \text{ and } \mathbf{b} \equiv \left\{ b_x \quad b_y \quad b_z \right\}; \mathbf{s}_{II} = \left\{ l_2 \quad m_2 \quad n_2 \right\} \tag{10.22}$$

In light of experimental facts on rocks, the following relations may be written among normal, sliding and neutral vectors and the maximum and minimum principal stresses:

$$\mathbf{s} \cdot \mathbf{s}_I = \cos(45 - \frac{\phi}{2}) \text{ ve } \mathbf{s} \cdot \mathbf{s}_{III} = \cos(135 - \frac{\phi}{2}) \tag{10.23a}$$

$$\mathbf{n} \cdot \mathbf{s}_I = \cos(45 + \frac{\phi}{2}) \text{ ve } \mathbf{n} \cdot \mathbf{s}_{III} = \cos(45 - \frac{\phi}{2}) \tag{10.23b}$$

$$\mathbf{b} \cdot \mathbf{s}_I = 0 \text{ and } \mathbf{s} \cdot \mathbf{s}_{III} = 0 \tag{10.23c}$$

Therefore the direction vectors $\mathbf{s}_I$ and $\mathbf{s}_{II}$ of the maximum and minimum principal stresses can be easily obtained from the preceding relations as:

$$\mathbf{s}_I = \begin{Bmatrix} l_1 \\ m_1 \\ n_1 \end{Bmatrix} = \begin{bmatrix} s_x & s_y & s_z \\ n_x & n_y & n_z \\ b_x & b_y & b_z \end{bmatrix}^{-1} \begin{Bmatrix} \cos(45 - \frac{\varphi}{2}) \\ \cos(45 + \frac{\varphi}{2}) \\ 0 \end{Bmatrix} \tag{10.24a}$$

$$\mathbf{s}_{III} = \begin{Bmatrix} l_3 \\ m_3 \\ n_3 \end{Bmatrix} = \begin{bmatrix} s_x & s_y & s_z \\ n_x & n_y & n_z \\ b_x & b_y & b_z \end{bmatrix}^{-1} \begin{Bmatrix} \cos(135 - \frac{\varphi}{2}) \\ \cos(45 - \frac{\varphi}{2}) \\ 0 \end{Bmatrix} \tag{10.24b}$$

When the Mohr-Coulomb yield criterion is used, the value of intermediate principal stress becomes indeterminate, although its value is bounded by the maximum and minimum principal stresses. Aydan (2000a) established a relation through the use of the Mohr-Coulomb criterion and Drucker-Prager yield criterion for frictional condition, which is commonly used in the numerical analysis of structures in geomaterials, and he derived the following inequality relation to obtain the value of intermediate principal stress:

$$\beta^2 - \frac{(1 + 6\alpha^2)(q + 1)}{(1 - 3\alpha^2)} \beta + \frac{(1 - 3\alpha^2)(q^2 + 1) - q(1 + 6\alpha^2)}{(1 - 3\alpha^2)} = 0 \tag{10.25}$$

where

$$\beta = \frac{\sigma_{II}}{\sigma_{III}} \tag{10.26}$$

The inequality relation yields two roots, and one of the roots is chosen so that the intermediate stress would have a value between the maximum and minimum principal stresses. When the peak friction angle is utilized, the stress state would correspond to

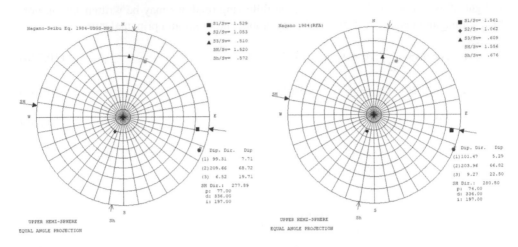

(a) Stress state for peak friction　　(b) Stress state for residual friction

Figure 10.10 Normalized stress states obtained for peak and residual friction angles for the 1984 Nagano Prefecture Seibu earthquake

Table 10.1 Stress and orientation changes for 1984 Nagano Prefecture Seibu earthquake

Normal Stress $\Delta\sigma_n / \sigma_v$	Shear Stress $\Delta\tau_d / \sigma_v$	Orientation $\Delta\theta$
0.14	0.114534	2.11

the stress state at the time slip. On the other hand, if the residual friction is utilized, it should correspond to the stress state at the equilibrium following the termination of the earthquake.

An example of this concept has been applied to the 1984 Nagano Prefecture Seibu earthquake with a moment magnitude of 6.2. The normalized stress states by the vertical stress for peak (30 degrees) and residual friction (26 degrees) angles are obtained and shown in lower-hemisphere stere-net projections in Figure 10.10. The computed normalized stress changes are given in Table 10.1. The average shear stress change on the fault plane would be about 14.88 MPa for a linear distribution of vertical stress.

10.4　Earthquake-induced waves

It is known that earthquakes cause fundamentally two types of earthquakes waves (Figure 10.11). The first type of waves, called body waves, are the P-wave and S-wave. P-waves, or primary waves, pass through all materials. S-waves, or secondary waves, arrive at the observation point after the P-wave. The second-type waves are called surface waves, and

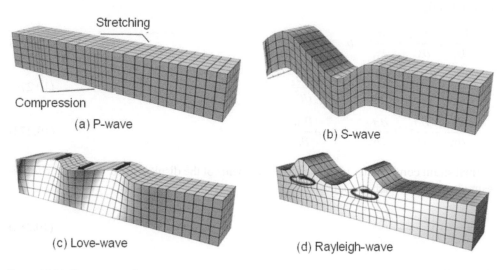

Figure 10.11 Illustration of wave types

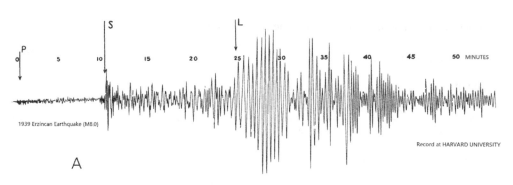

Figure 10.12 Seismogram of the 1939 Erzincan earthquake at Harvard University

Source: From Ketin (1973)

they are further subdivided into Rayleigh and Love waves. Figure 10.12 shows a record of 1939 Erzincan earthquake taken at Harvard University. It is known that shear waves are not transmitted through materials in the liquid phase. As the outer core of the Earth is in liquid phase, no shear waves are observed beyond 143 degrees from the earthquake focus. Surface waves (Rayleigh and Love) are observed near the Earth surface, and they disappear as the depth increases.

The equation of motion can be written in the following form:

$$\frac{\partial \sigma_{ij}}{\partial x_j} + b_i = \rho \frac{\partial^2 u_i}{\partial t^2} \tag{10.27a}$$

or specifically:

$$\frac{\partial \sigma_{11}}{\partial x_1} + \frac{\partial \sigma_{12}}{\partial x_2} + \frac{\partial \sigma_{13}}{\partial x_3} + b_1 = \rho \frac{\partial^2 u_1}{\partial t^2} \tag{10.27b}$$

$$\frac{\partial \sigma_{12}}{\partial x_1} + \frac{\partial \sigma_{22}}{\partial x_2} + \frac{\partial \sigma_{23}}{\partial x_3} + b_2 = \rho \frac{\partial^2 u_2}{\partial t^2} \tag{10.27c}$$

$$\frac{\partial \sigma_{13}}{\partial x_1} + \frac{\partial \sigma_{23}}{\partial x_2} + \frac{\partial \sigma_{33}}{\partial x_3} + b_3 = \rho \frac{\partial^2 u_3}{\partial t^2} \tag{10.27d}$$

Normal strain components are related to components of the displacement vector if the infinitesimal strain approach is adopted:

$$\varepsilon_{ij} = \frac{1}{2}\left(\frac{\partial u_i}{\partial x_j} + \frac{\partial u_j}{\partial x_i}\right) \tag{10.28a}$$

or specifically:

$$\varepsilon_{11} = \frac{\partial u_1}{\partial x_1}, \varepsilon_{22} = \frac{\partial u_2}{\partial x_2}, \varepsilon_{33} = \frac{\partial u_3}{\partial x_3} \tag{10.28b}$$

Engineering shear strains are related to the components:

$$\gamma_{ij} = 2\varepsilon_{ij} \text{ with } i \neq j \tag{10.28c}$$

or specifically:

$$\gamma_{23} = \frac{\partial u_2}{\partial x_3} + \frac{\partial u_3}{\partial x_2}, \gamma_{12} = \frac{\partial u_1}{\partial x_2} + \frac{\partial u_2}{\partial x_1}, \gamma_{13} = \frac{\partial u_1}{\partial x_3} + \frac{\partial u_3}{\partial x_1} \tag{10.28d}$$

Rotational strains are defined as:

$$\omega_1 = \frac{1}{2}\left(\frac{\partial u_3}{\partial x_2} - \frac{\partial u_2}{\partial x_3}\right), \omega_2 = \frac{1}{2}\left(\frac{\partial u_1}{\partial x_3} - \frac{\partial u_3}{\partial x_1}\right), \omega_3 = \frac{1}{2}\left(\frac{\partial u_2}{\partial x_1} - \frac{\partial u_1}{\partial x_2}\right) \tag{10.28e}$$

The constitute law between stress and strain can be expressed if material is an isotropic elastic body as:

$$\sigma_{ij} = \lambda \delta_{ij} \varepsilon_{kk} + 2\mu \varepsilon_{ij}; \varepsilon_{kk} = \varepsilon_{11} + \varepsilon_{22} + \varepsilon_{33}; \tag{10.29a}$$

or specifically:

$$\begin{Bmatrix} \sigma_{11} \\ \sigma_{22} \\ \sigma_{33} \\ \sigma_{12} \\ \sigma_{23} \\ \sigma_{13} \end{Bmatrix} = \begin{bmatrix} \lambda+2\mu & \lambda & \lambda & 0 & 0 & 0 \\ \lambda & \lambda+2\mu & \lambda & 0 & 0 & 0 \\ \lambda & \lambda & \lambda+2\mu & 0 & 0 & 0 \\ 0 & 0 & 0 & \mu & 0 & 0 \\ 0 & 0 & 0 & 0 & \mu & 0 \\ 0 & 0 & 0 & 0 & 0 & \mu \end{bmatrix} \begin{Bmatrix} \varepsilon_{11} \\ \varepsilon_{22} \\ \varepsilon_{33} \\ \gamma_{12} \\ \gamma_{23} \\ \gamma_{13} \end{Bmatrix} \tag{10.29b}$$

where λ and μ are Lamé coefficients, specifically given in the following form:

$$\lambda = \frac{Ev}{(1+v)(1-2v)} \text{ and } \mu = \frac{E}{2(1+v)} \tag{10.30}$$

Let us introduce the following:

$$\Delta = \frac{\partial u_1}{\partial x_1} + \frac{\partial u_2}{\partial x_2} + \frac{\partial u_3}{\partial x_3} \tag{10.31a}$$

$$\nabla^2 = \nabla \cdot \nabla = \frac{\partial^2}{\partial x_1^2} + \frac{\partial^2}{\partial x_2^2} + \frac{\partial^2}{\partial x_3^2} \tag{10.31b}$$

Equation (10.31a) corresponds to volumetric strain, while Equation (10.31b) is called Laplacian operator.

Inserting constitutive law given by Equation (10.29), together with relations between strain and displacement components given by Equation (10.28), into the equation of motion and differentiating Equations (10.28b), (10.28c) and (10.28d) with respect to x_1, x_2 and x_3, respectively, yields for each respective directions provided that elastic coefficients, density and body forces are constant as follows:

$$(\lambda + \mu)\frac{\partial^2 \Delta}{\partial x_1^2} + \mu \nabla^2 \frac{\partial u_1}{\partial x_1} = \rho \frac{\partial}{\partial x_1}\left(\frac{\partial^2 u_1}{\partial t^2}\right) \tag{10.32a}$$

$$(\lambda + \mu)\frac{\partial^2 \Delta}{\partial x_2^2} + \mu \nabla^2 \frac{\partial u_2}{\partial x_2} = \rho \frac{\partial}{\partial x_2}\left(\frac{\partial^2 u_2}{\partial t^2}\right) \tag{10.32b}$$

$$(\lambda + \mu)\frac{\partial^2 \Delta}{\partial x_3^2} + \mu \nabla^2 \frac{\partial u_3}{\partial x_3} = \rho \frac{\partial}{\partial x_3}\left(\frac{\partial^2 u_3}{\partial t^2}\right) \tag{10.32c}$$

Summing up Equations (10.32a), (10.32b) and (10.32c) results in the following equation:

$$(\lambda + \mu)\nabla^2 \Delta + \mu \nabla^2 \Delta = \rho \frac{\partial^2 \Delta}{\partial t^2} \text{ or } V_p^2 \nabla^2 \Delta = \frac{\partial^2 \Delta}{\partial t^2} \text{ or } V_p^2 \nabla^2 \varepsilon_v = \frac{\partial^2 \varepsilon_v}{\partial t^2} \tag{10.33}$$

where

$$\lambda + 2\mu = \frac{Ev}{(1+v)(1-2v)} + \frac{E}{1+v} = \frac{E(1-v)}{(1+v)(1-2v)} \tag{10.34a}$$

$$V_p = \sqrt{\frac{E(1-v)}{\rho(1+v)(1-2v)}} \tag{10.34b}$$

$$\varepsilon_v = \Delta \tag{10.34c}$$

Equation (10.33) is known as the governing equation of P-wave propagation in solids. As noted from this equation, P-wave propagation is directly related to volumetric straining. During the propagation of P-wave, solids will undergo dilatational and compressive volumetric straining.

Similarly inserting constitutive law given by Equation (10.33), together with relations between strain and displacement components given by Equation (10.28), into the equation of motion takes the following form specifically for each respective direction:

$$(\lambda+\mu)\frac{\partial\Delta}{\partial x_1}+\mu\nabla^2 u_1+b_1=\rho\frac{\partial^2 u_1}{\partial t^2}\tag{10.35a}$$

$$(\lambda+\mu)\frac{\partial\Delta}{\partial x_2}+\mu\nabla^2 u_2+b_2=\rho\frac{\partial^2 u_2}{\partial t^2}\tag{10.35b}$$

$$(\lambda+\mu)\frac{\partial\Delta}{\partial x_3}+\mu\nabla^2 u_3+b_3=\rho\frac{\partial^2 u_3}{\partial t^2}\tag{10.35c}$$

Differentiating Equations (10.35b) and (10.35c) with respect to x_3 and x_2 yields the following, provided that elastic coefficients, density and body forces are constant:

$$(\lambda+\mu)\frac{\partial^2\Delta}{\partial x_3\partial x_2}+\mu\nabla^2\frac{\partial u_2}{\partial x_3}=\rho\frac{\partial^2}{\partial t^2}\left(\frac{\partial u_2}{\partial x_3}\right)\tag{10.36a}$$

$$(\lambda+\mu)\frac{\partial^2\Delta}{\partial x_2\partial x_3}+\mu\nabla^2\frac{\partial u_3}{\partial x_2}=\rho\frac{\partial^2}{\partial t^2}\left(\frac{\partial u_3}{\partial x_2}\right)\tag{10.36b}$$

Subtracting Equations (10.36a) from (10. 36b) results in:

$$(\lambda+\mu)\left(\frac{\partial^2\Delta}{\partial x_2\partial x_3}-\frac{\partial^2\Delta}{\partial x_3\partial x_2}\right)+\mu\nabla^2\left(\frac{\partial u_3}{\partial x_2}+\frac{\partial u_2}{\partial x_3}\right)=\rho\frac{\partial^2}{\partial t^2}\left(\frac{\partial u_3}{\partial x_2}-\frac{\partial u_2}{\partial x_3}\right)\tag{10.37}$$

Using the rotational strain definition given by Equation (10.28e) and dividing Equation (10.37) gives:

$$\frac{\mu}{\rho}\nabla^2\omega_1=\frac{\partial^2\omega_1}{\partial t^2}\tag{10.38a}$$

Using the same procedure for other directions together with rotation strain components given by Equation (10.28e), one can easily derive the following:

$$\frac{\mu}{\rho}\nabla^2\omega_2=\frac{\partial^2\omega_2}{\partial t^2}\tag{10.38b}$$

$$\frac{\mu}{\rho}\nabla^2\omega_3=\frac{\partial^2\omega_3}{\partial t^2}\tag{10.38c}$$

Equation (10.38) is the governing equation of Rayleigh waves. The coefficient in Equation (10.38) is interpreted as the propagation velocity of rotational waves:

$$V_s=\sqrt{\frac{\mu}{\rho}}\text{ or }V_s=\sqrt{\frac{E}{2\rho(1+v)}}\tag{10.39}$$

If Δ is 0 and body force is negligible, one easily gets the following expressions from Equation (10.36):

$$\frac{\mu}{\rho}\nabla^2 u_1 = \frac{\partial^2 u_1}{\partial t^2} \text{ or } V_s^2\nabla^2 u_1 = \frac{\partial^2 u_1}{\partial t^2}$$ (10.40a)

$$\frac{\mu}{\rho}\nabla^2 u_2 = \frac{\partial^2 u_2}{\partial t^2} \text{ or } V_s\nabla^2 u_2 = \frac{\partial^2 u_2}{\partial t^2}$$ (10.40b)

$$\frac{\mu}{\rho}\nabla^2 u_3 = \frac{\partial^2 u_3}{\partial t^2} \text{ or } V_s\nabla^2 u_2 = \frac{\partial^2 u_2}{\partial t^2}$$ (10.40c)

Equation (10.40) is the fundamental equation of distortion (shear) waves known as S-waves. It should be noted that the propagation velocity of S-waves is the same as that of Rayleigh waves.

10.5 Inference of faulting mechanism of earthquakes

The striations and internal structure of these faults are evidence of what type of stress state caused them, and they may also indicate what type of earthquake they produced. The methodology for the inference of the possible stress state and focal plane solutions of earthquakes from the faults require data on dip, dip direction and striation orientation (Aydan, 2000a; Aydan and Kim, 2002). Figure 10.9 shows an illustration of how striation angle is measured. Figure 10.13 shows the focal mechanism of earthquakes estimated from the fault striations obtained for the 1891 Kiso-Beya earthquake in Japan and the 1999 Kocaeli earthquake in Turkey.

The focal mechanism of the earthquakes may also be inferred from the P-waves and S-waves induced by earthquakes. Figure 10.14 illustrates the concept of obtaining the focal mechanism of a vertical strike-slip and associated wave responses at the observation points around the focus of the earthquake. Figure 10.15(b) shows the focal mechanism solution for the motion of the fault plane shown in Figure 10.15(a), and the seismograms for this solution would look like those shown in Figure 10.15.

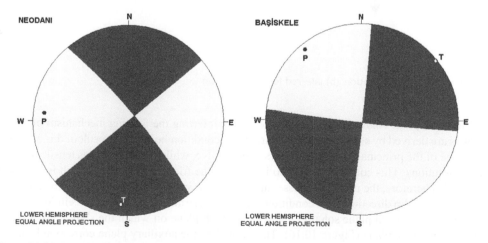

Figure 10.13 Inferred faulting mechanism for some earthquakes from fault striations

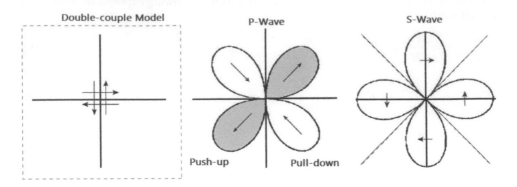

Figure 10.14 Illustration of fundamental concept to obtain focal mechanism solutions from seismic waves

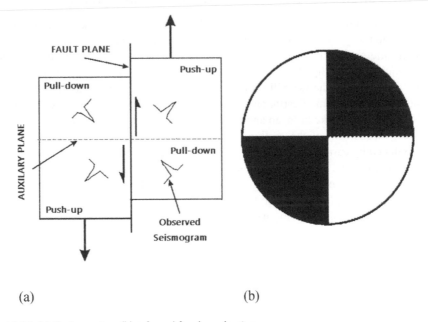

Figure 10.15 (a) Fault motion, (b) inferred focal mechanism

The focal plane solutions used in geoscience for inferring the faulting mechanism of earthquakes are derived by assuming that the pure-shear condition holds. As a result of this assumption, one of the principal stresses is always compressive, while the other one is tensile in focal plane solutions. This condition may also imply that the friction angle of the fault is assumed to be nil. Therefore, the principal stresses are inclined at an angle of 45 degrees with respect to the normal of slip direction. This condition is used to determine p-axis and t-axis in focal plane solutions. Each focal plane solution involves the fault plane on which the sliding takes place and the auxiliary plane (Figure 10.16). The normal of the auxiliary plane corresponds to the slip vector, and it is orthogonal to the neutral plane on which p-axis and t-axis exist.

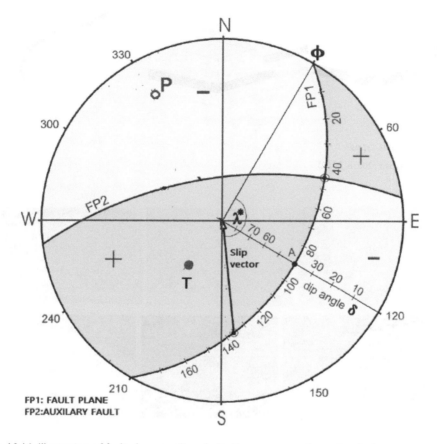

Figure 10.16 Illustration of fault plane, auxiliary fault, slip vector on a lower hemisphere stereo net

10.6 Characteristics of earthquake faults

The fault is geologically defined as a discontinuity in geological medium along which a relative displacement takes place. Faults are broadly classified into three big groups, namely normal faults, thrust faults and strike-slip faults, as seen in Figures 10.17 and 10.18. A fault is geologically defined as active if a relative movement took place in a period of less than 2 million years.

It is well-known that a fault zone may involve various kinds of fractures as illustrated in Figure 10.19(a), and it is a zone having a finite volume (Aydan *et al.*, 1997; Ulusay *et al.*, 2002). In other words, it is not a single plane. Furthermore, the faults may have a negative or positive flower structure as a result of their transtensional or transpressional nature and the reduction of vertical stress near the Earth surface (Aydan *et al.*, 1999). For example, even a fault having a narrow thickness at depth may cause broad rupture zones and numerous fractures on the ground surface during earthquakes (Figure 10.19(b)). Furthermore, the movements of a fault zone may be diluted if a thick alluvial deposit is found on the top of the fault (e.g. the 1992 Erzincan earthquake (Hamada and Aydan, 1992)). The appearance of ground breaks is closely related to geological structure, the characteristics of sedimentary deposits, their geometry, the magnitude of earthquakes and fault movements.

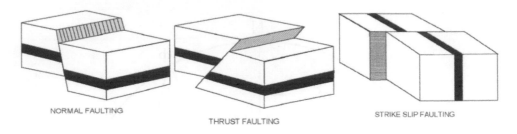

NORMAL FAULTING   THRUST FAULTING   STRIKE SLIP FAULTING

Figure 10.17 Fault types

Source: From Aydan (2003b, 2012)

(a) Normal faulting   (b) Strike-slip faulting   (c) Thrust faulting

Figure 10.18 Some examples of faulting

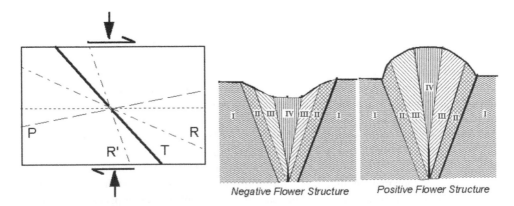

Negative Flower Structure   Positive Flower Structure

Figure 10.19 (a) Fractures in a shear zone or fault, (b) negative and positive flower structures due to transtension or transpression faulting and zoning

Sources: (a) From Aydan *et al.* (1997), (b) modified from Aydan *et al.* (1997, 1999), Aydan (2003b)

10.7 Characterization of earthquakes from fault ruptures

Turkey is one of the well-known earthquake-prone countries in the world, and most of her large earthquakes involve ground surface rupturing. The data from the past and present earthquakes of Turkey, as well as those of other countries, may be quite useful to establish and/or to revise empirical relations among the characteristics of earthquakes accompanying ground surface rupturing. The data compiled by the author come from the Turkish earthquake database (TEDBAS) developed by the author (Aydan, 1997) and additional inputs from recent earthquakes (Ambraseys, 1988; Hamada and Aydan, 1992; Aydan *et al.* 1991; Aydan and Kumsar, 1997; Aydan *et al.*, 1998, 1999, 2000a, 2003b, 2005a, 2005b, 2010; Ergin *et al.*, 1960; Soysal *et al.*, 1981; Eyidogan *et al.*, 1991; Gencoglu *et al.*, 1990; Ohta, 2010). The data for other countries is compiled by Wells and Coppersmith (1994), Matsuda (1975) and Sato (1989). The data on the source properties of earthquakes is gathered from the well-known seismological institutes such as the U.S. Geological Survey (USGS), Harvard, ERI of Tokyo University and Swiss Seismological Institute. The number of a data set varies depending upon the studied empirical relations. For example, the number of data for the relation between M_w and M_s is 2010. The following items are chosen as the characteristics of earthquakes:

1 Magnitude (moment and surface wave magnitudes, M_w, M_s)
2 Length of earthquake fault (L), which denotes the length of the source fault or that estimated by the ground surface trace observed in the field or aftershock distribution if the surface rupture is hindered by the thick sedimentary deposits (i.e. 1992 Erzincan, 1998 Adana-Ceyhan earthquake)
3 Depth of earthquake hypocenter (D)
4 Rupture area (S), which denotes the ruptured area of the earthquake fault inferred from aftershock distribution or the multiplication of surface rupture length produced by the earthquake by its hypocenter depth with the assumption of a rectangular source area
5 Net slip of the earthquake fault (U_{max}), which denotes the maximum slip along the slip direction (Whitten and Brooks, 1972)
6 Maximum ground acceleration and velocity (a_{max}, v_{max}) (hypocenter distance is mostly in the range of 15–25 km)
7 Rupture mode–striation orientation
8 Ratio of vertical maximum acceleration to the horizontal maximum acceleration (RVAHA)

It should be noted that the minimum value of M_w is assumed to be 0 in all-empirical relations presented hereafter.

10.7.1 Relation between surface wave magnitude and moment magnitude

Aydan (Aydan, 1997; Aydan *et al.*, 1996) selected surface wave magnitude M_s in developing his empirical relations for Turkish earthquakes since a lot of data based on surface wave magnitude M_s is available, and the magnitude of earthquakes did not exceed 8 so far. It is pointed out that surface wave magnitude M_s becomes unreliable if it exceeds the value of 8 (i.e. Fowler, 1990). Furthermore, it is becoming more popular to use the moment magnitude M_w in place of surface wave magnitude M_s since, recently, many seismological institutes release moment magnitude data rather than surface wave magnitude data. Nevertheless, the moment magnitude data determined by various institutes for the same earthquake is not always the same. Furthermore, the moment magnitude data must be assigned to previous earthquakes before the development of moment magnitude concept. Kanamori (1983) suggested that the surface wave and moment magnitudes of earthquakes can be taken as equal to each other within the range of 5–7.6 Aydan (Aydan, 1997; Aydan *et al.*, 1996) proposed the following relation between surface and moment magnitudes of earthquakes for Turkish earthquakes:

$$M_w = 1.044M_s \text{ or } M_s = 0.958M_w \qquad (10.41)$$

However, Ulusay *et al.* (2004) recently suggested the following formula for their data set on Turkish earthquakes:

$$M_w = 0.6798M_s + 2.0402 \qquad (10.42)$$

Figure 10.20 compares the data set of Turkish earthquakes compiled by the author, including all recent data on Turkish earthquakes and worldwide data, which is fitted to the following empirical relation:

$$M_w = 1.2M_s e^{-0.028M_s} \qquad (10.43)$$

As noted from the figure, all data sets generally support the suggestion of Kanamori (1983) for relating surface wave magnitude to moment magnitude for the magnitude range of 4–8. Therefore, it can be safe to adopt the previous empirical relations proposed by Aydan (1997, 2001) based on surface wave magnitude for the seismic characteristics of Turkish earthquakes, together with the replacement of surface wave magnitude with moment magnitude. However, the constants of functions will have to be recalculated if the independent variable is chosen as moment magnitude.

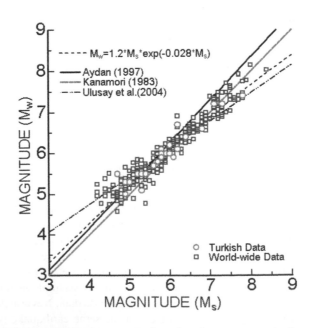

Figure 10.20 Relation between moment magnitude and surface wave magnitude

10.7.2   Relation between MMI intensity and moment magnitude

The magnitude of historical earthquakes is mainly inferred from the Modified Mercalli Seismic (MMI) intensity. Aydan *et al.* (1996) and Aydan (1997) proposed the following empirical relation between MMI intensity and surface wave magnitude Ms (see also Aydan, 2001):

$$I_o = 1.317 M_s \tag{10.44}$$

In this study, the following relation between moment magnitude and MMI intensity is proposed with the consideration of recent large earthquakes in Turkey and worldwide:

$$I_o = 1.32 M_w \tag{10.45}$$

Yarar *et al.* (1980) and Gürpınar *et al.* (1979) also studied the relation between MKS intensity and earthquake magnitude for Turkish earthquakes. Their proposed relations were essentially similar to those given by Equations (10.44) and (10.45) except for the constant with a minus sign. However, the magnitude in their formula is local magnitude. Furthermore, the coefficient for magnitude is about 1.10 times the one given in Equations (10.44) and (10.45). Also, Kudo (1983) studied the relation between the MKS intensity and the Intensity Scale of the Japan Meteorological Agency. As the maximum values of intensity were different from each other, it can be roughly said that two intensity values of MKS would be designated as one intensity value in the Intensity Scale of the Japan Meteorological Agency. However, the intensity scales of 4, 5 and 6 of the Japan Meteorological Agency have been recently revised and subdivided into weak and strong intensity levels. These days the broad intensity scale of the Japan Meteorological Agency discussed by Kudo (1983) is no longer used in Japan.

10.7.3   Relation between moment magnitude and rupture length, area and net slip of fault

It is natural to expect that the rupture sense of faulting of earthquakes greatly influences their seismic characteristics. In this study, the faulting sense is considered in empirical relations between moment magnitude and rupture length (L), rupture area (A) and net slip (U_{max}) of fault given here as an extension of the previous works of Aydan *et al.* (1996) and Aydan (1997, 2001):

$$L, S \text{ or } U_{max} = A \cdot M_w e^{M_w/B} \tag{10.46}$$

The functional form of the empirical relations is the same, while their constants A and B differ depending upon the faulting sense, which are given in Table 10.2. Kudo (1983) and

Table 10.2 Values of constants for Equation (10.46) for each fault parameter

	Rupture Length L (km)		Rupture Area S (km²)		Maximum Displacement U_{max} (cm)	
	A	B	A	B	A	B
Normal faulting	0.0014525	1.21	0.003	1.5	0.0003	1.6
Strike-slip faulting	0.0014525	1.25	0.001	1.7	0.00035	1.6
Thrust faulting	0.0014525	1.19	0.0032	1.5	0.0014	1.4

Ambraseys (1988) proposed similar empirical relations between rupture length and earth-quake magnitude for Turkish earthquakes. Most of them were caused by strike-slip faulting. Although their relations are good fits to the data set they used, they have limited applicability for the range of data set used in this study. Figure 10.21 shows the plot of data for several parameters previously listed together with empirical functions given by Equation (10.46) in Aydan (1997) and Wells and Coppersmith (1994). The horizontal axis of the plots is the moment magnitude of earthquakes. As seen from the figure, the data is somewhat scattered. Nevertheless, the proposed function, together with constants for each seismic parameter, is the best fit to observational data. The standard deviations of fitted equations to observational data were obtained. Nevertheless, they will not be presented in this article for the purpose of clarity. Furthermore, the relation proposed by Aydan (1997) without considering faulting sense is still valid as noted from Figure 10.21.

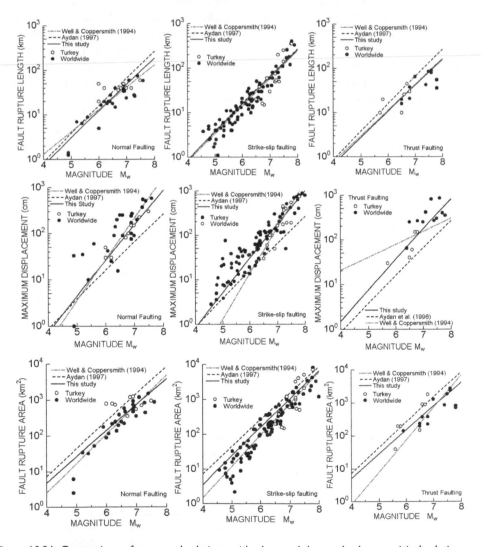

Figure 10.21 Comparison of proposed relations with observed data and other empirical relations

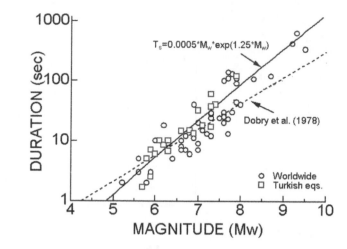

Figure 10.22 Comparison of proposed relation with observed data and other empirical relations

Another important source parameter is the duration of fault rupture. The duration is longer if the rupture propagation is unilateral. However, it is shortened if bilateral rupture propagation takes place. Figure 10.22 shows a compilation of data on Turkish earthquakes and worldwide including the most recent events such as the 2004 and 2005 Sumatra earthquakes and the 2005 Pakistan earthquake. The functional form of the empirical equation has the form of Equation (10.46), and it is shown in Figure 10.22 together with empirical relations proposed by Dobry *et al.* (1978). The proposed empirical relation holds for the data of Turkish earthquakes as well as worldwide data.

10.8   Strong motions and permanent deformation

10.8.1   Observations on strong motions and permanent deformations

10.8.1.1   Observations on maximum ground accelerations

It is observationally known that the ground motions induced by earthquakes could be much higher in the hanging wall block or mobile side of the causative fault as observed in the recent earthquakes such as the 1999 Kocaeli earthquake (strike-slip faulting), the 1999 Chi-chi earthquake (thrust faulting), the 2004 Chuetsu earthquake (blind thrust faulting) and the 2000 Shizuoka earthquake and l'Aquila earthquake (normal faulting) (Ohta, 2011; Chang *et al.*, 2004; Somerville *et al.*, 1997; Tsai and Huang, 2000; Aydan *et al.*, 2009; Abrahamson and Somerville, 1991; Aydan, 2003b; Aydan *et al.*, 2007) as seen in Figures 10.23–10.26.

Figure 10.26 illustrates the effect of the hanging wall effect on the attenuation of maximum ground accelerations observed in the 1999 Chi-chi earthquake (Taiwan), 1999 Düzce earthquake (Turkey) and 2001 Geiyo earthquake (Japan) with different faulting mechanisms (Ohta and Aydan, 2010).

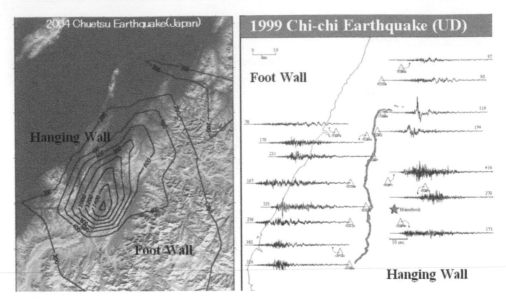

Figure 10.23 Footwall and hanging wall effects on the maximum ground accelerations (thrust faulting)

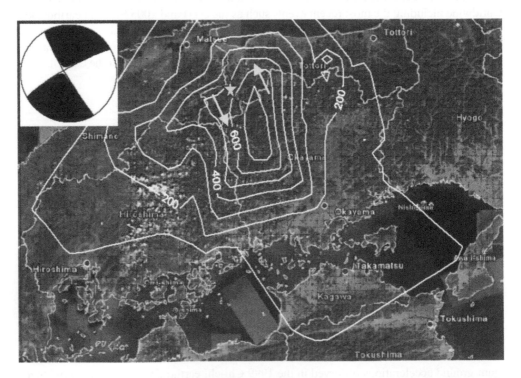

Figure 10.24 Mobile and stationary block effects on the maximum ground accelerations observed in 2000 Tottori Seibu earthquake (strike-slip faulting)

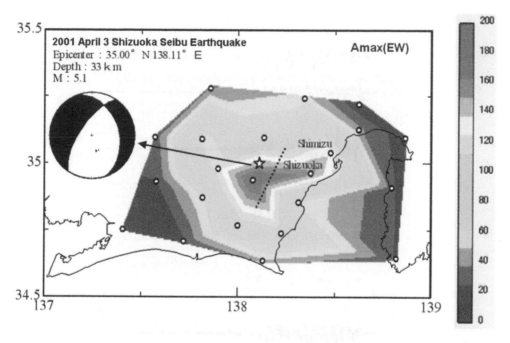

Figure 10.25 Footwall and hanging wall effects on the maximum ground accelerations (normal faulting)

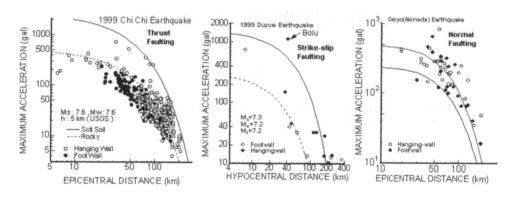

Figure 10.26 Attenuation of maximum ground accelerations for some earthquakes

Figure 10.27 shows the records of accelerations at the ground surface and at bedrock 260 m below at the Ichinoseki strong motion station (IWTH25) of the KiK-NET (2008) strong motion network of Japan measured during the 2008 Iwate-Miyagi earthquake. The strong motion station was located on the hanging wall side of the fault, and it was very close to the surface rupture. As noted from the figure, the ground acceleration of the UD component was amplified 5.67 times that at the bedrock, and the acceleration records are not symmetric with respect to time axis. This record is also the highest strong motion recorded in the world so far.

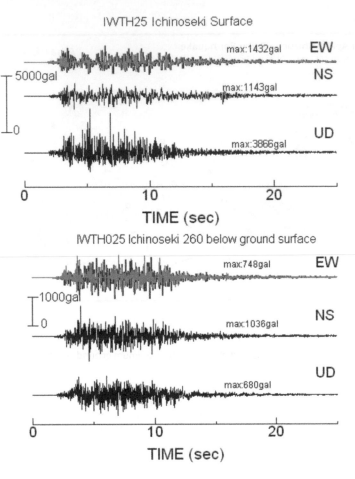

Figure 10.27 Acceleration records at ground surface and bedrock at Ichinoseki strong motion station IWTH25 of KiK-NET in Iwate-Miyagi earthquake

10.8.1.2   Permanent ground deformation

The recent global positioning system (GPS) also showed that permanent deformations of the ground surface occur after each earthquake (Figures 10.28 and 10.29). The permanent ground deformation may result from different causes such as faulting, slope failure, liquefaction and plastic deformation induced by ground shaking (Aydan *et al.*, 2010). These type of ground deformations have a limited effect on small structures as long as the surface breaks do not pass beneath those structures. However, such deformations may cause tremendous forces on long and/or large structures such as rock engineering structures. The ground deformation may induce large tensile or compression forces, as well as bending stresses in structures depending upon the character of permanent ground deformations. Blind faults and folding processes may also induce some peculiar ground deformations and associated folding of soft overlaying sedimentary layers. Such deformations caused tremendous damage on tunnels during the 2004 Chuetsu earthquake, although no distinct rupturing took place.

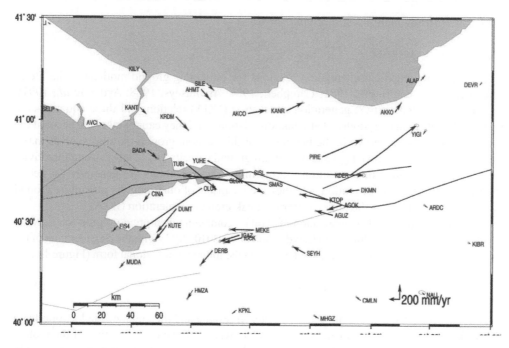

Figure 10.28 Permanent ground deformations and associated straining induced by the 1999 Kocaeli earthquake

Source: Reilinger *et al.* (2000)

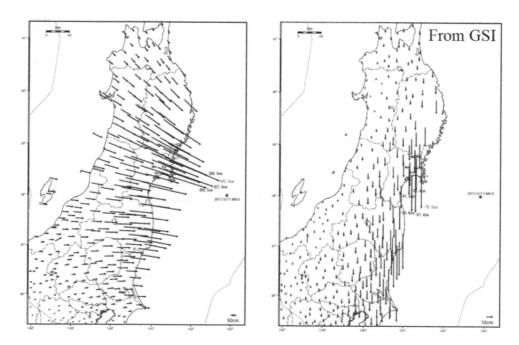

Figure 10.29 Ground deformation induced by the great East Japan earthquake

Source: Measured using GPS by GSI (2011)

10.8.1.3   Strong motion estimations

(A)   EMPIRICAL APPROACH

There are many empirical attenuation relations for estimating ground motions in the literature (i.e. Joyner and Boore, 1981; Campbell, 1981; Ambraseys, 1988; Aydan *et al.*, 1991). Including so-called next (?) generation attenuation (NGA) relations, all these equations are essentially spherical or cylindrical attenuation relations, and they cannot take into account the directivity effects. As shown in the beginning of this section, ground motions such as maximum ground acceleration (A_{max}) and maximum ground velocity (V_{max}) have strong directivity effects in relation to fault orientation. Furthermore, these relations are generally far below the maximum ground acceleration, and they are incapable of obtaining the maximum ground acceleration (AMAX) or the preferred term, "peak ground acceleration (PGA)."

Aydan (2012) proposed an attenuation relation by combining their previous proposals (Aydan *et al.*, 1997; Aydan, 1997, 2001, 2007; Aydan and Ohta, 2010, 2011) with the consideration of the inclination and length of the earthquake fault using the following functional form (Figure 10.30):

$$\alpha \max = F_1(V_s) * F_2(R, \theta, \varphi, L^*) * F_3(M) \tag{10.47}$$

where Vs, θ, φ, L^* and M are the shear velocity of ground and the angle of the location from the strike and dip of the fault (measured anticlockwise with the consideration of the mobile side of the fault) and earthquake magnitude. The following specific forms of functions in Equation (10.47) were put forward as:

$$F_1(V_s) = A e^{-v_s/B} \tag{10.48a}$$

$$F_2(R, \theta, \varphi, L^*) = e^{-R(1 - D\sin\theta + E\sin^2\theta)(1 + F\cos\varphi)/L^*} \tag{10.48b}$$

$$F_3(M) = e^{M/G} - 1 \tag{10.48c}$$

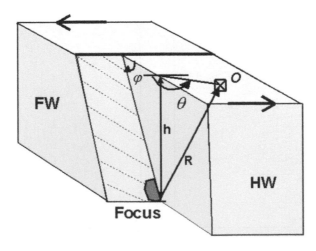

Figure 10.30 Illustration of geometrical fault parameters (R, θ, φ)

The same form is also used for estimating the maximum ground velocity (V_{max}). L^* (in km) is a parameter related to half of the fault length. And it is related to the moment magnitude in the following form:

$$L^* = a + be^{cM_w} \tag{10.49}$$

The specific values of constants of Equations (10.48)–(10.49) for this earthquake are given in Tables 10.3–10.5.

The most important parameter in this approach is the estimation of magnitude of the potential earthquake. If a very reliable database exists for a given region, one may estimate the magnitude of the most likely earthquake from such as database. Another approach may be the estimation from the characteristics (length, area, maximum relative slip) of active faults. Matsuda (1975), Sato (1989), Wells and Coppersmith (1994) and Aydan (1997) proposed empirical relations. Aydan (2007, 2012) recently established several relations between moment magnitude and rupture length (L), rupture area (A) and net slip (U_{max}) of fault given here and checked their validity with available data as well as the data from the most recent event of the 2011 Great East Japan earthquake:

$$L, S \, or \, U_{max} = A \cdot M_w e^{M_w/B} \tag{10.50}$$

The functional form of the empirical relations is the same, while their constants A and B differ depending upon the faulting sense, which are given in Table 10.4. If striation or sense of deformation of the potential active fault is known, it is also possible to infer its focal mechanism. Such a method is proposed by Aydan (2000a) and compared with the focal mechanism solutions inferred from fault striations or sense of deformation with those from telemetric wave solutions.

Table 10.3 Values of constants in Equation (10.48) for interplate earthquakes

	A	B(m s⁻¹)	D	E	F	G(Mw)
A_{max}	2.8	1000	0.5	1.5	0.5	1.05
V_{max}	0.4	1000	0.5	1.5	0.5	1.05

Table 10.4 Values of constants in Equation (10.48) for intraplate earthquakes

	A	B(m/s)	D	E	F	G(Mw)
A_{max}	2.8	1000	0.5	1.5	0.5	1.16
V_{max}	0.4	1000	0.5	1.5	0.5	1.16

Table 10.5 Values of constants in Equation (10.49) for earthquakes

Faulting Type	a	b	c
Normal faulting	30	0.002	1.35
Strike-slip faulting	20	0.002	1.40
Thrust faulting	20	0.002	1.27

Table 10.5 Values of constants for Equation (10.48) for each fault parameter

Fault Type	Parameter	L (km)	S (km²)	U_{max} (cm)
Normal	A	0.0014525	0.003	0.0003
Faulting	B	1.21	1.5	1.6
Strike-slip	A	0.0014525	0.001	0.00035
Faulting	B	1.19	1.7	1.6
Thrust	A	0.0014525	0.0032	0.0014
Faulting	B	1.25	1.5	1.4

L is rupture length, S is rupture area, U_{max} is maximum displacement.

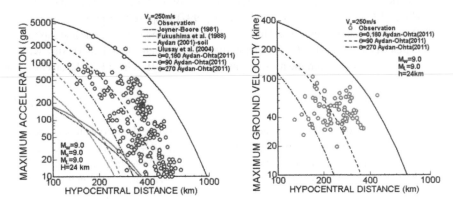

Figure 10.31 Comparison of estimated attenuation of maximum ground acceleration and ground velocity with observations for the 2011 Great East Japan earthquake

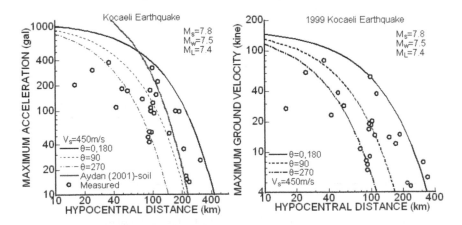

Figure 10.32 Comparison of estimated attenuation of maximum ground acceleration and ground velocity with observations for the 1999 Kocaeli earthquake

The attenuation relation given by (Equation 10.48) was used to evaluate the maximum ground acceleration and ground velocity of the 2011 Great East Japan earthquake (GEJE) and 1999 Kocaeli earthquake and compared with actual observation data in Figures 10.31–10.32. The same equation is used to evaluate the areal distribution of maximum ground

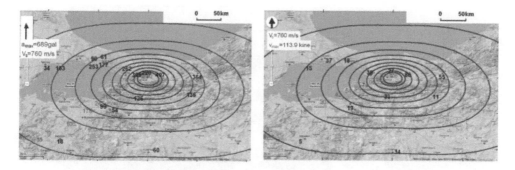

Figure 10.33 Comparison of estimated contours of maximum ground acceleration and ground velocity with observations for the 1999 Kocaeli earthquake

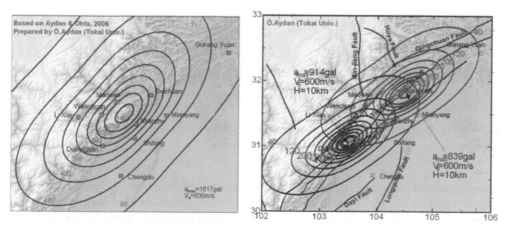

Figure 10.34 Comparison of single and double source models for maximum ground acceleration for the 2008 Wenchuan earthquake

acceleration and velocity for Kocaeli earthquake and compared with observational data in Figure 10.33. For large earthquakes, the use of Equation (10.48) and the estimations based on the segmentation of faults may be more appropriate. Figure 10.34 shows the single- and double-source models for the 2008 Wenchuan earthquake.

(B)   GREEN-FUNCTION-BASED EMPIRICAL WAVE FORM ESTIMATION

The empirical Green's function method was initially introduced by Hartzell (1978). Follow-up methods proposed by Hadley and Helmberger (1980) and Irikura (1983) are modifications of Hartzell's method of summing empirical Green's functions. In the empirical Green's function approach, rupture propagation and radiation pattern were specified deterministically, and the source propagation and radiation effects were included empirically by assuming that the motions observed from aftershocks contained this information (Somerville *et al.*, 1991). A semiempirical Green's function summation technique has been used by Wald *et al.* (1998), Cohee *et al.* (1991) and Somerville *et al.* (1991), which allows the

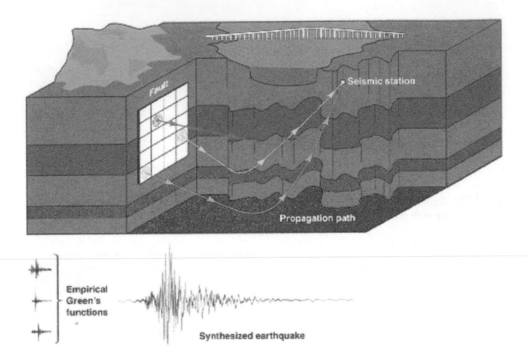

Figure 10.35 Illustration of the fundamental concept of the empirical Green's function method

Source: From Hutchings and Viegas (2012)

gross aspects of the source rupture process to be treated deterministically using a kinematic model based on first motion studies, teleseismic modeling and distribution of aftershocks. Gross aspects of wave propagation are modeled using theoretical Green's functions calculated with generalized rays (Figure 10.35). The empirical Green's function method can be used only for a region where small events (i.e., aftershocks or foreshocks) of the target event are available.

Ikeda *et al.* (2016) recently performed an analysis of strong motion induced by the 2014 Nagano-ken Hokubu earthquake using the empirical Green's function method. The earthquake fault was assumed to be 9.16 km long and 7.2 km wide with a dip angle of 50 degrees. The stress drop was about 12.10. MPa, and the rupture time was 2.7 km s^{-1} with a rise time of 0.16 s. Figure 10.36 shows the observed and simulated responses of acceleration, velocity and displacement for the north-south direction for the Hakuba strong motion station of K-NET. As seen from the figure, the simulated strong motions are close to the observations.

(C) NUMERICAL APPROACHES

There are several numerical techniques, which are known to be finite difference method (FDM), finite element method (FEM), boundary element method (BEM). The FDM is the earliest numerical model, while FEM and BEM have become available after the 1960s and 1970s, respectively. Therefore, the first application of the numerical methods for strong

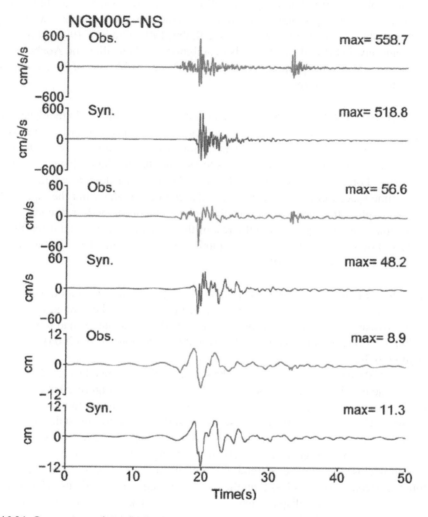

Figure 10.36 Comparison of simulations with observations for the north-south component of Hakuba record

motion estimation is related to the FDM. When this method is applied for strong motion estimation, one needs to solve Equation (10.27) together with appropriate constitutive laws for the medium and the assumption of a rupture plane. In particular, the geometrical definition of the rupture plane and its rupture velocity would be also the key parameters of the simulations. Furthermore, both the FDM and FEM consider the finite size domain; the prevention of reflections of waves from the boundaries would be necessary. This issue is generally dealt with by the introduction of the Lysmer-type viscous boundaries into the numerical model. Both the FDM and FEM would evaluate the wave propagation without any assumption on how waves generated at the source is transferred to any point of particular interest, which is a major issue in Green's function–based strong motion simulations. While the FEM can easily handle the irregular boundaries such as the surface of the model with its topography as a free boundary, the FDM has a severe restriction dealing with such boundaries with irregular geometry. Nevertheless, some procedures dealing with irregular surface topography have been proposed (i.e. Hestholm, 1999; Gravers, 1996)). For irregular surface topography, FDM and FEM are also combined (i.e. Ducellier and Aochi, 2012). In addition, there are also some proposals to combine FDM or FEM with BEM in order to deal with newly developing ruptures.

(i) Finite difference method (FDM)   The constitutive law for the medium adjacent to the rupture plane is generally assumed to be visco-elastic (i.e. Graves 1996). The most difficult aspect is the simulation of the fault plane associated with rupture process in the FDM schemes. The most conventional technique is to assume that the fault plane coincides with the grid planes. Forced displacement field is introduced at the domain where two points occupy the same space initially and can move relative to each other after the rupture. Many schemes also explore the incorporation of the finite element method or boundary integral model to simulate the fault plane, and the rest of the domain is discretized using the FDM. Figure 10.37 shows a simulation of strong motion induced by the 1995 Kobe earthquake using the FDM by Pitarka *et al.* (1998).

(ii) Finite element method (FEM)   Toki and Miura (1985) utilized Goodman-type joint elements in 2-D-FEM to simulate both the rupture process and the ground motions (Figure 10.38). Fukushima *et al.* (2010) utilized this method to simulate ground motions caused by the 2000 Tottori earthquake (Figure 10.39). Later Mizumoto *et al.* (2004) extended the same method to 3-D.

Iwata *et al.* (2016) recently investigated the strong motions induced by the 2014 Nagano-Hokubu earthquake. The model is based on 3-D FEM version. Figure 10.40(a) shows the fault parameters, and Figure 10.40(b) shows the 3-D mesh of the earthquake fault and its vicinity. Figure 10.41(a) shows the time histories of surface acceleration at distances of 1 km and 2 km from the surface rupture in the 3-D-FEM model. Rupture time is about 7–8 s. The maximum acceleration is higher in the east-side (hanging wall) than that in west-side (footwall), which is close to the general trend observed in strong motion records. Nevertheless, the computed acceleration was less than the measured accelerations. Figure 10.41(b) shows the time histories of surface displacement at distances of 1 km and 2 km from the surface rupture. The east side of the fault moves upward with respect to the footwall together with movement to the north, and the vertical displacement of the east side is larger than that of the footwall, and the computed results are close to the observations. However, it is necessary to utilize finer meshes for better simulations of ground accelerations, which requires use of the supercomputers.

(iii) GPS method   The recent global positioning system (GPS) also showed that permanent deformations of the ground surface occur after each earthquake (Figure 10.28–10.29). The permanent ground deformation may result from different causes such as faulting, slope failure, liquefaction and plastic deformation induced by ground shaking (Aydan *et al.*, 2010). These types of ground deformations have a limited effect on small structures as long as the surface breaks do not pass beneath those structures. However, such deformations may cause tremendous forces on long and/or large structures. The ground deformation may induce large tensile or compression forces, as well as bending stresses in structures depending upon the character of permanent ground deformations. As an example, the ground deformations

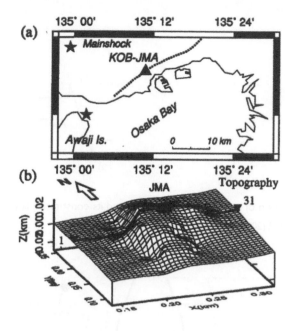

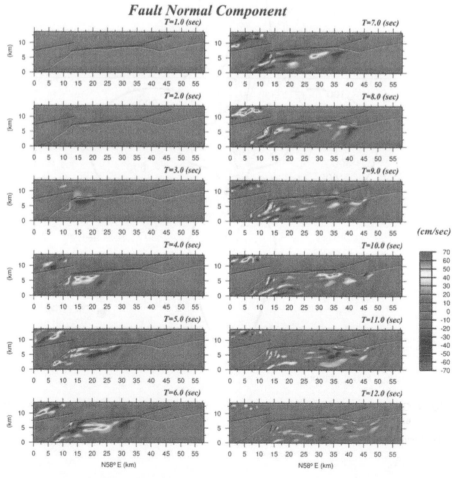

Figure 10.37 Fault normal ground velocity propagation induced by the 1995 Kobe earthquake

Source: From Pitarka *et al.* (1998)

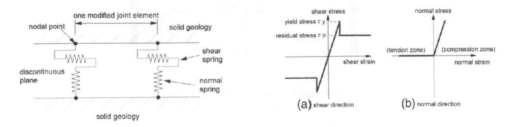

Figure 10.38 Representation of joint elements for faults and its constitutive law

Source: From Fukushima *et al.* (2010)

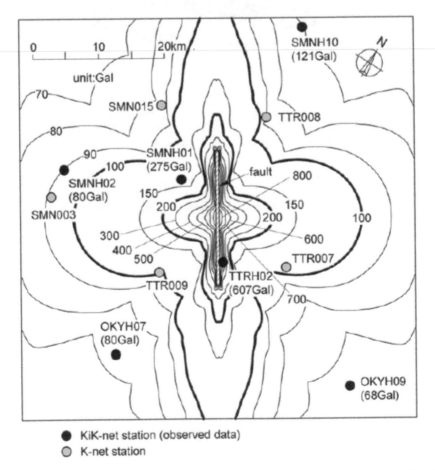

Figure 10.39 Comparison of computed and observed maximum ground acceleration for 2000 Tottori earthquake

Source: From Fukushima *et al.* (2010)

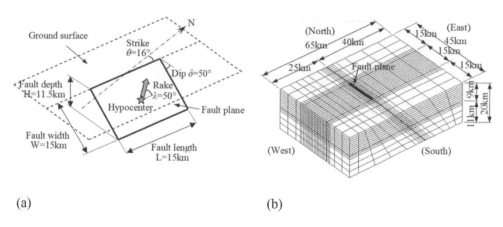

(a)   (b)

Figure 10.40 Fault model and 3-D FEM mesh

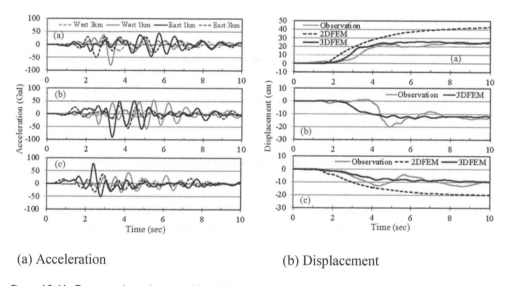

(a) Acceleration   (b) Displacement

Figure 10.41 Computed acceleration (a) and displacement (b) responses

reported by Reilinger *et al.* (2000) are shown in Figure 10.42, which were caused by a strike-slip fault during the 1999 Kocaeli earthquake in Turkey. Blind faults and folding processes may also induce some peculiar ground deformations and associated folding of soft overlaying sedimentary layers. Such deformations caused tremendous damage on tunnels during the 2004 Chuetsu earthquake although no distinct rupturing took place.

(iv) InSAR method   Interferometric synthetic aperture radar, abbreviated InSAR or IfSAR, is a radar technique used in geodesy and remote sensing. This geodetic method uses two or more synthetic aperture radar (SAR) images to generate maps of surface deformation or digital elevation, using differences in the phase of the waves returning to the satellite or aircraft.

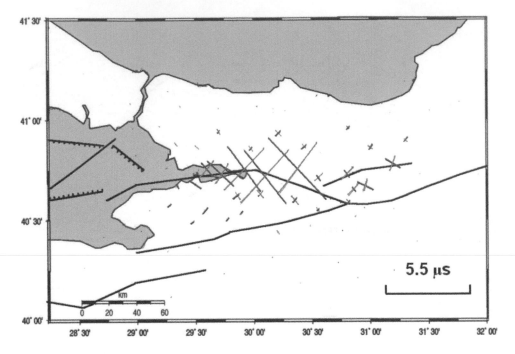

Figure 10.42 Computed ground straining from GPS measurements

Source: From Aydan *et al.* (2010)

The technique can potentially measure centimeter-scale changes in deformation over spans of days to years. It has applications for geophysical monitoring of natural hazards, for example earthquakes, volcanoes and landslides, and in structural engineering, in particular the monitoring of subsidence and structural stability. Figure 10.43 shows an application of the InSAR to estimate ground deformations induced by the 1999 Kocaeli earthquake.

(v) EPS method   Ohta and Aydan (2007) and Aydan and Ohta (2011) have recently showed that the permanent ground deformations may be obtained from the integration of acceleration records. The erratic pattern screening (EPS) method proposed by Ohta and Aydan (2007) and Aydan and Ohta (2011) can be used to obtain the permanent ground displacement with the consideration of features associated with strong motion recording. The duration of shaking should be naturally related to the rupture time t_r. Depending upon the arrival time difference of S-wave and P-wave, the shaking duration would be a sum of rupture duration and S-P arrival time difference Δt_{sp}. If the ground exhibits a plastic response due to yielding or ground liquefaction, the duration of shaking would be elongated. If coseismic crustal deformations are to be obtained, the integration duration should be restricted to the rupture duration with the consideration of the S-P arrival time difference. The existence of plastic deformation can be assessed by comparing the effective shaking duration and the sum of rupture duration t_r and S-P arrival time difference Δt_{sp}. If the effective shaking duration is longer than the sum of rupture duration t_r and S-P waves arrival time difference Δt_{sp}, the integration can be carried out for both durations, and the difference can be interpreted as the

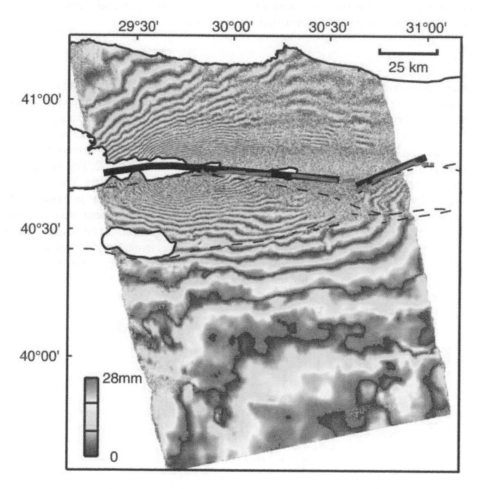

Figure 10.43 Interferogram produced using ERS-2 data from 13 August and 17 September 1999 for the 1999 Kocaeli earthquake

Source: NASA/JPL-Caltech

plastic ground deformation. The most critical issue is the information of rupture duration. The data on rupture duration is generally available for earthquakes with a moment magnitude greater than 5.10. worldwide. The effective shaking duration may be obtained from the acceleration records using the procedure proposed by Housner (1965). When such data is not available, the empirical relation proposed by Aydan (2007) may be used:

$$t_r = 0.005 M_w \exp(1.25 M_w)$$
(10.51)

S-P arrival time difference Δt_{sp} can also be easily evaluated from the acceleration record.

They divided an acceleration record into three sections and applied filters in Section 1 and Section 3, and the integration is directly carried in Section 2 without any filtering. The times to differentiate sections are t_1 and t_2. Time t_1 is associated with the arrival of the P-wave, while time t_2 is related to the arrival time of P-wave, rupture duration and S-P waves' arrival time difference for the crustal deformations as given here:

$$t_2 = t_1 + t_r + \Delta t_{sp} \tag{10.52}$$

Any deformation after time t_2 must be associated with deformations related to the local plastic behavior of the ground at the instrument location. We show one example for defining times t_1 and t_2 on a record taken at HDKH07 strong motion station in the 2003 Tokachi-oki earthquake, Japan (Figure 10.44). The estimated rupture time for this earthquake is about 40 s (Yamanaka and Kikuchi, 2003) with about 18 s S-P waves' arrival time difference (Δt_{sp}).

Another important issue is how to select filter values in Sections 1 and 2. This is somewhat a subjective issue, and it depends upon the sensitivity of the accelerometers. The filter value ε_1 is generally small, and this stage is associated with the pretrigger value of instruments. Our experience with the selection of ε_1 for K-NET and KiK-NET accelerometer records implies that its value be less than ±2 gals. As for the value of ε_2, higher values must be assigned. Again, our experiences with the records of K-NET and KiK-NET accelerometers imply that its value should be ±6 gals. The threshold values in the acceleration records of Turkey and Italy are much less than those from the highly sensitive accelerometers of networks in Japan.

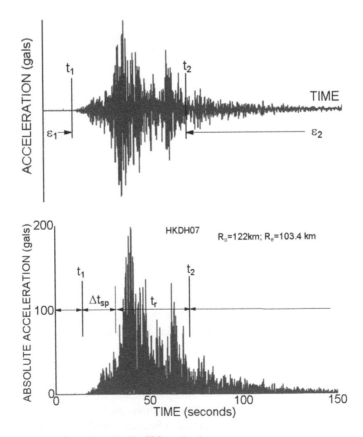

Figure 10.44 Definition of sections in the EPS method

This method is applied to the results of laboratory faulting and shaking table tests, in which shaking was simultaneously recorded using both accelerometers and laser displacement transducers. Furthermore, the method was applied to the strong motion records of several large earthquakes with measurements of ground movements by GPS as seen in Figure 10.45. The comparison of computed responses with actual recordings was almost the same, implying that the proposed method can be used to obtain actual recoverable as well as permanent ground motions from acceleration recordings. Figure 10.46 shows the application of the EPS method to the strong motions records of the 2009 L'Aquila earthquake to estimate

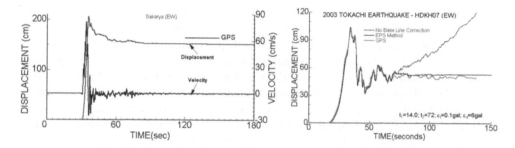

Figure 10.45 Comparison of the permanent ground deformation by the EPS method with measured GPS recordings

Source: From Ohta and Aydan (2007)

Figure 10.46 Estimated permanent ground displacements by EPS method

Source: From Aydan et al. (2009)

the coseismic permanent ground displacements. These results are very consistent with the GPS observations. However, it should be noted that the permanent ground deformations recorded by the GPS does not necessarily correspond to those of the crustal deformation. Surface deformations may involve crustal deformation as well as those resulting from the plastic deformation of ground due to ground shaking. The records at ground surface and 210 m below the ground surface taken at IWTH25 during the 2008 Iwate-Miyagi earthquake clearly indicated the importance of this fact in the evaluation of GPS measurements (KiKNET, 2008).

(vi) Okada's method   Okada (1992) proposed closed-form solutions for dislocation in a half-space isotropic medium (Figures 10.47 and 10.48). Closed-form analytical solutions are presented in a unified manner for the internal displacements and strains due to shear and tensile faults in a half space for both point and finite rectangular sources. These expressions evaluate deformations in an infinite medium, and a term related to surface deformation is obtained through multiplying by the depth of observation point. Stein (2003) utilized the solutions of Okada's method in his software to compute permanent ground deformation and associated stress changes. This method is also used to forecast earthquakes with the introduction of superposing displacement field and associated stress changes together with the use of the Mohr-Coulomb criterion (King *et al*). His method has been upgraded by his research group and applied to various earthquakes in recent years. Figure 10.49 shows an example of computations by Toda *et al*. (2002) for earthquake activity in the Izu islands.

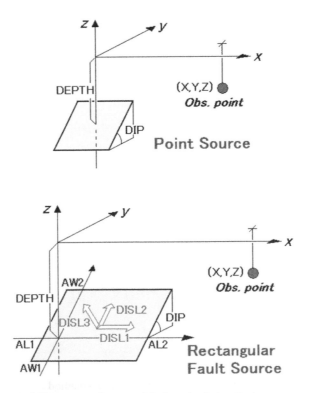

Figure 10.47 Geometrical illustration of assumed fault and relative displacements

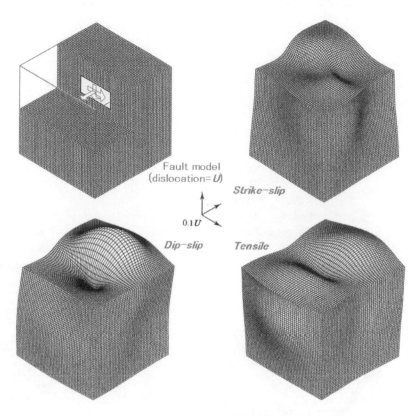

Fault model
(dislocation=U)

Strike-slip

0.1U

Dip-slip   *Tensile*

Figure 10.48 Illustration of ground deformation associated with faulting

Figure 10.49 Relations between static stress changes and seismicity
Source: From Toda *et al.* (2002)

(a)

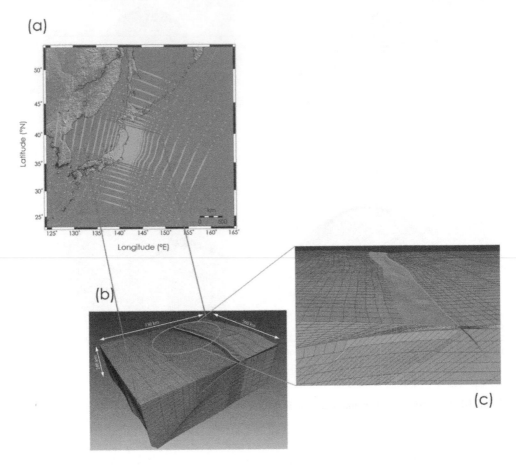

Figure 10.50 3D-FEM mesh for strong motion simulation of 2011 Great East Japan earthquake

Source: From Romano et al. (2014)

(vii) Numerical methods   FDM, FEM, BEM or combined FDM, FEM and BEM produce permanent ground deformation as a natural output of computations provided that the rupture process is well simulated. Figures 10.50 and 10.51 show an example of such a computation for 2011 Great East Japan earthquake (or Tohoku earthquake). Although, the displacement response could be more easily simulated, acceleration responses simulation requires fine meshes, which undoubtedly require the use of supercomputers. It should be also noted that FEM models could easily simulate both permanent displacements in addition to strong ground motions.

10.9   Effects of surface ruptures induced by earthquakes on rock engineering structures

In this section, typical examples of damage to various structures induced by the fault breaks observed in recent large earthquakes since 1995 are presented, and details can be found in the quoted references (e.g. Aydan *et al.*, 2011; Ohta et al., 2014).

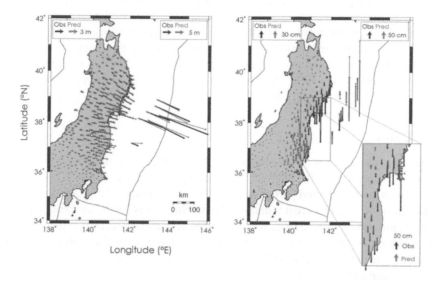

Figure 10.51 3-D-FEM simulations of ground motions associated with the 2011 Great East Japan earthquake

Source: From Romano et al. (2014)

(a) Arifiye   (b) Pefong Bridge

Figure 10.52 (a) Collapse of the overpass, (b) collapse of Pefong Bridge

(Note the uplifted ground on RHS.)

10.9.1   *Bridges and viaducts*

Along the damaged section of the TEM motorway previously mentioned, there were several overpass bridges. Among them, a four-span overpass bridge at Arifiye junction collapsed as a result of faulting (Figure 10.52(a)). The fault rupture passed between the northern abutment and the adjacent pier. The overpass was designed as a simply supported structure according to the modified AASHTO standards, and girders had elastometric bearings. However, the girders were connected to one another through prestressed cables. The angle between the motorway and the strike of the earthquake fault was approximately 15 degrees, while the angle between the axis of the overpass bridge and the strike of the fault was 65 degrees. The measurements of the relative displacement in the vicinity of the fault range between 330 and 450 cm. Therefore an average value of 390 cm could be assumed for the relative displacement between the pier and the abutment of the bridge. The Pefong bridge collapsed due to thrust faulting in the 1999 Chi-Chi earthquake, which passed between the piers near its southern abutment, as seen in Figure 10.52(b).

10.9.2 Dams

The Shihkang dam, which is a concrete gravity dam with a height of 25 m, was ruptured by thrust-type faulting during the 1999 Chi-Chi earthquake (Figure 10.53). The relative displacement between the uplifted part of the dam was more than 980 cm. Liyutan rockfill dam with a height of 90 m and a crest width of 210 m, which was on the overhanging block of Chelongpu fault, was not damaged, even though the acceleration records at this dam showed that the acceleration was amplified 4.5 times that at the base of the dam (105 gals). The deformation zone of faulting during the 2008 Wenchuan earthquake caused some damage at the Zipingpu dam with concrete facing.

10.9.3 Tunnels

The past experience on the performance of tunnels through active fault zones during earthquakes indicates that the damage is restricted to certain locations. Portals and the locations where the tunnel crosses the fault may be damaged, as occurred in the 2004 Chuetsu, 2005 Kashmir and 2008 Wenchuan earthquakes (Figure 10.54). A section nearby Elmalık portal of Bolu Tunnel collapsed (Figure 10.55). This section of the tunnel was excavated under very heavy squeezing conditions. The well-known examples of damage to tunnels at locations where the fault rupture crossed the tunnel are mainly observed in Japan. The Tanna fault ruptured during the 1930 Kita-Izu earthquake, causing damage to a railway tunnel; the relative displacement was about 100 cm. The 1978 Izu-Oshima Kinkai earthquake induced damage to the Inatori railway tunnel. Similar type of damage to the tunnels of Shinkansen and subway lines through the Rokko mountains with a small amount of relative displacement due to motions of the Rokko, Egeyama and Koyo faults was also observed. During the

Figure 10.53 Failure of Shihkang dam due to thrust faulting

Figure 10.54 Examples of damaged portals of tunnels

Figure 10.55 Collapse of Bolu Tunnel during the 1999 Düzce earthquake

1999 Chi-Chi earthquake, the portal of the water intake tunnels was ruptured for a distance of 10 m as a result of thrust faulting. Except for this section, the tunnel was undamaged for its entire length.

Jiujiaya Tunnel is a 2282 m long double-lane tunnel that is 226.6 km away from the earthquake epicenter and that is about 3–5 km away from the earthquake fault of the Wenchuan earthquake. The tunnel face was 983 m from the south portal at the time of the earthquake. The concrete lining follows the tunnel face at a distance of approximately 30 m. Thirty workers were working at the tunnel face, and one worker was killed by the flying pieces of rock bolts, shotcrete and bearing plates caused by the intense deformation of the tunnel face during the earthquake. The concrete lining was ruptured and fallen down at several sections (Figure 10.56). However, the effect of the unreinforced lining rupturing was quite large and intense in the vicinity of the tunnel face. The rupturing of the concrete lining generally occurred at the crown sections, although there was rupturing along the shoulders of the tunnel at several places. Furthermore, the invert was uplifted due to buckling at the middle sections.

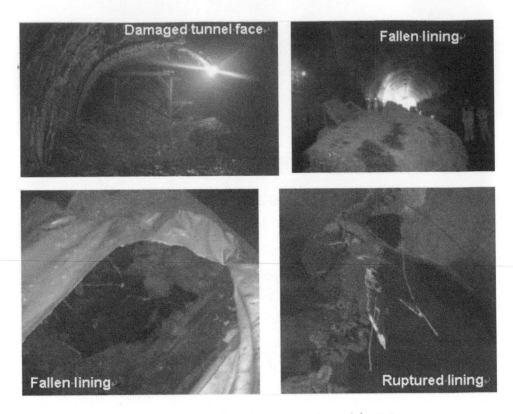

Figure 10.56 Earthquake damage at Jiujiaya Tunnel due to permanent deformations

The Kumamoto earthquake on 16 April 2016 caused heavy damage to several tunnels in the vicinity of Tateno and Minami-Aso villages. Damage to the Tawarayama Roadway Tunnel and Aso Railway Tunnel and Minami-Aso Tunnel was publicized. The damage to Tawarayama Tunnel occurred at two locations (Figure 10.57). The first damage occurred approximately 50–60 m from the west portal of the tunnel, and the concrete lining was displaced by about 30 cm almost perpendicularly to the tunnel axis. The heaviest damage occurred for a length of 10 m about 1600 m from the west portal and about 460 m from the east portal. The angle between the relative movement and tunnel axis was about 20–30 degrees. At this location, the nonreinforced concrete lining collapsed for a length of about 5 m. Although the tunnel is located about 2 km away from the main fault, the tunnel was damaged by secondary faults associated with the transtension nature of the earthquake fault.

10.9.4   Landslides and rockfalls

The recent large earthquakes caused mega-scale slope failures and rockfalls particularly along the surface ruptures on the hanging wall side of the fault. The slope failure induced in Beichuan town during the 2008 Wenchuan earthquake is of great interest. In association

Figure 10.57 Views of damage and their locations at Tawarayama Tunnel

with the sliding motion of the earthquake fault, northwest- or southeast-facing slopes failed during this earthquake. There were two large-scale slope failures (landslides) in Beichuan town, which destroyed numerous buildings and facilities. The northwest-facing landslide (Jingjiashan) involved mainly limestone while the southeast-facing landslide (Wangjiaya) involved phyllite (mudstone, according to some) rock unit (Figure 10.58). Limestone layers dipped toward the valley-side with an inclination of about 30 degrees. Furthermore, there are several faults dipping parallel to the failure surface within the rock mass. The angles of the lower and upper parts of the failed slope are 10.0 degrees and 30–35 degrees, respectively. The existence of several faults dipping parallel to the slope with an inclination of about 60–65 degrees creates a stepped failure surface.

The SE facing slope (Wangjiaya landslide) may involve a slippage along the steeply dipping bedding plane (fault plane?) and shearing through the layered rock mass. In other words, it may be classified as a combined sliding and shearing sliding (Aydan et al., 1992). The angles of the lower and upper parts of the failed slope are 40–45 degrees and 30–35 degrees, respectively. The layers dip at an angle of 40 degrees toward the valley, and the shearing plane is inclined at an angle of 20 degrees.

Figure 10.58 Views of landslides in Beichuan

10.10 Response of Horonobe underground research laboratory during the 20 June 2018 Soya region earthquake and 6 September 2018 Iburi earthquake

10.10.1 Characteristics of Soya region earthquake

The Soya region earthquake occurred on 20 June 2018 at 5:28 a.m. The moment magnitude of the earthquake was 4.0 according to F-NET of NIED. The focal mechanism of the earthquake was estimated to be thrust fault (Figure 10.59a), which is a consistent mechanism in view of the tectonics of the Soya region. Figure 10.59b. shows the inferred stress state for the earthquake. The maximum horizontal stress acts in almost in an east-west (EW) direction.

10.10.2 Characteristics of Iburi earthquake

Another major earthquake with a moment magnitude of 6.6 (Mj 6.7) occurred on 6 September 2018 at 3:08 in Iburi Region of Hokkaido island, which is about 216 km away from Horonobe. The focal mechanism of this earthquake was due to the blind steeply dipping thrust fault. The earthquake was felt in Horonobe as recorded by the Kik-Net network. However, the maximum ground acceleration was 3.3 gals.

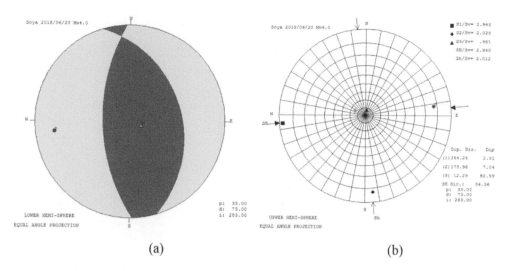

(a)   (b)

Figure 10.59 (a) Redrawn focal mechanism obtained by F-NET, (b) inferred stress state for the focal mechanism obtained by F-NET by Aydan's method (2000a)

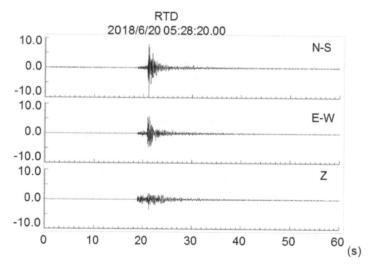

Figure 10.60 Seismic records of the 20 June 2018 Soya region earthquake (20 June 2018, M = 4.8) observed in Horonobe URL (maximum acceleration is 8.0, 5.8 and 2.0 gals for N-S, E-W and Z directions.)

10.10.3 Acceleration records at Horonobe URL

Accelerometers are set at the ground surface, GL.-250 m and GL.-350 m galleries in Horonobe URL. Figure 10.60 shows the seismic records of the 20 June 2018 Soya Region earthquake observed at GL.-350 m gallery and shaft bottoms (Figure 10.61). Table 10.6

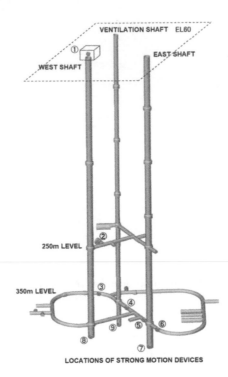

Figure 10.61 Locations of strong motion observation stations

Table 10.6 Maximum acceleration and amplification

Locations	NS (gals)	EW (gals)	UD (gals)
Surface (+10.0 m)	8.0	5.8	2.0
250 m level	8.72	10.04	3.8
350 m level	10.4	8.710.	3.45
West shaft (310.5 m)	7.79	5.410.	2.74
Ventilator shaft (380 m)	0.87	0.10.4	0.94

Source: Kik-NET and K-NET data.

compares the maximum acceleration at each strong motion station installed at various depths. The ground motions are amplified toward the ground surface. The data even in the same level is scattered, which may imply some local effects such as the geological conditions and the geometry of the opening where the devices are installed.

The National Research Institute for Earth Science and Disaster Prevention of Japan (NIED) has been operating the Kik-Net and K-Net strong motions networks. There is a strong motion station of the Kik-Net in Horonobe town that recorded the accelerations at the ground surface and at a depth of 100 m from the ground surface (-70 m). Figure 10.62 shows the acceleration records taken at the ground surface and at the base (100 m below the

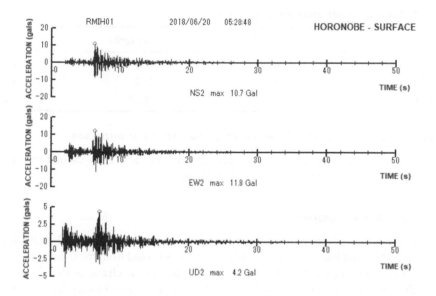

(a) Ground surface (30m)

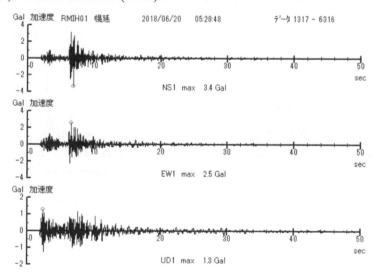

(b) Base (100m below the ground surface)

Figure 10.62 Acceleration records taken at the ground surface and at the base

ground). Table 10.7 gives the maximum ground accelerations and their amplifications. Theoretically, the amplification is expected to be greater than 2 for elastic ground (Nasu, 1931). The comparison indicates that the amplification is more than 3 times. Compared to data from the Kik-NET, the measurements at the Horonobe URL are somewhat scattered.

Table 10.7 Maximum acceleration and amplification

	NS (gals)	EW (gals)	UD (gals)
Surface (+30 m)	3.4	2.5	1.3
Base (−70 m)	10.7	11.8	4.2
Amplification	3.6	4.72	3.23

10.10.4 Fourier and acceleration response spectra analyses

The Fourier and accelerations response spectra analyses have been carried out for each strong motion stations. We report some of them here.

(a) Fourier spectra analyses

The Fourier spectra analyses of acceleration records measured by the Kik-NET (RMIH01) at the ground surface and base are shown in Figure 10.63. As noted from the figure, the dominant frequency ranges between 4 Hz and 8 Hz, and the Fourier Spectra characteristics do not change with depth, although the amplitude of the ground surface is at least 3 times that at the base.

Figure 10.64 shows the FFT of records taken at the ground surface and at a depth of 380 m at the shaft bottom in Horonobe URL. The FFT amplitude of the shaft bottom records are

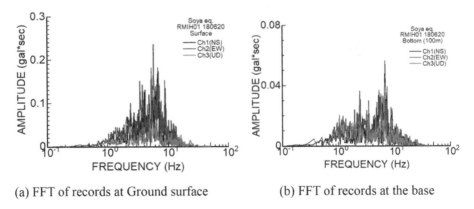

(a) FFT of records at Ground surface   (b) FFT of records at the base

Figure 10.63 FFT of records at the ground surface and base

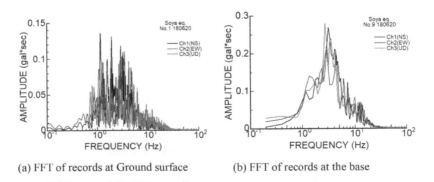

(a) FFT of records at Ground surface   (b) FFT of records at the base

Figure 10.64 FFT of records at the ground surface and base

almost the same as that of the ground surface. The frequency characteristics are also quite similar, resembling those of the Kik-Net records.

Figure 10.65 shows the FFT spectra of the record taken at No.9 observation station during the 2018 Iburi earthquake. Except for the UD component, the other components, except amplitude, are quite similar to those of the Soya region earthquake shown in Figure 10.64. The normalized amplitude may be useful for comparison purposes.

(b)   Acceleration response spectra analyses

A series of acceleration response analyses are carried out. Figure 10.66 shows the acceleration response spectra for RHIM01 and Horonobe URL No. 1 strong ground motion stations. The amplitude and frequency characteristics are somewhat different. The ground conditions at the RHIM01 may be softer than those at the Horonobe URL site.

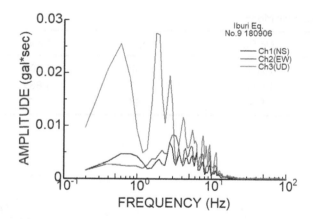

Figure 10.65 FFT of records at the ground surface for the records due to 2018 Iburi earthquake of 6 September 2018

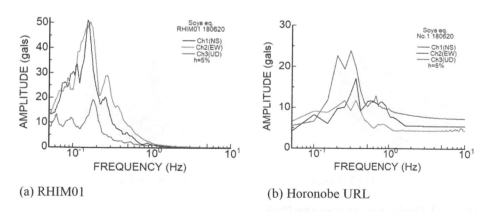

(a) RHIM01   (b) Horonobe URL

Figure 10.66 Comparison of acceleration response spectra for RHIM01 and Horonobe URL site

10.10.5   Modal analyses

A series of 3-D finite element modal analyses were carried out for four conditions: no shafts, single shaft, double shafts and triple shafts. The software used was 3-D MIDAS-FEA. Table 10.8 gives the material properties used in numerical analyses, while Table 10.9 compares the eigenvalues for four different conditions; Figure 10.67 shows the displacement response for Mode 1.

Table 10.8 Material properties

Material	UW (kN m^{-3})	E (GPa)	Poisson's ratio
Rock mass	26.5	0.600	0.37
Concrete	23.5	11.042	0.20

Table 10.9 Eigenvalues for Mode I

	No Shaft	Single	Double	Triple
Mode I(s)	1.763	1.752	1.203	1.199
Mode 2(s)	1.645	1.635	1.889	1.172
Mode 3(s)	1.564	1.554	1.117	1.111

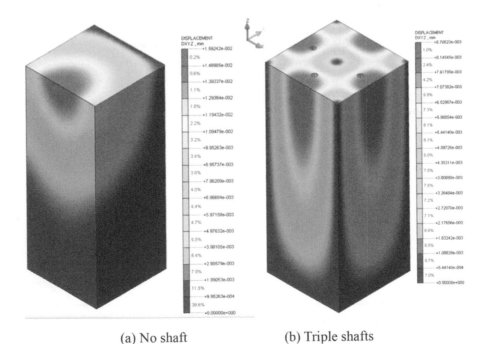

(a) No shaft   (b) Triple shafts

Figure 10.67 Displacement response for Mode I

10.11   Global positioning method for earthquake prediction

As stated previously, if the stress state and the yielding characteristics of the earth's crust are known at a given time, one may be able to predict earthquakes with the help of some mechanical, numerical and instrumental tools. The GPS method may be used to monitor the deformation of the Earth's crust continuously with time. From these measurements, one may compute the strain rates and probably the stress rates. The stress rates derived from the GPS displacement rates can be effectively used to locate the areas with high seismic risk, as proposed by Aydan *et al.* (2000b). Thus, daily variations of derived strain–stress rates from dense continuously operating GPS networks in Japan and the United States may provide high-quality data to understand the behavior of the Earth's crust preceding earthquakes.

10.11.1   Theoretical background

First we describe a brief outline of the GPS method proposed by Aydan (2000b, 2004, 2008). The crustal strain rate components can be related to the displacement rates at an observation point (x, y, z) through the geometrical relations (i.e. Eringen, 1980) as given here:

$$\dot{\varepsilon}_{xx} = \frac{\partial \dot{u}}{\partial x}; \dot{\varepsilon}_{yy} = \frac{\partial \dot{v}}{\partial y}, \dot{\varepsilon}_{zz} = \frac{\partial \dot{w}}{\partial y}, \dot{\gamma}_{xy} = \frac{\partial \dot{v}}{\partial x} + \frac{\partial \dot{u}}{\partial y}, \dot{\gamma}_{yz} = \frac{\partial \dot{w}}{\partial y} + \frac{\partial \dot{v}}{\partial z}, \dot{\gamma}_{zx} = \frac{\partial \dot{w}}{\partial x} + \frac{\partial \dot{u}}{\partial z} \quad (10.53)$$

where $\dot{u}$, v and $\dot{w}$ are displacement rates in the direction of x, y and z, respectively. $\dot{\varepsilon}_{xx}$, $\dot{\varepsilon}_{yy}$ and $\dot{\varepsilon}_{zz}$ are strain rates normal to the x, y and z planes and $\dot{\gamma}_{xy}$, $\dot{\gamma}_{yz}$, $\dot{\gamma}_{zx}$ are engineering shear strain rates. The GPS measurements can only provide the displacement rates on the Earth's surface (x (EW) and y (NS) directions), and it does not give any information on displacement rates in the z-direction (radial direction). Therefore, it is impossible to compute normal and shear strain rate components in the vertical (radial) direction near the Earth's surface. The strain rate components in the plane tangential to the Earth's surface would be $\dot{\varepsilon}_{xx}$, $\dot{\varepsilon}_{yy}$ and $\dot{\gamma}_{xy}$. Additional strain rate components $\dot{\gamma}_{yz}$ and $\dot{\gamma}_{zx}$, which would be interpreted as tilting strain rate in this article, are defined by neglecting some components in order to make the utilization of the third component of displacement rates measured by GPS as follows:

$$\dot{\gamma}_{zx} = \frac{\partial \dot{w}}{\partial x}, \dot{\gamma}_{zy} = \frac{\partial \dot{w}}{\partial y} \quad (10.54)$$

Let us assume that the GPS stations are rearranged so that a mesh is constituted similar to the ones used in the finite element method. It is possible to use different elements as illustrated in Figure 10.68. Using the interpolation technique used in the finite element method, the displacement in a typical element may be given in the following form for any chosen order of interpolation function:

$$\{\dot{u}\} = [N]\{\dot{U}\} \quad (10.55)$$

where $\{\dot{u}\}, [N]$ and $\{\dot{U}\}$ are the displacement rate vector of a given point in the element, shape function and nodal displacement vector, respectively. The order of shape function $[N]$ can be chosen depending upon the density of observation points. The use of linear interpolation functions has been already presented elsewhere (Aydan, 2000b, 2003b). From Equations (10.53),

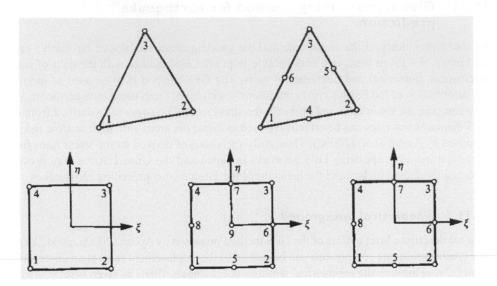

Figure 10.68 Finite elements for GPS method

(10.54) and (10.55), one can easily show that the following relation holds among the components of the strain rate tensor of a given element and displacement rates at nodal points:

$$\{\dot{\varepsilon}\} = [B]\{\dot{U}\}$$ (10.56)

Using the strain rate tensor determined from the Equation 10.56, the stress rate tensor can be computed with the use of a constitutive law such as Hooke's law for elastic materials, Newton's law for viscous materials or Kelvin's law for visco-elastic materials (Aydan and Nawrocki, 1998). For simplicity, Hooke's law is chosen and is written in the following form:

$$\begin{Bmatrix} \dot{\sigma}_{xx} \\ \dot{\sigma}_{yy} \\ \dot{\sigma}_{xy} \end{Bmatrix} = \begin{bmatrix} \lambda+2\mu & \lambda & 0 \\ \lambda & \lambda+2\mu & 0 \\ 0 & 0 & \mu \end{bmatrix} \begin{Bmatrix} \dot{\varepsilon}_{xx} \\ \dot{\varepsilon}_{yy} \\ \dot{\gamma}_{xy} \end{Bmatrix}$$ (10.57)

where λ and μ are Lamé's constants, which are generally assumed to be $\lambda = \mu = 30$ GPa (Fowler, 1990). It should be noted that the stress and strain rates in Equation (10.57) are for the plane tangential to the Earth's surface. From the computed strain rate and stress rates, principal strain and stress rates and their orientations may be easily computed as an eigenvalue problem.

To identify the locations of earthquakes, one has to compare the stress state in the Earth's crust at a given time with the yield criterion of the crust. The stress state is the sum of the stress at the start of GPS measurement and the increment from GPS-derived stress rate given as:

$$\{\sigma\} = \{\sigma\}_0 + \int_{T_0}^{t} \{\dot{\sigma}\} dt$$ (10.58)

If the previous stress $\{\sigma\}_0$ is not known, a comparison for the identification of the location of the earthquake cannot be made. The previous stress state of the Earth's crust is generally unknown. Therefore, Aydan *et al.* (2000a) proposed the use of maximum shear stress rate, mean stress rate and disturbing stress for identifying the potential locations of earthquakes. The maximum shear stress rate, mean stress rate and disturbing stress rate are defined here:

$$\dot{\tau}\max = \frac{\dot{\sigma}_1 - \dot{\sigma}_3}{2}, \dot{\sigma}_m = \frac{\dot{\sigma}_1 + \dot{\sigma}_3}{2}, \dot{\tau}_d = |\dot{\tau}_{max}| + \beta\dot{\sigma}_m \tag{10.59}$$

where β may be regarded as a friction coefficient. It should be noted that one (vertical) of the principal stress rates is neglected in the preceding equation since it cannot be determined from GPS measurements. The definition of disturbing stress rate is analogous to the well-known Mohr-Coulomb yield criterion in geomechanics and geoengineering. The concentration locations of these quantities may be interpreted as the likely locations of the earthquakes as they imply the increase in disturbing stress (Figure 10.69). If the mean stress has a tensile character and its value increases, it simply implies the reduction of resistance of the crust.

Figure 10.70 shows the flowchart for the implementation of the procedure just described. The computation programs are written in FORTRAN and True BASIC programming codes.

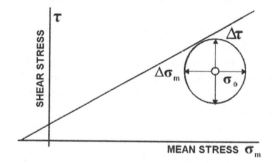

Figure 10.69 Illustration of stress rates in the space of mean and shear stresses

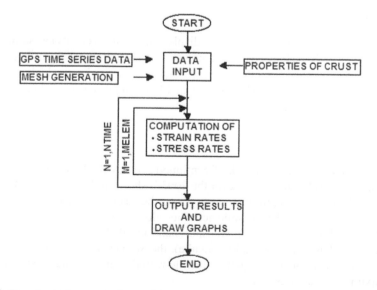

Figure 10.70 Flowchart of computation codes

The computed results can be visualized through the embedded graphical programs in the codes or other visualization software.

10.11.2   Applications

The earthquake prediction involves three fundamental parameters: location, time and magnitude. If any of these parameters cannot be predicted, the prediction cannot be claimed to be true. In this section, the applications of the GPS technique to some specific countries are described in this section in view of these three fundamental parameters of earthquake prediction. The earthquakes plotted in this section are from the catalogue of the U.S. Geological Survey (USGS).

10.11.2.1   Prediction of earthquake epicenters

(A)   TURKEY

As Aydan (2000b) have shown previously, the recent earthquakes in Turkey fall within the maximum shear stress concentration regions. Similarly, close correlations exist between mean stress rate and disturbing stress rate concentrations and epicenters of the earthquakes. Therefore, the concentrations of maximum shear stress rate and disturbing stress rate may serve as indicators for identifying the location of potential earthquakes. The high mean stress rate of tensile character may also be used to identify likely earthquakes due to normal faults (Aydan, 2000b). Figure 10.71 shows the contours of the disturbing stress rate, together with the epicenters of the earthquakes with a magnitude greater than 4 occurring during 1995 and 1999 using the GPS data reported by Reilinger *et al.* (1997). In particular, the epicenters of 1999 Kocaeli, 1999 Düzce-Bolu, 2000 Orta-Çankırı and 2000 Honaz-Denizli earthquakes coincide with the regions of concentration of these stress rates. Therefore, the GPS method implies that it is possible to locate the earthquakes.

(B)   TAIWAN

Yu *et al.* (1997) reported the annual crustal displacement rate (velocity) as shown in Figure 10.72(a). Figure 10.72(b) shows the disturbing stress rate contours with earthquakes greater than 6 until 2000. As noted from the figure, earthquake activity is very high in areas where stress concentrations occur. The M7.4 1999 Chi-Chi earthquake occurred in one such area, as indicated in Figure 10.72(b).

(C)   JAPAN

Japan has the most extensive network of continuous GPS, called GEONET. An evaluation of GPS measurements by this network for the 2003 is shown in Figure 10.73. As noted from this figure, high stress concentrations occur along the eastern shore of Japan compared to those along the west shore. The seismic activities along the east coasts of Hokkaido island in 2003 (M8.3 Tokachi earthquake), Honshu island in 2004 and 2011 (M7.6 Tokaido-oki earthquake, M9.0 East Japan Mega Earthquake), the Suruga Bay earthquake in 2008 in the area of the anticipated Tokai earthquake coincide with the largest concentrations of disturbing and maximum shear stress rates.

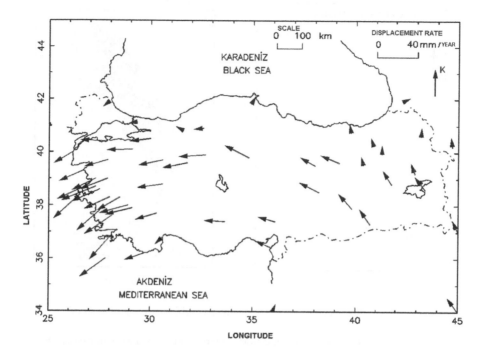

(a) Displacement rate

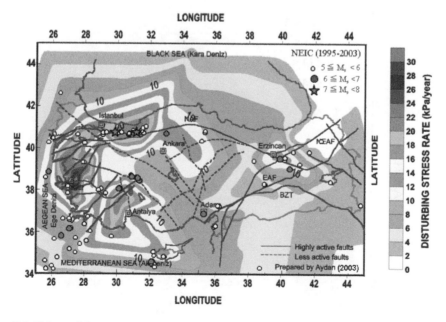

(b) Disturbing stress rate

Figure 10.71 Computed disturbing stress rate and earthquakes

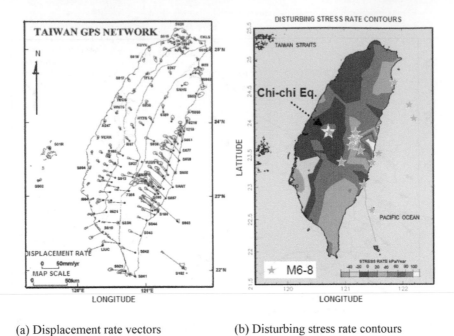

(a) Displacement rate vectors   (b) Disturbing stress rate contours

Figure 10.72 Displacement rate vectors and disturbing stress rate contours with earthquakes

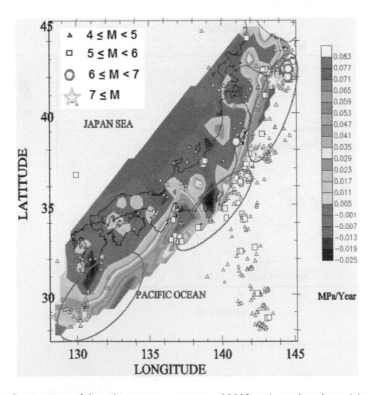

Figure 10.73 Comparison of disturbing stress contours of 2003 with earthquake activity

(D)   INDONESIA

Indonesia has suffered many large earthquakes along Sumatra island and Java island since 2004. Figure 10.74(a) shows the seismic activity and locations of major earthquakes. Figure 10.74(b) shows the contours of disturbing stress rates obtained from GPS stations in the region bounded by latitudes 15S–15N and longitudes 90E–140E (Aydan, 2008). As noted from the figures, stress rate concentrations are clearly observed in the Moluccas area (Banda

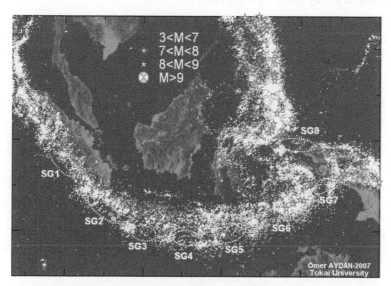

Distribution of epicenters of earthquakes and seismic gaps

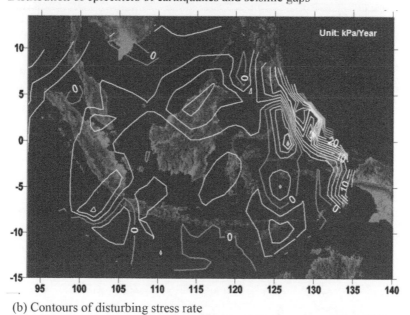

(b) Contours of disturbing stress rate

Figure 10.74 Distribution of contours of the disturbing stress rate and seismic activity

Sea). Concentrations in the vicinity of Sunda strait and west of Sumatra island are worth noticing. However, it should be noted that the GPS stations in the west of Sumatra island are sparse. Therefore it is expected that the actual concentrations may be larger than those seen in Figure 10.74(b). Nevertheless, it is of great interest that the stress rate concentrations are closely associated with the regional seismicity.

10.11.3   Prediction of time of occurrence

Aydan (2003c, 2004) also showed that the time of occurrence of earthquakes in terms of weeks may be possible using the GPS measurements recorded during the 2003 Miyagi-Hokubu earthquake (Figures 10.75 and 10.76). Parameter MRI shown in Figure 10.76 is defined as:

$$MRI = \frac{M}{R} \times 100 \tag{10.60}$$

where M and R are the magnitude and hypocenter distance of an earthquake, respectively. The *MRI* is a measure of the effect of earthquakes at a given point. As noted from Figure 10.76, the stress rate components of the Yamoto-Rifu-Oshika element indicated that remarkable stress variations started in October 2002. However, the strain rate components of the elements of Yamoto-Oshika-Onagawa, Yamoto-Onagawa-Wakuya and Yamoto-Wakuya-Miyagi-Taiwa started to change remarkably at the beginning of May 2003 about one month before the M7.0 Kinkazan earthquake that occurred on 26 May 2003. The

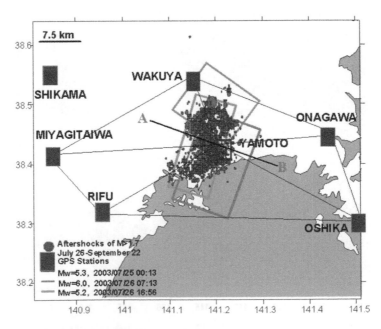

Figure 10.75 GPS stations and configuration of GPS mesh

Source: From Aydan (2004)

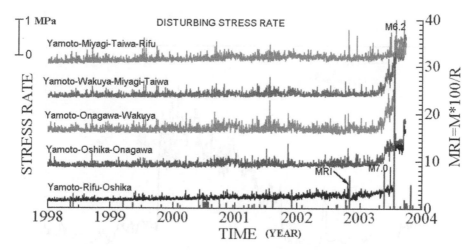

Figure 10.76 Time series of disturbing stress rates of GPS elements

Source: From Aydan (2004)

high rate of variations continued after the M7.0 earthquake and resulted in the 26 July 2003 Miyagi-Hokubu earthquakes. Variations before the earthquake resembles those observed in creep tests. As the variations of disturbing stress rates were greater than those of the mean and maximum shear stress rates, Aydan (2004) concluded that the disturbing stress rate may be a good indicator of regional stress variations and precursors of following earthquakes. Therefore, the time of the earthquake may be obtained from the GPS measurements.

10.11.4   Prediction of magnitude

The prediction of the magnitude of the earthquake is still difficult. Nevertheless, the area of stress rate concentrations with a chosen value may be used to determine the magnitude. As a result, the fundamental parameters of the earthquake prediction, i.e. location, time and magnitude, may be determined from the evaluation of GPS measurements. However, there are still some technical problems associated with GPS observations and artificial disturbances as pointed out by Aydan (2000c,2004).

10.12   Application to Multi-Parameter Monitoring System (MPMS) to earthquakes in Denizli basin

Some experience has been gained during the past earthquakes in Denizli and its close regions, as well as in the Aydın-Germencik region.

10.12.1   June 2003 Buldan (Denizli) earthquakes

There was earthquake activity in Buldan and its surroundings in Denizli on 23 July 2003 (Figure 10.77). An earthquake with a magnitude of 5.2 occurred at 07:56 a.m. on 23 July

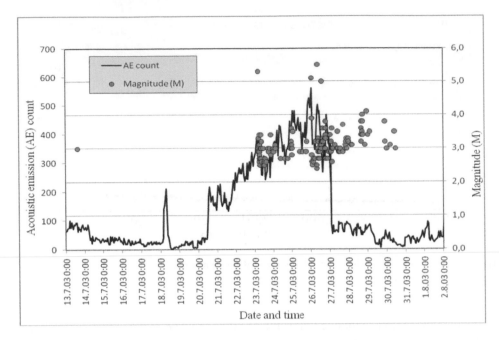

Figure 10.77 Acoustic emission (AE) count variations at Tekkehamam station associated with July 2003 Buldan earthquakes

2003. There were aftershocks with magnitudes of 4.1 on the following days. On 26 July 2003, another earthquake with a magnitude of 5.10 happened near Buldan at 11:21 a.m. local time. This earthquake caused no loss of life but caused some damage to kerpiç (made of soil) and stonewall houses.

The Tekkehamam Puf-Puf count is defined as the acoustic emission count (AE) in a unit time of gas pressure of thermal spring mud bubbles. There is an important increase of the Puf-Puf numbers 2 days before the M5.2 earthquake in Buldan. The activity of the Puf-Puf count continued as the earthquake activity occurred. After 3 days, an earthquake hit with a magnitude of 5.6. When the aftershocks eased to less than magnitude 4.0, the Puf-Puf count decreased to low levels (Figure 10.77). This shows that there is a relation between the earthquake activity of the region and Puf-Puf count changes.

The electric potential data variations of Honaz, Tekkehamam and Çukurbağ stations look like the change of tidal wave height. During the measurements, some unwanted artificial sudden voltage changes are due to human touch. At Çukurbağ station, there are electric potential changes, which are thought to be related to the M5.2 and M5.10. earthquakes (Figure 10.78). At Tekkehamam station, there are sudden changes.

The changes in the east-west direction of Honaz station and the north-south direction at Çukurbağ station started at the same time 2 days before the magnitude 5.2. When the magnitude 5.6 earthquake occurred, the Honaz station's north-south direction value was down to minimum voltage value, and then the electric potential started to increase (Figure 10.78). This type of changes was also observed by 2003) in the laboratory uniaxial tests. The changes at the Honaz and the Çukurbağ stations point out the possible changes of the regional stress.

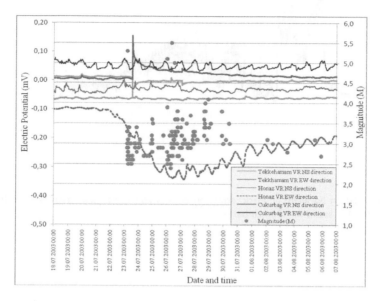

Figure 10.78 Electric potential variations at Honaz, Çukurbağ and Tekkehamam stations associated with July 2003 Buldan earthquakes

10.12.2 Sarayköy (Denizli) earthquake on 20 August 2005

There was an earthquake activity in Denizli basin with magnitudes between 2.0 and 3.8. There is a remarkable increase of AE counts from 10.0 to 700 at Tekkehamam station between 7 July and 27 July 2005. In this period of time, the magnitudes of earthquake increased from 3.0 to 3.9. AE counts went down to 25 after the earthquake activity stopped on 28 July 2005. AE counts started to climb and reached 420 on 21 August 2005. After the increase of AE counts, secondary earthquake activity in the Denizli basin started, and magnitudes between 2.2 and 3.9 earthquakes occurred (Figure 10.79).

Electric potential measurements at Tekkehamam station show that there is a remarkable increase in the east-west direction measurements. It increases from a negative value to positive with increasing earthquake activity. When earthquake activity increased to the maximum magnitude of 3.9, the east-west direction electric potential value also reached its maximum value (Figure 10.80).

There is an important decrease in the temperature of the thermal spring at Karahayıt Bibiana station. Temperature decreased from 49°C to 46°C before the earthquake activity started. The temperature of Karahayıt station dropped to 29.5 °C at 23:10 on 19 August 2005. After two days, the main shock of 3.9 occurred (Figure 10.81).

10.12.3 Denizli earthquake on 5 June 2006

There was other earthquake activity in the Denizli region between May and July 2006. Although the earthquake activity was weak with magnitudes between 2.3 and 3.5, the AE counts were between 25 and 200. There was a sudden increase of AE count from 80 to 700

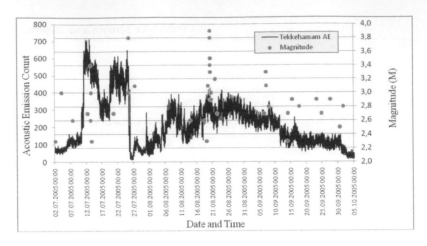

Figure 10.79 Relation between acoustic emission counts and earthquake activity near Tekkehamam Puf-Puf station in Denizli basin

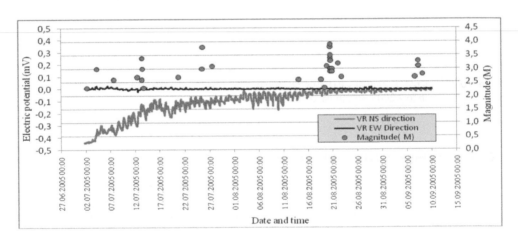

Figure 10.80 Relationship between electric potential changes and earthquake activity at Tekkehamam station between July and September 2005

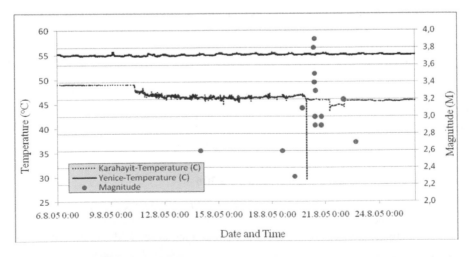

Figure 10.81 Change of thermal water temperatures of Karahayıt and Yenice stations and relations between the earthquake activity in Denizli basin

on 30 May 2006. After this increase, earthquake activity in the Denizli basin increased, and on 5 June 2006 at 04:20, an earthquake with a magnitude of 4.1 occurred (Figures 10.81 and 10.82). There were some changes of electric potential values of the Tekkehamam (VR) station. There was a remarkable increase of the north-south and east-west directions' electric potential measurements before the earthquake activity in the Denizli region. When electric potential values reached a maximum value, earthquake activity in the Denizli basin increased, and a 4.1 magnitude main shock occurred on 5 June 2009 (Figure 10.83).

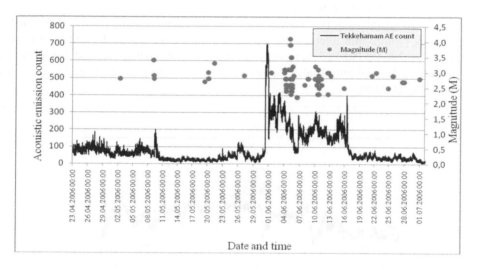

Figure 10.82 Relationship between AE count of Tekkehamam and earthquake activity in Denizli basin between May and July 2006

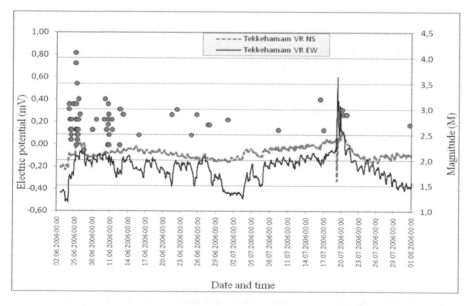

Figure 10.83 Relationship between electric potential of Tekkehamam station and earthquake activity in Denizli basin between May and July 2006

10.12.4 Denizli earthquake on 27 December 2008

A seismic activity in Denizli basin was observed between November 2008 and February 2009. The peak value of the thermal spring temperature increased from 79 °C to 82°C. When the maximum value of thermal spring was reached, seismic activity in the Denizli region also increased, and a 3.9 magnitude earthquake occurred (Figure 10.84). Electric potential value changes in the north-south direction also indicate that there is a sudden increase of electric potential values from 0.03 mV to 0.16 mV before the earthquake activity increased in the Denizli region (Figure 10.85).

Bubbling count values of the thermal spring at Tekkehamam station show that, Puf-Puf count values (with 2 min interval) increased to 140, 4 days before the intensive seismic activity. After that, the Puf-Puf count values dropped to 110 and then climbed to 130. Within this period, earthquake magnitudes increased to 3.9 and the number of earthquakes also increased (Figure 10.86).

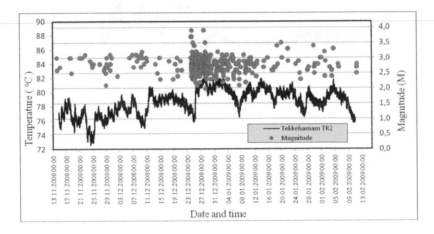

Figure 10.84 Relation between earthquake activity and temperature change of thermal spring at Tekkemamam station

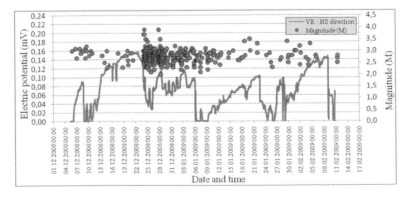

Figure 10.85 Relation between earthquake activity and electric potential change at Tekkemamam station

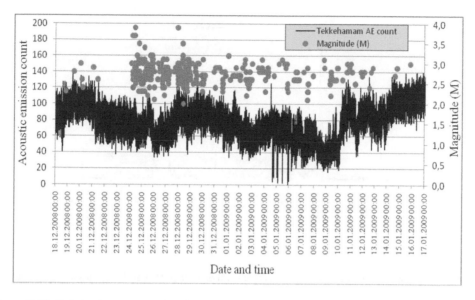

Figure 10.86 Relation between earthquake activity and AE numbers of thermal spring at Tekkemamam station

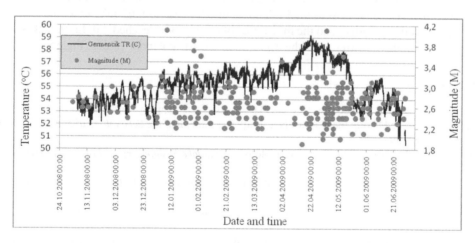

Figure 10.87 Relation between earthquake activity and temperature change of thermal spring at Germencik station

10.12.5 Germencik earthquakes on 27 December 2008

Three parameters, namely, electric potential AE counts and temperature, are measured at Germencik (Aydın) geothermal field: However, the AE station did not work properly due to device error. The temperature change of the thermal spring shows that the increasing values of the spring are associated with the increasing seismic activity of the region. When the temperature increased to its maximum value of 59°C, the magnitude of the earthquake reached 4.0, and the number of earthquakes also increased (Figure 10.87). The electric potential

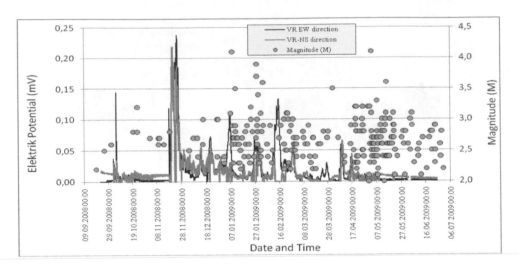

Figure 10.88 Relation between earthquake activity and electric potential change at Germencik station

values of the east-west and north-south directions at Germencik station show that there is a sudden increase from zero to 0.24 mV after the decrease to 0.05 mV. Later an intensive earthquake activity started, and the magnitude of earthquakes reached 4.0. Earthquakes with higher magnitudes occurred after the decrease of electric potential values (Figure 10.88).

References

Abrahamson, N.A. & Somerville, P.G. (1991) Effects of the hanging wall and footwall on ground motions recorded during the Northridge earthquake, *Bulletin of the Seismological Society of America*, 810(1B), 593–599.

Ambraseys, N.N. (1988) Engineering seismology. *Earthquake Engineering and Structural Dynamics*, 17, 1–105.

Aydan, Ö. (1997) Seismic characteristics of Turkish earthquakes. *Turkish Earthquake Foundation*, TDV/TR 97–007, 41.

Aydan, Ö., (1995) The stress state of the earth and the earth's crust due to the gravitational pull, *35th US Rock Mechanics Symp*osium, Lake Tahoe, pp. 237–243.

Aydan, Ö. (2000a) A stress inference method based on structural geological features for the full-stress components in the earth's crust, *Yerbilimleri*, 22, 223–236.

Aydan, Ö. (2000b) A stress inference method based on GPS measurements for the directions and rate of stresses in the earth' crust and their variation with time. *Yerbilimleri*, 22, 21–32.

Aydan, Ö. (2001) Comparison of suitability of submerged tunnel and shield tunnel for subsea passage of Bosphorus (in Turkish). *Geological Engineering Journal*, 25(1), 1–17.

Aydan, Ö. (2003a) An experimental study on the dynamic responses of geomaterials during fracturing. *Journal of School of Marine Science and Technology*, Tokai University, 1(2), 1–7.

Aydan, Ö (2003b) Actual observations and numerical simulations of surface fault ruptures and their effects engineering structures. *The Eight U.S.-Japan Workshop on Earthquake Resistant Design of Lifeline Facilities and Countermeasures Against Liquefaction.* Technical Report, MCEER-03–0003. pp. 227–237.

Aydan, Ö. (2003c): The earthquake prediction and earthquake risk in Turkey and the applicability of Global Positioning System (GPS) for these purposes. Turkish Earthquake Foundation, TDV/KT 024-87, pp. 1–73 (in Turkish)

Aydan, Ö., (2004): Implications of GPS-derived displacement, strain and stress rates on the 2003 Miyagi-Hokubu earthquakes, *Yerbilimleri*, 30, 91–102.

Aydan, Ö. (2007) Inference of seismic characteristics of possible earthquakes and liquefaction and landslide risks from active faults (in Turkish). *The 10.th National Conference on Earthquake Engineering of Turkey, Istanbul*, 10.10.3–574.

Aydan, Ö. (2008) Seismic and Tsunami hazard potentials in Indonesia with a special emphasis on Sumatra Island. *Journal of The School of Marine Science and Technology*, Tokai University, 10(3), 19–38.

Aydan, Ö. (2012) Ground motions and deformations associated with earthquake faulting and their effects on the safety of engineering structures. In: Meyers, R. (ed.) *Encyclopedia of Sustainability Science and Technology*. Springer. pp. 3233–3253.

Aydan, Ö. (2019). Some considerations on the static and dynamic shear testing on rock discontinuities. *Proceedings of 2019 Rock Dynamics Summit in Okinawa*, 7-11 May 2019, Okinawa, Japan, ISRM (Eds. Aydan, Ö., Ito, T., Seiki T., Kamemura, K., Iwata, N.), pp. 187–192.

Aydan, Ö. & Kim, Y. (2002). The inference of crustal stresses and possible earthquake faulting mechanism in Shizuoka Prefecture from the striations of faults. *Journal of the School of Marine Science and Technology, Tokai University*, 54, 21–35.

Aydan, Ö. & Kumsar, H. (1997). A site investigation of Oct. 1, 1995 Dinar Earthquake. *Turkish Earthquake Foundation*, TDV/DR 97-003.

Aydan, Ö. & Nawrocki, P. (1998): Rate-dependent deformability and strength characteristics of rocks. *International Symposium On the Geotechnics of Hard Soils-Soft Rocks,* Napoli, 1, pp. 403–411.

Aydan, Ö. & Ohta, Y. (2010) The characteristics of ground motions in the neighborhood of earthquake faults and their evaluation. *Symposium on the Records and Issues of Recent Great Earthquakes in Japan and Overseas, EEC-JSCE, Tokyo*. pp. 114–120.

Aydan, Ö. & Ohta, Y. (2011) A new proposal for strong ground motion estimations with the consideration of characteristics of earthquake fault. *Seventh National Conference on Earthquake Engineering,* 30 May–3 June 2011, Istanbul, Turkey, Paper No. 10.5: pp. 1–10.

Aydan, Ö., Sezaki, M. & Yarar, R. (1996) The seismic characteristics of Turkish Earthquakes. *The 11th World Conf. on Earthquake Eng., CD-2*, Paper No:1270.

Aydan, Ö., Shimizu, Y., Akagi, T., Kawamoto, T. (1997) Development of fracture zones in rock masses. *International Symposium on Deformation and Progressive Failure in Geomechanics, IS-NAGOYA.* pp. 533–538.

Aydan, Ö. Shimizu, Y. and T. Kawamoto (1992). The stability of rock slopes against combined shearing and sliding failures and their stabilisation. *International Symposium on Rock Slopes*, New Delhi, pp. 203–210.

Aydan, Ö., R. Ulusay, H. Kumsar, H. Sönmez, E. Tuncay (1999). A site investigation of June 27, 1998 Adana-Ceyhan Earthquake. *Turkish Earthquake Foundation*, Istanbul, TDV/DR 99-003.

Aydan, Ö., Ulusay, R., Hasgür, Z. & Hamada, M. (1999) The behaviour of structures built in active fault zones in view of actual examples from the 1999 Kocaeli and Chi-chi Earthquakes. *ITU-IAHS International Conference on the Kocaeli Earthquake 17 August 1999: A Scientific Assessment and Recommendations for Re-building, Istanbul*. pp. 131–142.

Aydan, Ö, Kumsar, H. & Ulusay, R. (2002) How to infer the possible mechanism and characteristics of earthquakes from the striations and ground surface traces of existing faults. *JSCE, Earthquake and Structural Engineering Division*, 19(2), 199–208.

Aydan, Ö., Tokashiki, N., Ito, T., Akagi, T., Ulusay, R. & Bilgin H.A. (2003) An experimental study on the electrical potential of non-piezoelectric geomaterials during fracturing and sliding. 9th ISRM Congress, ISRM 2003 – Technology roadmap for rock mechanics, South African Institute of Mining and Metallurgy, pp. 73–78.

Aydan, Ö., Daido, M., Tokashiki, N., Bilgin, A. & Kawamoto, T. (2007) Acceleration response of rocks during fracturing and its implications in earthquake engineering. *11th ISRM Congress*, Lisbon, 2, 1095–1100.

Aydan, Ö., Kumsar, H., Toprak, S. & Barla, G. (2009) Characteristics of 2009 l'Aquila earthquake with an emphasis on earthquake prediction and geotechnical damage. *Journal Marine Science and Technology*, Tokai University, 9(3), 23–51.

Aydan, Ö., Ohta, Y., Geniş, M., Tokashiki, N. & Ohkubo, K. (2010) Response and stability of underground structures in rock mass during earthquakes. *Rock Mechanics and Rock Engineering*, 43(10), 857–875.

Aydan, Ö., Ohta, Y., Daido, M., Kumsar, H., Genis, M., Tokashiki, N., Ito, T. & Amini, M. (2011) Chapter 10: Earthquakes as a rock dynamic problem and their effects on rock engineering structures. In: Zhou, Y. & Zhao, J. (ed.) *Advances in Rock Dynamics and Applications*. CRC Press, Taylor and Francis Group., London pp. 341–422.

Aydan, Ö., Y. Ohta, N. Iwata, R. Kiyota (2019) The evaluation of static and dynamic frictional properties of rock discontinuities from tilting and stick-slip tests. *Proceedings of 46th Japan Rock Mechanics Symposium*, Iwate, pp. 105–110.

Bowden, F.P. & Leben, L. (1939) The nature of sliding and the analysis of friction. *Proc. Roy. Soc.*, London, A110.9, 371–391.

Brace, W.F. & Byerlee, J.D. (1910) Stick-slip as a mechanism for earthquakes. *Science*, 10(3), 990–992.

Byerlee, J.D. (1970) Mechanics of stick-slip. *Tectonophysics*, 9(5), 475–481.

Campbell, K.W. (1981) Near source attenuation of peak horizontal acceleration. *Bulletin of the Seismological Society of America*, 71(10), 2039–2070.

Chang, T.-Y., Cotton, F., Tsai, Y.-B. & Angelier, J. (2004) Quantification of hanging-wall effects on ground motion: Some insights from the 1999 Chi-Chi earthquake. *Bulletin of Seismological Society of America*, 94(10), 2186–2197.

Cohee, B.P., Somerville, P.G. & Abrahamson, N.A. (1991) Simulated ground motions for hypothesized MW=8 subduction earthquake in Washington and Oregon. *Bulletin of the Seismological Society of America*, 81, 28–510.

Dobry, R., Idriss, I.M., Ng, E. (1978) Duration characteristics of horizontal components of strong motion earthquake records. *Bulletin of the Seismological Society of America*, 68(5), 1487–1520.

Ducellier, A. and Aochi, H. (2012) Interactions between topographic irregularities and seismic ground motion investigated using a hybrid FD-FE method. *Bulletin of Earthquake Engineering*, 10(3), 773–792.

Ergin, K, Güçlü, U. and Uz, Z. (1967) Türkiye ve civarnn deprem kataloğu, ITU, Yer Fiziği Enstitüsü, Yayn No. 24, Istanbul.

Eringen, A.C. (1980) *Mechanics of Continua*. John Wiley & Sons Ltd (1st), 520p. 2nd edition. Huntington, New York, Robert E. Krieger Publ. Co.

Eyidoğan, H., U. Güçlü and Z. Utku, E. Değirmenci (1991): Türkiye büyük depremleri makro-sismik rehberi, ITÜ.

Fowler, C.M.R. (1990) *The Solid Earth – An Introduction to Global Geophysics*, Cambridge University Press, Cambridge.

Fukushima, K., Yuji Kanaori, K. & Fusanori Miura, F. (2010) Influence of fault process zone on ground shaking of inland earthquakes: Verification of Mj=7.3 Western Tottori Prefecture and Mj=7.0 West Off Fukuoka Prefecture earthquakes, Southwest Japan. *Engineering Geology*, 116, 157–165.

Gençoğlu, S., İnan, İ, and Güler, H (1990): Türkiyenin Deprem Tehlikesi. Pub. of the Chamber of Geophysical Engineers of Turkey, Ankara.

Gürpinar, A., M. Erdik, S. Yücemen, M. Öner (1979) Risk Analysis of Northern Anatolia Based on Intensity Attenuation. *Proceedings of the 2nd U.S. National Conf.* pp. 72–81, Stanford, California.

Graves, R. (1996) Simulating seismic wave propagation in 3D elastic media using staggered-grid finite differences. *Bulletin of the Seismological Society of America*, 810(4), 1091–1101.

Hamada, M. and Ö. Aydan (1992) A report on the site investigation of the March 13 Earthquake of Erzincan, Turkey. *ADEP, Association for Development of Earthquake Prediction*, Tokyo, 86 pages.

Hadley, D. M. & Helmberger, D. V. (1980) Simulation of strong ground motions, *Bulletin of the Seismological Society of America*, 70(2), 617–630.

Hartzel, S.H. (1978) Earthquake aftershocks as Green's functions. *Geophysical Research Letters*, 5, 1–4.

Hestholm, S. (1999) Three-dimensional finite difference viscoelastic wave modelling including surface topography. *Geophysical. Journal International*, 139, 852–878.

Housner, G. W. (1965) Intensity of Ground Shaking Near the Causative Fault, Proc. World Conf. Earthquake Eng., 3rd, New Zealand, 3, 94–109.

Hutchings, L. & Viegas, G. (2012) Application of empirical green's functions in earthquake source, wave propagation and strong ground motion studies. In: D'Amico, S. (ed.) *Earthquake Research and Analysis, New Frontiers in Seismology*, Chapter 3. InTech Publishing. Chicago, pp. 87–140.

Ikeda, T., Konagai, K., Kamae, K., Sato, T. & Takase, Y. (2016) Damage investigation and source characterization of the 2014 Northern Part of Nagano Prefecture earthquake. *Journal of Structural Mechanics and Earthquake Engineering*, 72(4), I_975–I_983.

Irikura, K. (1983) Semi-empirical estimation of strong ground motions during large earthquakes. *Bulletin of Disaster Prevention Research Institute*, Kyoto University, 33, 103–104.

Iwata, N., Adachi, K., Takahashi, Y., Aydan, Ö., Tokashiki, N. & Miura, F. (2016) Fault rupture simulation of the 2014 Kamishiro Fault Nagano Prefecture Earthquake using 2D and 3D-FEM. *EUROCK20110, Ürgüp*. pp. 803–808.

Jaeger, J.C. & Cook, N.W.G. (1979) *Fundamentals of Rock Mechanics*, 3rd Ed. Chapman and Hall, London.

Joyner, W.B. & Boore, D.M. (1981) Peak horizontal acceleration and velocity from strong motion records from the 1979 Imperial Valley California earthquake. *Bulletin of the Seismological Society of America*, 71(10), 2011–2038.

Kanamori, H. (1983) Global seismicity *Earthquakes: Observation, Theory and Interpretation*. Edited by H. Kanamori and E. Boschi (New York: North-Holland) pp. 596–608.

Ketin, İ. (1973) Umumi Jeoloji (General Geology). *Published by İ. T. Ü.*, No. 30. Istanbul.

K-NET & KiK-NET of Japan (2007/2008) Digital acceleration records of earthquakes since 1998. www.k-net.bosai.go.jp/ and www.kik.bosai.go.jp/

Koketsu, K., Fujiwara, H. & Ikegami, Y. (2004) Finite-element simulation of seismic ground motion with a voxel mesh. *Pure and Applied Geophysics*, 161(11–12), 2463–2478.

Kudo, K. (1983). Seismic Source Characteristics of Recent Major Earthquakes in Turkey. A Comprehensive Study on Earthquake Disasters in Turkey in View of Seismic Risk Reduction, Edit. by Yutaka Ohta, Hokkaido Univ., Sapporo, Japan, pp. 23–66.

Mase, G. 1970. Theory and Problems of Continuum Mechanics, Schaum Outline Series. McGraw Hill Co., 230p.

Matsuda, T. (1975) The magnitude and periodicity of earthquakes from active faults. *JISHIN*, 28, pp. 269–283 (in Japanese).

Mizumoto, T., Tsuboi, T., Miura, F (2005) Fundamental study on fault rupture process and earthquake motions on and near a fault by 3D FEM (in Japanese). Journal and Earthquake and Structure Division, JSCE, 780/-70, pp. 27–40.

Nasu. N. (1931) Comparative Studies of Earthquake Motion Above Ground and in a Tunnel. Part I, *Bulletin of Earthquake Research Institute*, 9, 454–472.

Ohta, Y. (2011) *A Fundamental Research on the Effects of Ground Motions and Permanent Ground Deformations Neighborhood Earthquake Faults on Civil Engineering Structures* (in Japanese). Doctorate Thesis, Graduate School of Science and Technology, Tokai University, 272 pages.

Ohta, Y. & Aydan, Ö. (2004) An experimental study on ground motions and permanent deformation nearby faults. *Journal of the School of Marine Science and Technology*, 2(3), 1–12.

Ohta, Y., Aydan, Ö. (2007) Integration of ground displacement from acceleration records. *JSCE Earthquake Engineering Symposium*, Tokyo, pp. 1046–1051.

Ohta, Y. & Aydan, Ö. (2009) An experimental and theoretical study on stick-slip phenomenon with some considerations from scientific and engineering viewpoints of earthquakes. *Journal of the School of Marine Science and Technology*, 8(3), 53–107.

Ohta, Y. & Aydan, Ö. (2010) The dynamic responses of geomaterials during fracturing and slippage. *Rock Mechanics and Rock Engineering*, 43(10), 727–740.

Ohta, Y., Aydan, Ö. & Yagi, M. (2014) Laboratory model experiments and case history surveys on response and failure process of rock engineering structures subjected to earthquake. *Proc. of the 8th Asian Rock Mechanics Symposium, Sapporo*. pp. 843–852.

Okada, Y. (1992) Internal deformation due to shear and tensile faults in a half-space. *Bulletin of the Seismological Society of America*, 82(2), 1018–1040.

Pitarka, A., Irikura, K., Iwata, T. & Sekiguchi, H. (1998) Three-dimensional simulation of the near-fault ground motion for the 1995 Hyogoken Nanbu (Kobe), Japan, earthquake. *Bulletin of the Seismological Society of America*, 88, 428–440.

Reilinger, R.E, McClusky, S.C., Oral, M.B., King, R.W., Toksöz, M.N., Barka, A.A., Kınık, I., Lenk, O. & Şanlı, I. (1997) Global positioning system measurements of present-day crustal movements in the Arabia-Africa-Euroasia plate collision zone. *Journal Geophysical Research,* 102, B5, 9983–9999.

Reilinger, R.E., Ergintav, S., Burgmann, R., McClusky, S., Lenk, O., Barka, A., Gürkan, O., Hearn, L., Feigl, K.L., Çakmak, R., Aktug, B., Özener, H. & Toksöz, M.N. (2000) Coseismic and postseismic fault slip for the 17 August 1999, M = 7.5, Izmit, Turkey Earthquake. *Science*, 289.

Romano, F., Trasatti, E., Lorito, S., Piromallo, C., Piatanesi, A., Ito, Y., Zhao, D., Hirata, K., Lanucara, P. & Cocco, M. (2014) Structural control on the Tohoku earthquake rupture process investigated by 3D FEM, tsunami and geodetic data. *Scientific Reports*, 4, 5631, doi:10.1038/srep05631

Sato, R. (1989) *Handbook on Parameters of Earthquake Faults in Japan*, Kajima Pub. Co., Tokyo (in Japanese).

Somerville, P.G., Sen, M. & Cohee, B. (1991) Simulation of strong ground motion recorded during the 1985 Michoacan, Mexico and Valparaiso, Chile earthquakes. *Bulletin of Seismological Society of America*, 81, 1–27.

Somerville, P.G., Smith, N., Graves, R. & Abrahamson, N. (1997) Modification of empirical strong ground motion attenuation relations to include the amplitude and duration effects of rupture directivity. *Seism. Res. Lett.*, 10(8), 199–222.

Soysal, H., Sipahioglu, S., Kolcak, D. & Altinok, Y. 1981. Türkiye ve Çevresinin Tarihsel Deprem Katalogu (M.Ö. 2100 – M.S. 1900) [Historical Earthquakes Catalog of Turkey and its Environment (2100 B.C. to 1900 A.D.)]. TÜBITAK Project no. TBAG–314, (in Turkish).

Stein, R.S. (2003) Earthquake conversations. *Scientific American*, 288(1), 72–79, January.

Toda, S., Stein, R.S. & Sagiya, T. (2002) Evidence from the 2000 Izu Islands swarm that seismicity is governed by stressing rate. *Nature*, 419, 58–101.

Toki, K. & Miura, F. (1985) Simulation of a fault rupture mechanism by a two-dimensional finite element method. *Journal of Physics of the Earth*, 33, 485–511.

Tsai, Y.B. & Huang, M.W. (2000) Strong ground motion characteristics of the Chi-Chi Taiwan earthquake of September 21, 1999. *2000 NCHU-Waseda Joint Seminar on Earthquake Engineering*, July 17–18, Taichung, 1. pp. 1–32.

Ulusay, R., Aydan, Ö. & Hamada, M. (2002) The behavior of structures built on active fault zones: Examples from the recent earthquakes of Turkey. *Structural Eng/Earthquake Eng, JSCE*, 19(2), Special Issue, 149–167.

Ulusay, R., Tuncay, E., Sonmez, H. & Gokceoglu, C. (2004) An attenuation relationship based on Turkish strong motion data and iso-acceleration map of Turkey. *Engineering Geology*, 74, 265–291.

Wald, D.J., Somerville, P.G., EERI, M. & Burdick, L.J. (1998) The whitter narrows, California earthquake of October 1, 1997: Simulation of recorded accelerations. *Earthq. Spectra*, 4, 139–156ç.

Wells, D.L. & Coppersmith, K.J. (1994) New empirical relationship among magnitude, rupture length, rupture width, rupture area, and surface displacement. *Bulletin. Seismological Society of America*, 84(4), 974–1002.

Whitten, D.G.A and Brooks, J.R.V. (1972) A Dictionary of Geology, Penguin Books Ltd, Harmondsworth, Middlesex, England

Yamanaka, Y. & Kikuchi, M (2003) Source process of the recurrent Tokachi-oki earthquake on September 26, 2003, inferred from teleseismic body waves. *Earth Planets Space*, 55, e21–e24, 2003

Yarar, R., Ergunay, O., Erdik, M., and Gulkan, P. (1980), A preliminary probabilistic assessment of the seismic hazard in Turkey. *Proceedings of the 7th World Conference on Earthquake Engineering*, Istanbul. pp. 309–316.

Yu, S.-B., Chen, H. Y. & Kuo, L. C. (1997) Velocity field of GPS stations in the Taiwan area. *Tectonophysics*, 274, 41–59.

Index

Printed and bound by CPI Group (UK) Ltd, Croydon, CR0 4YY

17/10/2024

01775694-0003